R. O'DONNELL (1979).

ADRENERGIC NEURONS

ADRENERGIC NEURONS

THEIR ORGANIZATION, FUNCTION
AND DEVELOPMENT IN THE
PERIPHERAL NERVOUS SYSTEM

Geoffrey Burnstock
Ph.D., D.Sc., F.A.A.
Head of Department of Anatomy and Embryology,
Professor of Anatomy,
University College,
University of London,
England

and

Marcello Costa
Laurea in Medicine and Surgery
Lecturer, Department of Human Physiology,
Flinders University,
South Australia

LONDON
CHAPMAN AND HALL

*First published 1975
by Chapman and Hall Ltd.,
11 New Fetter Lane, London EC4 4EE*
© *1975 G. Burnstock and M. Costa*

*Printed in Great Britain by
Richard Clay (The Chaucer Press) Ltd
Bungay, Suffolk*

ISBN 0 412 14060 8

All rights reserved. No part of this book
may be reprinted, or reproduced or utilized in
any form or by any electronic, mechanical or
other means, now known or hereafter invented,
including photocopying and recording, or in
any information storage and retrieval system,
without permission in writing from the
Publisher.

Distributed in the U.S.A.
by Halsted Press, a Division
of John Wiley & Sons Inc., New York
Library of Congress Catalog Card Number 75-2022

CONTENTS

1	**Introduction and History**	1
2	**General Organization and Function of Adrenergic Nerves**	4
	2.1 Distribution of adrenergic neurons	4
	2.2 Input of adrenergic neurons	5
	2.2.1 Cholinergic input from the CNS	5
	2.2.2 Cholinergic input from peripheral neurons	8
	2.2.3 Intraganglionic input	8
	2.3 Output of adrenergic neurons	11
	2.4 General functions of adrenergic neurons	12
	2.5 Relation between adrenergic and non-adrenergic neurons	14
	2.5.1 Adrenergic and cholinergic neurons	14
	2.5.2 Adrenergic and purinergic neurons	16
	2.6 Summary	17
3	**Structure of Adrenergic Neurons and Related Cells**	19
	3.1 Adrenergic neurons	19
	3.1.1 General structure	19
	3.1.2 Transmitter storage	22
	3.1.3 Axoplasmic transport of NA and related enzymes	31
	3.1.4 Origin, life-span and fate of adrenergic vesicles	33
	3.2 Other catecholamine-containing cells	35
	3.2.1 Adrenal chromaffin cells	35
	3.2.2 Extra-adrenal chromaffin cells	36
	3.3 Summary	37
4	**Biosynthesis and Metabolic Degradation of NA**	39
	4.1 Biosynthesis	39
	4.1.1 Biosynthetic pathway and enzymes	39
	4.1.2 Site of NA synthesis	39
	4.1.3 Regulation of NA synthesis in adrenergic neurons	41
	4.2 Metabolic degradation of neuronal NA	47
	4.2.1 Localization and properties of MAO	47
	4.2.2 Localization and properties of COMT	48
	4.2.3 Metabolic routes of neuronal NA inactivation	48
	4.3 Summary	50

CONTENTS

5 Adrenergic Neuroeffector Transmission — 51
- 5.1 Relation between adrenergic terminals and muscle effectors — 51
  - 5.1.1 Varicose terminal axons — 51
  - 5.1.2 Junctional cleft — 52
  - 5.1.3 Post-junctional specialization — 53
  - 5.1.4 Muscle effector bundle and nexuses — 54
- 5.2 Release of NA — 54
  - 5.2.1 Mechanism of release — 55
  - 5.2.2 Role of calcium in NA release — 58
  - 5.2.3 Physiological control of NA release — 58
  - 5.2.4 Quantal hypothesis and quantitative estimation of NA release — 61
- 5.3 Postjunctional action of NA — 64
  - 5.3.1 Adrenergic receptors — 64
  - 5.3.2 Chemical nature of adrenergic receptors — 66
  - 5.3.3 Effect of NA on membrane conductivity and calcium activity — 67
- 5.4 Inactivation of NA — 68
  - 5.4.1 Neuronal uptake of NA — 68
  - 5.4.2 Extraneuronal uptake of NA — 69
  - 5.4.3 Metabolic degradation of NA — 71
  - 5.4.4 Diffusion of NA from the neuroeffector junction into the bloodstream — 71
- 5.5 Electrophysiology of adrenergic transmission — 72
  - 5.5.1 Adrenergic excitatory transmission — 72
  - 5.5.2 Adrenergic inhibitory transmission — 76
- 5.6 Pharmacology of adrenergic transmission — 76
  - 5.6.1 Drugs which affect synthesis of NA — 76
  - 5.6.2 Drugs which affect storage of NA — 77
  - 5.6.3 Drugs which affect release of NA — 79
  - 5.6.4 Drugs which affect uptake of NA — 81
  - 5.6.5 Drugs which affect enzyme inactivation of NA — 83
  - 5.6.6 Drugs which affect postsynaptic receptors — 84
  - 5.6.7 False adrenergic transmitters — 86
- 5.7 General discussion and models of adrenergic transmission — 88
  - 5.7.1 Transmitter mechanisms at adrenergic neuromuscular junctions — 88
  - 5.7.2 Maintenance of adrenergic transmission — 90
  - 5.7.3 Adrenergic responses and neuromuscular geometry — 96
  - 5.7.4 Dual role of adrenergic nerves — 100
  - 5.7.5 Models of neuromuscular junctions — 101
- 5.8 Summary — 104

6	**Growth and Degeneration of Adrenergic Neurons**		107
	6.1 Growth of adrenergic neurons and related cells in tissue culture		107
		6.1.1 Adrenergic neurons	107
		6.1.2 Interactions between adrenergic neurons and effector tissues	108
		6.1.3 Chromaffin cells	109
	6.2 Ontogenesis of the sympatho-adrenal system		109
		6.2.1 Development of adrenergic neurons	110
		6.2.2 Development of adrenergic transmission	112
		6.2.3 Factors which control the maturation of adrenergic neurons	114
		6.2.4 Development of chromaffin cells	117
	6.3 Growth and regeneration of adrenergic neurons in the adult		119
	6.4 Degeneration of adrenergic neurons		121
		6.4.1 Surgical sympathectomy	121
		6.4.2 Drug-induced long term structural changes in adrenergic neurons and chemical sympathectomy	122
		6.4.3 Immunosympathectomy	124
		6.4.4 Secondary effects accompanying sympathectomy	125
	6.5 Summary		127
7	**General Discussion and Conclusions about Adrenergic Neurons**		129
	7.1 Heterogeneity of peripheral adrenergic neurons		129
	7.2 Modulating factors in the heterogeneity of adrenergic neurons		130
	7.3 Functional requirements of different adrenergic neurons		131
	7.4 Comparison of neurons with different functions		132
	7.5 Speculations on the functional organization of the sympatho-adrenal system		133
	7.6 Summary		136
	References		137
	Index		219

ACKNOWLEDGEMENTS

The authors are greatly indebted to their colleagues C. Bell, G. Campbell, B. Dumsday, J. B. Furness, Marian McCulloch, R. Purves, D. H. Jenkinson, G. Gabella and Betty Twarog for criticism. Special thanks go to Jay McKenzie, Lynne Chapman, Daniel Costa, Kerrie Dawson and Susan Livermore for their invaluable help in the preparation of the manuscript and figures, to Gail Liddell, Bron Robinson and Marian Rubio for typing the manuscript, and to G. Mucznik and B. Pump for their expert assistance.

Most of the work on this book was carried out in the Department of Zoology, Melbourne University, and was supported by Grants from the Australian Research Grants Committee, the National Health and Medical Research Council and the National Heart Foundation of Australia.

ABBREVIATIONS

CA	=	Catecholamine(s)
A	=	Adrenaline
NA	=	Noradrenaline
DA	=	Dopamine
DOPEG	=	3,4-dihydroxyphenylglycol
NMN	=	Normetanephrine
5-HT	=	5-hydroxytryptamine
5-OHDA	=	5-hydroxydopamine
6-OHDA	=	6-hydroxydopamine
ACh	=	Acetylcholine
AChE	=	Acetylcholinesterase
ATP	=	Adenosinetriphosphate
PGS	=	Prostaglandins
PGE	=	Prostaglandin E
PGF	=	Prostaglandin F
T-OH	=	Tyrosine hydroxylase
DβH	=	Dopamine-β-hydroxylase
PNMT	=	Phenylethanolamine-N-methyl transferase
COMT	=	Catechol-*o*-methyl transferase
MAO	=	Monoamine oxidase
EM	=	Electron microscopy
EJP	=	Excitatory junction potential
IJP	=	Inhibitory junction potential
SEJP	=	Spontaneous excitatory junction potential
EPP	=	End plate potential
EPSP	=	Excitatory postsynaptic potential
IPSP	=	Inhibitory postsynaptic potential
S.C.G.	=	Superior cervical ganglion
CNS	=	Central nervous system
NGF	=	Nerve growth factor
SMPS	=	Sulphomucopolysaccharides
PBA	=	Phenoxybenzamine
DMI	=	Desmethylimipramine
DCI	=	Dichloroisoproterenal
ACTH	=	Corticotrophic hormones
DMPP	=	Dimethyl phenyl piperazinium

1 INTRODUCTION AND HISTORY

The object of this monograph is to present an overall multi-disciplinary picture of the mammalian peripheral adrenergic neuron and of adrenergic neuroeffector transmission. Readers are referred to reviews by leading workers for further discussion on adrenergic physiology, biochemistry, pharmacology and morphology.* In addition, some topics have been included which have not been reviewed recently, such as the organization and development of adrenergic nerves. A discussion of the heterogeneity of peripheral adrenergic neurons, their functional requirements and the factors which affect them is included in the last section and some speculations made concerning the functional organization of the sympatho-adrenal system.

The term 'adrenergic nerve' was proposed by Dale in 1933 'to assist clear thinking, without committing us to precise chemical identification, which may be long in coming'. The similarities between the effects of sympathetic nerve stimulation and those of directly applied adrenaline (A) (Lewandowsky, 1900; Langley, 1901) led to the assumption (Elliot, 1904) that the transmitter liberated at sympathetic endings was A (A was discovered following its isolation from adrenal glands, Oliver & Schafer, 1895; Aldrich, 1901; Takamine, 1901). Loewi (1921) and Cannon & Uridil (1921) provided the first experimental confirmation of this hypothesis. Barger & Dale (1910) showed that the action of noradrenaline (NA) corresponded more closely with that of sympathetic nerves than that of A, and it was suggested that it might be a sympathetic transmitter (Bacq, 1933; Stehle & Ellsworth, 1937; Melville, 1937; Greer, Pinkston, Baxter and Brannon, 1938; Tiffeneau, 1939). However, it was not until NA was shown to be the predominant catecholamine (CA) in mammalian adrenergic nerves in the late 1940s by Von Euler (von Euler, 1946, 1956), that NA rather than A was recognized as the transmitter released from adrenergic nerves. It is

* Banks & Mayor, 1972; Bennett, 1973a; Bevan & Su, 1973; Bloom, 1972; Burnstock, 1970; Burnstock & Iwayama, 1971; Campbell, 1970; Coupland, 1972; Dahlström, 1971; De Potter, Chubb & De Schaepdryver, 1972; Folkow & Neil, 1971; Furchgott, 1972; Furness & Costa, 1974; Geffen & Livett, 1971; Haefely, 1972; Holman, 1970; Iversen, 1973; Jacobowitz, 1970; Jenkinson, 1973; Kosterlitz & Lee, 1972; Ljung, 1970; Molinoff & Axelrod, 1971; Shore, 1971; Skok, 1973; Smith, 1972b; Smith & Winkler, 1972; Thoenen, 1972; Thoenen & Tranzer, 1973; Trendelenburg, 1972; Triggle, 1972; Verity, 1971.

interesting that Loewi's (1921) original proposition that A was the sympathetic neurotransmitter substance released in frog heart was correct, since the neurotransmitter in adrenergic nerves in most anuran amphibians (unlike that in most other vertebrates) was found to be A rather than NA (see Burnstock, 1969). Catecholamines occur in the tissues of many species of invertebrates and vertebrates and even some plants. Tabulated summaries of the distribution of catecholamines in different tissues and species can be found in several reviews (see for example Iversen, 1967; Burnstock, 1969; Holzbauer & Sharman, 1972).

In 1939, both Blaschko and Holtz proposed a hypothetical series of reactions for the biosynthesis of A from tyrosine, which have been largely confirmed by later workers (see reviews: Axelrod, 1966; Udenfriend, 1966a, 1968; Geffen & Livett, 1971).

Although the general organization and distribution of the sympathetic nervous system was described by Gaskell (1916) and Langley (1921), little was known of the relationship of sympathetic nerves to smooth muscle. In 1946, Hillarp reviewed the theories of the innervation of autonomic effectors and introduced the concept of the 'autonomic ground plexus', which is 'a plexus of axon ramifications embedded in a fine-meshed network of anastomosing strands formed by the terminal Schwann plasmodium and directly superimposed on and probably contacting all effector cells' (Hillarp, 1946).

In the late 1940's it was proposed that the different responses of effector tissues to catecholamines were due to interaction with two different receptor sites, termed α and β (Ahlquist, 1948) rather than to the formation of two different substances, excitatory (E) and inhibitory (I) simpathin, as suggested by Cannon and Rosenblueth (1933, 1937).

Significant advances in our knowledge of the adrenergic nerves have been made following these early studies, largely due to the development of new techniques. Some examples of these developments are: isolation of the storage particles of noradrenaline from adrenergic nerves (see von Euler, 1956; Geffen & Livett, 1971); electronmicroscopic, histochemical and autoradiographic localization of NA and associated enzymes (Taxi & Droz, 1969; Burnstock, 1970); quantitative assay of CA by biochemical and fluorometric methods and the use of radioactive labelled compounds to study the biosynthesis and inactivation of CA released from adrenergic nerves (see Iversen, 1967); techniques for the determination of the enzymes involved in NA metabolism (see Axelrod, 1972); the use of drugs which interfere with adrenergic mechanisms (see Carlsson, 1966a, b; von

Euler, 1972); immunoflorescence methods to localize macroproteins associated with NA transport, synthesis, storage and release (Livett, Geffen & Rush, 1971; Fuxe, Goldstein & Hökfelt 1971; Goldstein, Fuxe and Hökfelt, 1972; Hartman & Udenfriend, 1972; Hartman, 1973; Hökfelt, Fuxe, Goldstein & Joh, 1973); electronmicroscopic studies of the structure of the adrenergic neuron and its relation to effector tissues (see Hökfelt, 1971; Burnstock & Iwayama, 1971); the histological localization of adrenergic nerves by florescence microscopy, a technique also widely used to study the properties of the adrenergic neuron (Falck, 1962; Hamberger & Norberg, 1963; Malmfors, 1965; see Fuxe & Jonsson, 1973); analysis of adrenergic transmission with electrophysiological techniques (see Bennett & Burnstock, 1968; Holman, 1970; Bennett, 1973a, b); techniques for destroying adrenergic nerves by immunosympathectomy (Levi-Montalcini & Angeletti, 1966) and by chemical sympathectomy with 6-hydroxydopamine (6-OHDA) (Thoenen & Tranzer, 1968) or guanethidine (Burnstock, Evans, Gannon, Heath & James, 1971); the use of sympathetic nerves grown in tissue culture (see Chamley, Mark, Campbell & Burnstock, 1972; Chamley, Mark & Burnstock, 1972) and of anterior eye chamber transplants (Olson & Malmfors, 1970), which have opened new possibilities for studying the properties of adrenergic neurons and their relationship with effector organs.

A great stimulus to the development of adrenergic pharmacology (see Goodman & Gilman, 1971) came from the proposal of Burn and Rand (Burn & Rand, 1959, 1965) that a cholinergic link is involved in adrenergic transmission. The relation of adrenergic nerves to other components in the autonomic nervous system, namely cholinergic and purinergic nerves is discussed briefly in Chapter 2. The reader is referred to a more detailed review of purinergic nerves (Burnstock, 1972) and various aspects of cholinergic autonomic nerves (Bell & Burnstock, 1971; Burnstock & Bell, 1974). The relation of adrenergic nerves to metabolism and to hormones has not been treated in this book (Axelrod, 1972; Himms-Hagen, 1972; Wurtman, 1973). A systematic account of the role of adrenergic nerves in different effector systems has not been included, although this would be a valuable area for future reviews.

2 GENERAL ORGANIZATION AND FUNCTIONS OF ADRENERGIC NERVES

2.1 Distribution of adrenergic neurons

The term 'sympathetic' was originally used for all visceral nerves, but Gaskell (1886) showed that there were gaps in the autonomic outflow in the cranial and sacral regions, and in 1905, Langley demonstrated that the cranial and sacral outflows were pharmacologically different from the thoracolumbar (sympathetic) system. In his classical book, Langley (1921) considered the autonomic nervous system to be composed of two main subsystems, the sympathetic and the parasympathetic, and he regarded the enteric system to be separate from these. Both the sympathetic and parasympathetic efferent pathways are composed of a chain of at least two neurons, a preganglionic neuron which is located in the neuraxis and a postganglionic neuron which is located in a peripheral ganglion. The preganglionic neurons of both parasympathetic and sympathetic systems are cholinergic (see Dale, 1937). The postganglionic parasympathetic neurons are mainly cholinergic, while the postganglionic sympathetic neurons are mainly adrenergic (see Bacq, 1934; von Euler, 1956; Campbell, 1970). In the enteric system, cell bodies of cholinergic neurons, sensory neurons and non-adrenergic, non-cholinergic inhibitory (purinergic) neurons appear to be present, but with few exceptions, no cell bodies of adrenergic neurons (see Burnstock, 1969, 1972; Burnstock & Costa, 1973; Furness & Costa, 1974).

Although some sympathetic postganglionic fibres are cholinergic (see Burnstock, 1969; Campbell, 1970), all peripheral adrenergic neurons are sympathetic postganglionic. Therefore to discuss the function of the adrenergic nerves is equivalent to discussing the function of the sympathetic system.

Most postganglionic (adrenergic) sympathetic neurons are located in the paravertebral and prevertebral ganglia (see Pick, 1970). This has been confirmed by using the fluorescence histochemical method for biogenic amines (Hamberger & Norberg, 1963; Eränkö & Härkonen, 1963; Jacobowitz, 1970). However, this technique has also led to the confirmation or discovery of adrenergic cell bodies in more peripheral locations such as the bladder (Hamberger & Norberg, 1965) in the proximal colon of the guinea-pig (Furness & Costa, 1971a) and in the pelvic ganglia (see Falck, Owman & Sjöstrand, 1965; Norberg, 1967; Kanerva, 1972; Costa & Furness, 1973). There is no evidence for the existence of peripheral dopaminergic neurons in mammals with the possible exception of those extra-adrenal chromaffin cells in sympathetic ganglia of some species which contain dopamine (DA) and may have interneuronal functions (see Figure 2 and Section 3.2.2). Nevertheless, detectable amounts of the intermediates dopa and DA have been found in peripheral adrenergic neurons (Snider, Almgren & Carlsson, 1973). 5-Hydroxytryptamine (5-HT) has been found in the adrenergic neurons to the pineal gland together with NA but its presence is due to uptake of extraneuronal 5-HT which is synthesized and stored in the pinealocytes (Owman, 1964, 1965; Falck, Owman & Rosengren, 1966).

2.2 Input of adrenergic neurons

Adrenergic neurons receive excitatory cholinergic input from the preganglionic sympathetic nerves arising in the central nervous system; some also receive excitatory cholinergic input from neurons with their cell bodies in the viscera, while some may originate from neurons within the sympathetic ganglion itself. Inhibitory input is probably also present resulting from intraneuronal release of catecholamine (NA or DA). Figures 1 and 2 show schemas of the possible connections of adrenergic neurons.

2.2.1 *Cholinergic input from the CNS*

The central input arises from cholinergic neurons located in the intermediolateral nuclei of the thoracolumbar spinal cord (Pick, 1970; Petras & Cummings, 1972), the classical preganglionic neurons of the sympathetic outflow. The axons of these neurons, which in most animals are predominantly myelinated and emerge from the spinal cord via the anterior roots, form the white rami communicantes and contribute to the splanchnic nerves. Three types of preganglionic fibres, with different speeds of conduction appear to be present (see Dunant, 1967; Perri, Sacchi & Casella, 1970a) although four or a wider range of fibres were described by earlier workers (see Eccles, 1935). Preganglionic fibres usually lose their myelin sheaths on entering the sympathetic ganglia and end mainly on dendrites, establishing several synapses 'en passage' with each adrenergic neuron (Gibson, 1940; Botar, 1966; Grillo, 1966; Elfvin, 1963, 1971c; Hamori, Lang & Simon, 1968; Figure 2). The preganglionic fibres form synapses with more than one adrenergic neuron, so that there is divergence in the input to sympathetic ganglia (de Castro, 1932; Bishop & Heinbecker, 1932; Gibson, 1940; Botar, 1966). Thus, the number of fibres emerging from a sympathetic ganglion is greater than the number of fibres entering,

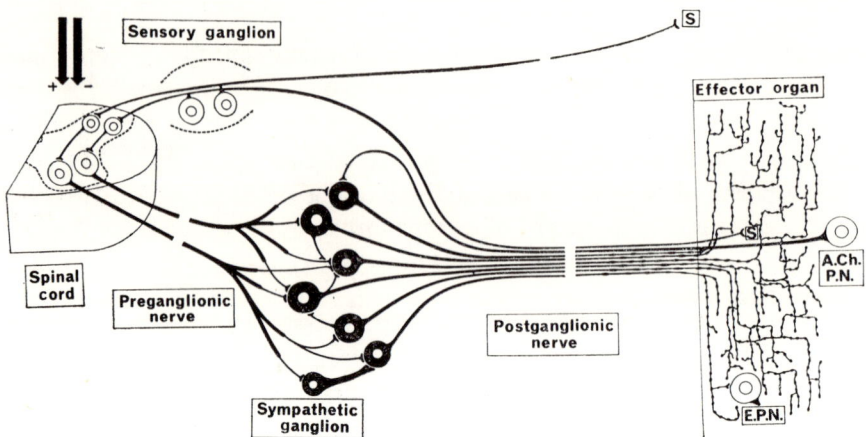

Figure 1. Schematic drawing showing the main features of the adrenergic innervation of an effector organ and summarizing the possible connections of adrenergic neurons (in black). The figure shows the neurons involved in a sympathetic reflex: the sensory information (S) from the innervated organ or from other sources reaches the spinal cord through sensory neurons located in the spinal dorsal root sensory ganglia. In the spinal cord, the afferent stimulus passes through at least one interneuron before stimulating the preganglionic sympathetic neuron. The preganglionic fibres form synapses with postganglionic sympathetic neurons with both convergence and divergence of signals. An inhibitory interneuron, which is believed to be a chromaffin cell, is also shown as well as the possible interconnections between adrenergic neurons. In the same ganglia, adrenergic neurons also receive synapses from peripheral cholinergic neurons (AChPN.). The axons of adrenergic neurons form the postganglionic sympathetic nerves. They become varicose shortly before entering the organs, branch in the effector organs forming the adrenergic 'ground plexus' and innervate different tissues including other peripheral excitatory neurons (E.P.N.). The two arrows on the spinal cord represent inhibitory and excitatory supraspinal influences on the spinal sympathetic reflex.

as was observed originally by Meckel (1751). The ratio of pre- to postganglionic fibres varies from 1:11 in cat ganglia to 1:196 in human superior cervical ganglia (see Ebbesson, 1968). Convergence of several preganglionic fibres on the same adrenergic neuron also occurs (Sala, 1893; Eccles, 1935; Bronk, 1939; Eccles, 1955; Erulkar & Woodward, 1968; Blackman & Purves, 1969; Perri, Sacchi & Casella, 1970b; Crowcroft & Szurszewski, 1971; Skok, 1973).

In the guinea-pig pelvic ganglia, adrenergic neurons do not have long dendrites. The preganglionic fibres end in invaginations of the cell bodies and receive inputs from only one or a few preganglionic fibres (Blackman, Crowcroft, Devine & Holman, 1969; Crowcroft & Szurszewski, 1971;

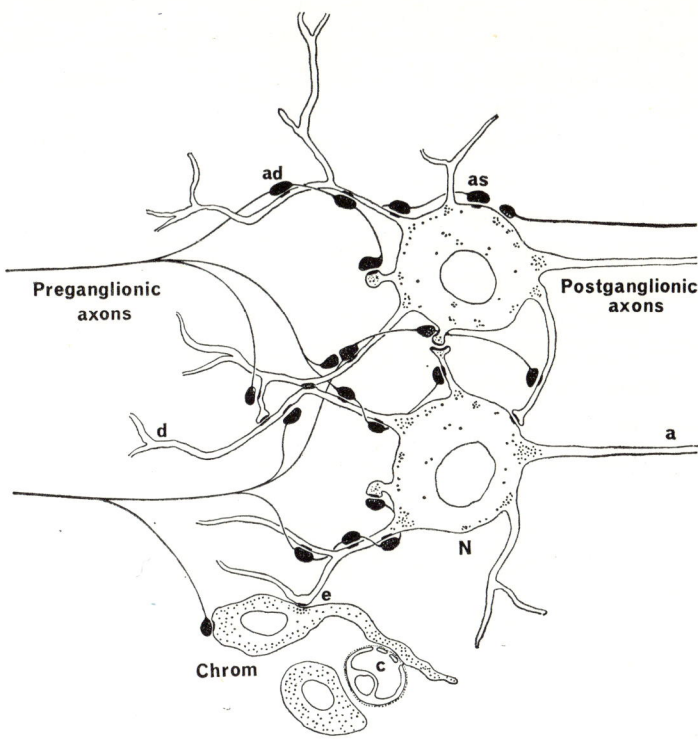

Figure 2. Schematic drawing of the intraganglionic connections of adrenergic neurons (slightly modified from Elfvin, 1971c and Tamarind and Quilliam, 1971). The preganglionic cholinergic axons branch and end on several neurons (*divergence*) by establishing several axodendritic (ad) synapses 'en passage' on the dendrites (d) and axosomatic synapses (as) on the cell body.

Each neuron has synapses from more than one preganglionic neuron (*convergence*). Dendritic processes of one cell form specialized contacts with processes of other nerve cells.

A nerve fibre of peripheral origin is shown to form synapses with a postganglionic neuron. Extra-adrenal chromaffin cells (Chrom) with possible interneuronal functions are also shown. The elongated chromaffin cell receives a preganglionic fibre and establishes an efferent synapse (e) with a dendrite of a neuron (N). The chromaffin cells are also in close proximity to fenestrated capillaries (c).

The black spots in the cytoplasm of the neurons and of the chromaffin cells represent the amine storing vesicles.

Watanabe, 1971; Costa & Furness, 1973). These ganglia probably have low integrative properties (Haefely, 1972).

Preganglionic sympathetic neurons receive excitatory and inhibitory impulses from higher centres and excitatory and inhibitory impulses from peripheral afferents, i.e. from chemoreceptors, baroreceptors, somatic and other visceral receptors (see Polosa, 1967, 1968; Folkow & Neil, 1971). Most of the preganglionic fibres (70 to 80 per cent) are silent during resting conditions (see also Section 2.3). Single fibres which are spontaneously active, discharge at up to 10 Hz and do not appear to show any definite rhythm either synchronized with heart or with respiratory activity. However, when the activities of single fibres are superimposed, definite rhythms become apparent which are comparable to those recorded in whole preganglionic nerves (Polosa, 1967, 1968; Mannard & Polosa, 1973). Spontaneous activity in preganglionic fibres represents the tonic activity of centres such as the supraspinal vasomotor centre and is little affected by peripheral spinal inputs. Silent preganglionic neurons can be activated by peripheral stimuli and they then discharge at frequencies which vary from 1 to 100 Hz. Tonic activity and reflex activity seems to be mediated by different preganglionic sympathetic neurons (Bronk, 1939; Iggo & Vogt, 1960; Koizumi & Suda, 1963; Beacham & Pearl, 1964; Widdicombe, 1966; Polosa, 1967, 1968; Coote, Downman & Weber, 1969; Cohen & Gootman, 1970; Janig & Schmidt, 1970; Folkow & Neil, 1971).

2.2.2 *Cholinergic input from peripheral neurons*

Less is known of input of peripheral origin. Afferent fibres synapsing with the neurons in sympathetic ganglia have been postulated to explain the persistence of reflex activity after interruption of the central connections (Schwartz, 1934; see Kuntz, 1953). In some cases the cell bodies of afferent fibres may be located in sympathetic ganglia (Schwartz, 1934). In other cases the cell bodies appear to be located in the wall of visceral organs (Kuntz, 1953; Úngvary & Leránth, 1970; Crowcroft, Holman & Szurszewski, 1971; see Figure 1); cholinergic fibres from the gut are activated by intestinal distension (Crowcroft, Holman & Szurszewski, 1971) and mediate reflex inhibition of gut motility (see Furness & Costa, 1974).

2.2.3 *Intraganglionic input*

Electrophysiological and pharmacological experiments suggest that there is an inhibitory system in the sympathetic ganglia which is activated by high intensity stimulation of preganglionic fibres and is due to release of a catecholamine of intraganglionic origin (Bronk, 1939; Eccles & Libet, 1961; Trendelenburg, 1961, 1967; Norberg & Sjöqvist, 1966; Dunant, 1967; Erulkar & Woodward, 1968; Volle, 1969; Libet, 1970; Volle & Hancock, 1970; Haefely, 1972). In extracellular recordings, stimulation of the preganglionic fibres in a curarized ganglion produces a positive wave (P-wave) which follows the negative wave (N-wave). In intracellular recordings, the P-wave appears as an inhibitory postsynaptic potential (IPSP) which follows the excitatory postsynaptic potential (EPSP) (see Haefely, 1972 for a detailed discussion).

There is also some morphological evidence for the presence of inhibitory inputs of intraganglionic origins on the adrenergic neurons. A network of adrenergic terminals has been described amongst the adrenergic neurons from fluorescence histochemistry (Hamberger, Norberb & Sjöqvist, 1965; Jacobowitz, 1970). These adrenergic endings on adrenergic neurons were suggested to be collaterals or dendritic processes of other postganglionic adrenergic neurons of the same ganglion (see Taxi, Gautron & L'Hermite, 1969; and Jacobowitz, 1970; Figure 2). Nevertheless, it seems unlikely that interactions between postganglionic neurons represent the mechanism of the intraganglionic inhibitory inputs, since antidromic stimulation does not elicit inhibitory responses in the ganglia (Libet, 1964). The mechanism might involve true interneurons. It is interesting that some adrenergic synapses on the nerve cells of the prevertebral ganglia survive following section of the pre- and postganglionic nerves (Joo, Lever, Ivens, Mottram & Presley, 1971; Tamarind and Quilliam, 1971) as well as the network of adrenergic terminals amongst the adrenergic neurons (Hamberger, Norberg & Sjöqvist, 1965). Furthermore some neurons do not show signs of retrograde degeneration following pre- and postganglionic denervation (Samuel, 1953; Yoshikawa, 1970). A type of chromaffin cell has been reported in the superior cervical ganglion of the rat (Siegrist, Dolivo, Dunant, Foroglou-Kerameus, Ribeupierre & Rouiller, 1968; Matthews & Raisman, 1969; Williams & Palay, 1969; see also Section 3.2). These cells have been reported to contain DA in several species (Björklund, Segrell, Falck, Ritzen & Rosengren, 1970; Fuxe, Goldstein & Hökfelt, 1971) and it has been suggested that DA is the amine involved in intraganglionic inhibition (Libet & Tosaka, 1970; Kebabian & Greengard, 1971). Preganglionic stimulation may stimulate the dopamine-containing cells, which in turn may release dopamine to activate a dopamine-sensitive adenyl cyclase located on the membrane of the adrenergic neuron producing the IPSP (Kebabian and Greengard, 1971; McAfee & Greengard, 1972;

Greengard, Kebabian & Nathanson, 1973; Machova & Kristofova, 1973; see Figures 2 and 3). Despite the fact that in many ganglia similar cells do not establish synapses with adrenergic neurons (Elfvin, 1968; Van Orden, Burke, Geyer & Lodoen, 1970; Eranko & Eranko, 1971a; Mustonen & Teravainen, 1971; Watanabe, 1971; Kanerva, 1972), they may still affect

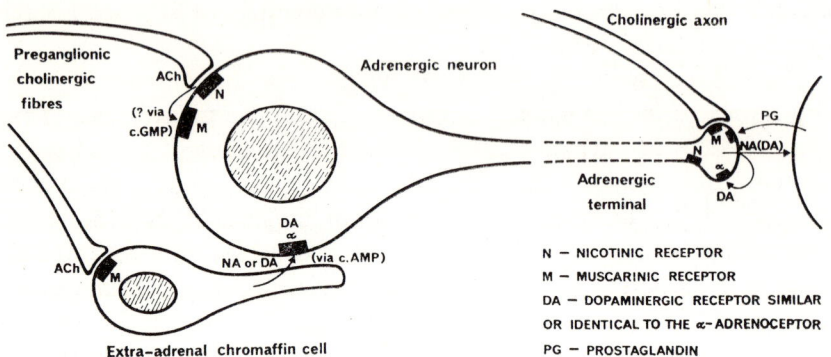

Figure 3. The schema shows the hypothetical types of receptors involved in adrenergic transmission and located on adrenergic cell bodies, adrenergic terminal axons and extra-adrenal chromaffin cells in sympathetic ganglia.

Preganglionic cholinergic fibres excite adrenergic neurons through nicotinic receptors (N). Acetylcholine diffuses away from the nicotinic synapse and activates muscarinic receptors (M), their physiological role being unclear. Adrenergic neurons may receive inhibitory inputs of intraganglionic origins from extra-adrenal chromaffin cells. These cells appear to receive an excitatory preganglionic input mediated through muscarinic receptors (M) and inhibit adrenergic neurons through α-adrenoceptors which are highly sensitive to dopamine (DA).

Both nicotinic and muscarinic receptors are present on adrenergic terminal axons, but acetylcholine released from neighbouring cholinergic fibres acts mostly through muscarinic receptors to inhibit noradrenaline release. Noradrenaline and dopamine released from adrenergic axons act on α-adrenoceptors located on the adrenergic terminal, inhibiting further release of transmitter. Prostaglandin (PG) released from the effector tissue may affect transmitter release from adrenergic terminals.

adrenergic neurons by releasing their amine content into the local circulation. The presence of extra-adrenal chromaffin cells in sympathetic ganglia has been known from the last century (see Coupland, 1965), and the hypothesis mentioned above does not differ substantially from that proposed by Eccles and Libet (1961) (see Figures 2 and 3).

It is not known whether 'adrenergic' inhibition represents a physiological mechanism in sympathetic ganglia. In the superior cervical ganglion of the cat, no evidence for an adrenergic inhibitory modulation of the transmission was found (Haefely, 1970). It is interesting in this respect that few adrenergic synapses (Tamarind & Quilliam, 1971) and few adrenergic terminal varicosities (Jacobowitz, 1970) were found in the cat superior cervical ganglion.

2.3 Output of adrenergic neurons

Adrenergic neurons have morphological and functional properties which enable them to integrate the inputs and give a modified output, as suggested originally by Cannon (1914).

With the exception of some neurons in the pelvic ganglia (Crowcroft & Szurszewski, 1971), a single pulse to a preganglionic fibre produces only a localized depolarization of the membrane of the neuron, i.e. the excitatory postsynaptic potential (EPSP) (Haefely, 1972), and cannot elicit discharge of the adrenergic neuron. Only the repetitive firing of one preganglionic fibre or the simultaneous firing of several excitatory fibres converging on the same neuron are able to elicit an action potential in the adrenergic neuron (Eccles, 1935; Eccles, 1955; Dunant, 1967; Erulkar & Woodward, 1968; Blackman & Purves, 1969; Libet & Tosaka, 1969; Perri, Sacchi & Casella, 1970a; Mirgorodsky & Skok, 1973; Skok, 1973). The frequency of discharge of an adrenergic neuron therefore depends on the temporal and spatial pattern of the inputs. Many adrenergic neurons are silent (Chalazonitis & Gonella, 1971; Skok, 1973), suggesting that they receive inputs from those preganglionic fibres which are silent (see above). The adrenergic neurons which are silent probably discharge only when a number of excitatory fibres are activated by sensory stimulation of sufficient intensity. A wide-spread and intense sensory stimulation, able to activate all the silent preganglionic fibres at their maximal frequency, would similarly activate the postganglionic (adrenergic) fibres (Green & Heffron, 1968).

Other neurons have a spontaneous activity with frequencies of discharge and rhythms similar to those of the preganglionic fibres (Knoefel & Davis, 1933; Pitts, Larrabee & Bronk, 1941; Koizumi & Suda, 1963; Bower, 1966; Green & Heffron, 1968; Vasalle, Levine & Stuckey, 1968; Chalazonitis & Gonella, 1971), indicating that the preganglionic fibres send impulses grouped in volleys which are transmitted by the adrenergic neuron without marked modification (Bishop & Heinbecker, 1932; Bronk, 1939; Skok, 1973). The postganglionic discharge remains synchronized with the preganglionic stimulation up to 20 to 30 Hz and becomes desynchronized at higher frequencies (Bishop & Heinbecker, 1932; Bronk, 1939; Eccles,

1944). These adrenergic neurons act as a filter by transmitting only those impulses which arrive arranged in a certain pattern (Volle & Hancock, 1970; Skok, 1973), such as the regular impuses from the tonic neurons in the central nervous system (Folkow & Neil, 1971). Little is known about the actual pattern of central and peripheral excitation of the adrenergic neuron in the living organism (see Skok, 1973). Plate 1 shows intracellular recording of tonically active neurons of the rabbit superior cervical ganglion (Skok, 1973).

2.4 General functions of adrenergic neurons

Adrenergic neurons innervate various tissues including smooth muscle, cardiac muscle, glands, adipose tissue, other peripheral neurons, central nervous structures and sensory receptors. The sympatho-adrenal system contributes to the internal stability of the organism, i.e. it has a homeostatic function (Cannon & Rosenblueth, 1937). Adrenergic nerves, as part of this system, contribute to homeostasis, but they also have specific local functions in particular organs or tissues.

An example of adrenergic neurons involved in the homeostasis is in the cardiovascular system. These neurons are driven by supraspinal centres and by increasing and decreasing the tonic input, the blood circulation is redistributed according to the needs of the organism. For instance during exercise, the blood supply to skeletal muscles increases by vasodilatation due to reduction of adrenergic vasoconstrictor tone and by increase of the cardiac output due to the activation of cardiostimulatory adrenergic nerves, while the blood in the viscera is diverted to other areas by increase of visceral vasoconstrictor adrenergic tone (Folkow & Neil, 1971). At the same time, the cerebral and coronary flow is maintained to guarantee an adequate blood supply to the brain and the heart. In species with brown adipose tissue, tonic discharge of adrenergic nerves innervating adipose cells and blood vessels increases during cold exposure and produces a thermogenic effect, contributing to homeostasis (Himms-Hagen, 1972).

An example of adrenergic nerves with both a general homeostatic function and a protective local function are those which supply the gastrointestinal tract. Adrenergic vasoconstrictor nerves to the gut reduce the proportion of the cardiac output entering the intestinal circulation when a redistribution of blood to other areas is required (see Folkow & Neil, 1971; Furness & Burnstock, 1975). It has been suggested that as a consequence of a selective ischaemia of the gut, the adrenergic nerves which supply the gut musculature and enteric nerve plexuses are stimulated and inhibit the activity of the enteric neurons, with consequent inhibition of peristalsis, and that in the absence of this adrenergic nerve mechanism, damage to enteric neurons would occur (Furness & Costa, 1974). Other adrenergic nerves with a protective function in certain critical situations appear to be those which supply the kidneys. Their activation produces a release of renin with consequent reduction of loss of water and sodium in situations of altered blood volume, extracellular fluid volume or intrathoracic pressure (Folkow & Neil, 1971; Walsner, 1971; Ganong, 1972), although it seems likely that adrenergic activity is less significant in this regard than a fall in arterial pulse pressure.

Adrenergic neurons, which do not seem to be involved in a general homeostatic function nor in a local protective reaction, include those which provide the excitatory innervation to the reproductive organs in males. They control the function of the vas deferens and seminal vesicles in emission and thus appear to play only a local role representing the final link in a complex behavioural pattern.

There is morphological and physiological evidence for adrenergic innervation of various sensory receptors, supporting the original suggestion of Claude Bernard (1851) that sympathetic nerves may modulate the sensory input. Adrenergic fibres innervate some mechanoreceptors (Loewenstein & Altamirano-Orrego, 1956; Fuxe & Nilsson, 1965; Santini, 1969; Andres, 1971; Nilsson, 1972), muscle spindles (Hines & Tower, 1928; Barker, 1948; Eldred, Schnitzlein & Buchwald, 1960; Hunt, 1960; Paintal, 1964), pain receptors in the tooth pulp (Edwall & Scott, 1971), taste receptors (Gabella, 1969), acoustic receptors (Spoendlin & Liechtensteiger, 1966; Wenzel, 1972), baroreceptors (Aars, 1971; Belmonte, Simon, Gallego & Baron, 1972), chemoreceptors (Duncan & Yates, 1967; Mills, 1968; Biscoe, 1971; Sampson, 1972), olfactory receptors (Alcocer, Aréchiga, Aguilar & Guevara, 1972), and visual receptors (Mascetti, 1972).

There is also some evidence that peripheral adrenergic neurons innervate some central nervous structures. Adrenergic fibres supply the pineal gland (Owman, 1964, 1965; Wolfe, 1965), the pituitary nervous lobe (Björklund, 1968), and the medial habenular nucleus and adjacent structures (Herbert, 1971; Björklund, Owman & West, 1972). It has been suggested that some peripheral adrenergic fibres penetrate the cerebral cortex (Falck, Owman & Rosengren, 1966; Owman, Edvinsson & Nielsen, 1974). Thus, adrenergic nerves are the final component in neuronal pathways of different complexity and function.

Adrenal chromaffin tissue also represents a final link in sympathetic function (see Section 2.2). For example, gastrointestinal function is influenced by CA released into the circulation from medullary and extramedullary chromaffin cells as well as NA released locally from sympathetic nerves. Conditions which lead to increased discharge of CA into the circulation also activate sympathetic nerves. CA has been shown to influence motility, blood flow, secretion, absorption and the activity of enteric neurons and these functions are interdependent. A detailed analysis of the sites and interdependence of CA action on the gut can be found in a recent review by Furness and Burnstock (1975).

2.5 Relation between adrenergic and non-adrenergic neurons

2.5.1 *Adrenergic and cholinergic neurons*

A review of the antagonistic or synergistic interactions between the postganglionic sympathetic and parasympathetic neurons at the level of the effector organs is beyond the scope of the present review. A brief account will be presented on the possible direct interactions between adrenergic and cholinergic neurons.

Cholinergic effects on adrenergic neurons. Adrenergic neurons are under excitatory control by preganglionic cholinergic sympathetic neurons and in some ganglia by peripheral cholinergic fibres (see Section 2.2). The excitatory effect is mediated by receptors on the cell body and on dendrites of adrenergic neurons which are predominantly nicotinic, although muscarinic receptors are also present, which can be activated by ACh diffusing from the synaptic junction and which mediate inhibition (see Trendelenburg, 1967; Haefely, 1972; Skok, 1973; Figure 3).

The possibility that ACh is involved in adrenergic neuroeffector transmission was postulated by Burn and Rand in 1959 to explain several pharmacological inconsistencies in the traditional view. Following their theory, ACh is present together with NA in adrenergic axons, and forms an intermediate link between nerve impulses and release of NA from the nerve terminals (see Burn & Rand, 1965; for opposing view see Ferry, 1966). This original proposition has not been supported by most recent workers (see Jaju, 1969; Campbell, 1970). Adrenergic fibres contain neither measurable ACh nor the synthetic enzyme choline-acetyltransferase (Nordenfeldt, 1965; Ehinger, Falck & Sporrong, 1970; Ehinger, Falck, Persson, Rosengren & Sporrong, 1971; Consolo, Garattini, Ladinski & Thoenen, 1972). While the enzyme acetylcholinesterase has been found in association with adrenergic axons (Imagawa, 1969; Eränkö, Rechardt, Eränkö &

Cunningham, 1970; Matsuda, 1970; Waterson, Hume & de la Lande, 1970; Eränkö & Eränkö, 1971d), this is not strong evidence for the cholinergic link theory, since this enzyme is widely distributed in non-cholinergic nerve tissues. The presence of acetylcholinesterase can be a misleading guide for the identification of cholinergic axons in the autonomic nervous system unless levels of the enzyme are taken into consideration (Koelle, 1963; Furness & Iwayama, 1972; Robinson, 1971).

In 1968, Burn (1968b) put forward a modification of the theory which, in accordance with the hypothesis of Leaders (1963), suggested that the source of ACh is a cholinergic axon in close association with the adrenergic terminal. Cholinergic and adrenergic fibres have been shown to run in the same nerve bundle (see Campbell, 1970; Furness & Iwayama, 1972; Kosterlitz & Lees, 1972), and there is convincing evidence that cholinergic and adrenergic fibres can be closely associated in effector tissues often within the same Schwann cell sheath (Richardson, 1964; Burnstock & Robinson, 1967; Robinson & Bell, 1967; Hökfelt, 1967; Tranzer & Thoenen, 1967c; Graham, Lever & Spriggs, 1968; Hökfelt & Jonsson, 1968; Ochi, Konishi, Yoshikawa & Sano, 1968; Ehinger, Falck & Sporrong, 1970; Iwayama, Furness & Burnstock 1970; Nelson & Rennels, 1970; Nielsen, Owman & Sporrong, 1971; Edvinsson, Nielsson, Owman & Sporrong, 1972; Furness & Iwayama, 1972; Hervönen & Kanerva, 1972b). Both nicotinic and muscarinic cholinoreceptors are present in the adrenergic axon terminals (Figure 3). High concentrations of cholinomimetic drugs activate nicotinic receptors which mediate depolarization of the adrenergic terminals with consequent release of NA; low concentrations of cholinomimetic drugs activate muscarinic receptors which appear to mediate hyperpolarization with inhibition of NA release (Haeusler, Thoenen, Haefely & Hürlimann, 1968; Malik & Ling, 1969; Muscholl, 1970; Rand & Varma, 1970; Hume, de la Lande & Waterson, 1972; Kirpekar, Prat, Puig & Wakade, 1972; Kosterlitz & Lees, 1972; Muscholl, Lindmar, Löffelholz & Fozard, 1973; Steinsland, Furchgott & Kirpekar, 1973; Vanhoutte & Shepherd, 1973). However, while there is no clear evidence for an excitatory effect of cholinergic fibres on release of NA from adrenergic fibres (see Campbell, 1970), there is some evidence for an inhibition of NA release by cholinergic parasympathetic fibres (Löffelholz & Muscholl, 1970; Figure 3). Thus, cholinergic fibres appear to excite adrenergic neurons when ACh released from preganglionic fibres reaches a sufficiently high concentration necessary to activate excitatory nicotinic receptors, as is the case in the cell body and dendrites; cholinergic fibres inhibit adrenergic neurons when ACh reaches adrenergic terminals in low concentrations by diffusion and activates inhibitory muscarinic receptors (Figure 3). The physiological significance of muscarinic inhibition of adrenergic nerve terminals is not known (Kosterlitz & Lees, 1972). However, this interaction does not always occur since adrenergic axons are often not associated with cholinergic axons, as for example in the longitudinal musculature of the vas deferens (Furness & Iwayama, 1972).

Adrenergic effects on cholinergic neurons. There is convincing evidence that postganglionic adrenergic sympathetic neurons exert an inhibitory effect on some cholinergic neurons in parasympathetic and enteric ganglia (see Figure 1).

The cholinergic postganglionic parasympathetic neurons located in the bladder of the cat receive adrenergic fibres (Hamberger & Norberg, 1965; Norberg & Sjöqvist, 1966) which, by acting on α-adrenoceptors, inhibit parasympathetic transmission (de Groat & Saum, 1971a, b, 1972; Saum & de Groat, 1972). The adrenergic nerves which supply the intestine end mainly around the enteric ganglion cells (see Read & Burnstock, 1969b; Costa & Gabella, 1971), and by acting through α-adrenoceptors, they reduce the ACh output from the enteric cholinergic neurons with consequent inhibition of intestinal peristalsis (Beani, Bianchi & Crema, 1969; Paton & Vizi, 1969; Kosterlitz & Lees, 1972; Furness & Costa, 1974). Adrenergic fibres have also been reported in the ciliary ganglion (Ehinger & Falck, 1970), in the intracardiac cholinergic ganglia (Jacobowitz, 1967) and in cholinergic ganglia of dog and calf lung (Jacobowitz, Kent, Fleisch & Cooper, 1973), but no experimental information is yet available about their possible function. It has been suggested that adrenergic sympathetic neurons exert an inhibitory effect on preganglionic cholinergic terminals (Christ & Nishi, 1971; Kosterlitz & Lees, 1972; see Section 2.2), and that adrenergic axons running close to cholinergic axons in terminal nerve bundles (see p. 15) may inhibit ACh release from them (Kosterlitz & Lees, 1972, and see Section 5.2).

In conclusion it appears that most direct interactions between adrenergic and cholinergic neurons are antagonistic.

2.5.2 *Adrenergic and purinergic neurons*

Non-adrenergic, non-cholinergic inhibitory nerves have been clearly demonstrated in the gastrointestinal tract in mammals (see Burnstock, 1969). Evidence has been presented that they release a purine compound, probably ATP, and have therefore been tentatively termed 'purinergic' (see Burnstock, 1972). The physiological relation between these inhibitory purinergic nerves and inhibitory adrenergic nerves is of interest. Purinergic nerves form the efferent link in a cascade of descending reflexes extending from the oesophagus to the anal sphincter. This 'descending inhibition', an essential part of the peristaltic reflex, prepares the way for material advancing through the alimentary tract by increasing gastric capacity, by opening sphincters and by dilating the intestine ahead of a bolus (Furness & Costa, 1973). In contrast adrenergic nerves exert reflex inhibition of peristalsis in situations in which the organism requires interruption of digestive function (Furness & Costa, 1974). It appears clear therefore that the functions of these nerves are different; purinergic nerves mediate the propulsive function of the gastrointestinal tract, while the adrenergic nerves are not involved in propulsive function, but rather in the prevention of it (Burnstock & Costa, 1973). There appears to be no direct interaction between these two kinds of neurons (Jansson & Martinson, 1966; Gershon, 1967, see Furness & Costa, 1974), although more recent work (Hirst & McKirdy, 1974) suggests that adrenergic nerves might antagonize descending inhibition in guinea-pig intestine by presynaptic inhibition of cholinergic terminals on purinergic neurons. In addition to purinergic neurons in the wall of the alimentary tract, nerves which are neither adrenergic nor cholinergic have been shown to supply a variety of other organs (see Burnstock, 1972). Some of these are excitatory (for example, those in the urinary bladder and intestine), while others are inhibitory (for example, those to parts of the vascular system). Whether these nerves release ATP or other neurotransmitters is not known and their relation to adrenergic nerve action has not been explored (see Burnstock, 1972; Furness & Costa, 1973).

2.6 Summary

All peripheral adrenergic neurons are sympathetic postganglionic neurones. Most of them are located in paravertebral and prevertebral sympathetic ganglia, although some are found more peripherally. All adrenergic neurons receive cholinergic excitatory inputs from the central nervous system. Some of the adrenergic neurons in the prevertebral ganglia also receive cholinergic excitatory inputs from visceral organs and in some ganglia there is evidence for inhibitory adrenergic inputs of intraganglionic origin. Adrenergic neurons have morphological and functional properties which enable them to integrate inputs and give a modified output; they are activated by inputs in specific temporal and spatial patterns. Many adrenergic neurons of the sympathetic ganglia are silent but can be reflexly activated. Others are tonically active, driven by supraspinal centres. Volleys of preganglionic impulses can be transmitted by the adrenergic neurons without marked modification. The physiological frequency of the tonically active adrenergic neuron varies from less than 1 Hz to 10 Hz, while during reflex activity higher frequencies occur.

The physiological function of adrenergic neurons, as part of the sympatho-adrenal system, is to contribute to body homeostasis. In addition to this general function, adrenergic neurons are involved in specific control of some organs.

There may be some interaction of terminal axons of adrenergic nerves with terminals of cholinergic nerves. In the bladder of some species and the intestine, adrenergic nerves form inhibitory synapses with the cell bodies of cholinergic neurons. Purinergic nerves are concerned with propulsive activity in the gut, while adrenergic nerves prevent this activity by inhibiting intramural cholinergic excitatory neurons; however, there appears to be no direct interaction between adrenergic and purinergic nerves.

3 STRUCTURE OF ADRENERGIC NEURONS AND RELATED CELLS

3.1 Adrenergic neurons

3.1.1 *General structure*

Most of the neurons in sympathetic ganglia are adrenergic. Therefore the morphology of adrenergic neurons studied by the fluorescence histochemical technique for catecholamines and by electronmicroscopy corresponds essentially to the descriptions of sympathetic ganglia based on classical histological techniques (Cajal, 1905; de Castro, 1932; Kuntz, 1953; Norberg & Hamberger, 1964; Botar, 1966; Jacobowitz, 1970; Plate 2).

Adrenergic neurons have cell bodies with diameters ranging widely from 15 to 60 μm and the majority are multipolar, with several short or long dendritic processes and one axon. Bipolar and monopolar adrenergic neurons have been reported in few situations (Botar, 1966; Pick, 1970; Costa & Furness, 1973). Axons extend to the effector organ where they branch and form a varicose terminal network which, together with the terminal varicose fibres of other neurons, form the 'autonomic ground plexus' (Hillarp, 1946, 1959; Malmfors & Sachs, 1965a; Dahlström & Häggendal, 1966a; Figures 1 and 21).

Several authors have described various neuron types on the basis of size, distribution of processes and relationship with other neurons (see De Castro, 1932; Kuntz, 1953; Botar, 1966), but no correlation between morphological types and function is known. Adrenergic cell bodies differ in degree of CA specific fluorescence, which varies from very intense to barely detectable indicating different content of CA (Härkonen, 1964; Hamberger, Norberg & Sjöqvist, 1965; Norberg, 1967; Jacobowitz, 1970; Plate 3). The degree of fluorescence intensity, which reflects the amount of CA in each cell body, appears to be related to the level of nerve activity (Costa & Eränkö, 1973 and see Section 4.1). A small percentage of neurons show an intense fluorescence and few processes; these resemble cell type V as described by De Castro (1932) and probably correspond to intensely fluorescent neurons described in tissue culture (Section 4.1). Adrenergic neurons which supply some internal reproductive organs in both sexes are located near to the effector tissues and therefore have short axons. These neurons respond differently to some drugs compared to typical adrenergic neurons with their cell bodies located in para- and prevertebral ganglia. These features have led to the suggestion that two classes of adrenergic neuron can be distinguished namely 'short' and 'long' (Owman & Sjöstrand, 1965; Sjöberg, 1967; Owman, Sjöberg, Sjostrand & Swedin, 1970; Owman, Sjöberg & Sjöstrand, 1974).

A more detailed analysis of the various components of the adrenergic neurons will now be described.

Cell body of adrenergic neurons. The ultrastructure of the cell bodies of adrenergic neurons appears to differ little between different ganglia in the same animal or between species (Elfvin, 1963, 1971a, b, c; Forssman, 1964; Norberg & Hamberger, 1964; Taxi, 1965; Grillo, 1966; Devine, 1967; Hökfelt, 1968, 1969; Geffen & Ostberg, 1969; Watanabe, 1971; Pick, 1970; Heath, Evans, Gannon, Burnstock & James, 1972; Plate 2). The cytoplasm contains a well developed endoplasmic reticulum with abundant ribosomes associated with it or free in the cytoplasm, numerous mitochondria, scattered lysosome-like particles, pigment particles, neurotubules and neurofilaments, and sometimes centrioles and cilia (Plate 2). Adrenergic vesicles have been demonstrated in the cell body of adrenergic neurons when specific fixation procedures have been applied (see Section 3.1) and make it possible to distinguish them from cell bodies of other autonomic neurons (Plates 3 and 4; Taxi, 1965; Grillo, 1966; Geffen & Livett, 1971; Hökfelt, 1969; Taxi, Gautron & L'Hermite, 1969; Van Orden, Burke, Geyer & Lodoen, 1970; Hervonen, 1971; Hökfelt & Dahlström, 1971; Tamarind & Quilliam, 1971; Elfvin, 1971a, b, c; Banks & Mayor, 1972; Eränkö, 1972a; Hökfelt & Ljungdahl, 1972; Kanerva, 1972). It has been suggested that under optimal conditions, the fluorescence histochemical method is able to detect vesicles storing NA in the cell bodies as bright spots of fluorescence mainly localized at the periphery of the cytoplasm where clusters of adrenergic vesicles have been described by electron microscopists (Eränkö, 1972a; Plate 5).

Non-terminal axons. Non-terminal axons have for the most part a uniform diameter of 0·3 to 2 μm and are enveloped by Schwann cell processes (Elfvin, 1958; Merrillees, Burnstock & Holman, 1963; Garrett, 1971). They contain neurotubules and/or neurofilaments, mitochondria, a smooth tubular system which appears swollen at intervals, and adrenergic vesicles

(Elfvin, 1961; Merrillees, Burnstock & Holman, 1963; Geffen & Ostberg, 1969; Kapeller & Mayor, 1969a, b; Pellegrino de Iraldi & De Robertis, 1970; Garrett, 1971; Thureson-Klein, Klein & Yen, 1973; Plate 6). All these organelles are also present in the dendritic processes, but it has not been possible yet to differentiate with certainty between dendritic processes and axons by light and electron microscopy in sympathetic ganglia.

The length of non-terminal axons depends on the size of the animal and on the distance of the effector tissue from the cell bodies. For instance, axons of the adrenergic neurons in the wall of the guinea-pig proximal colon are certainly shorter than those of the neurons which supply the blood vessels of the hind legs. However, the division of adrenergic neurons into 'short' and 'long' (see page 20) does not seem to be a sufficient reason *per se* for such a firm division in view of the side spectrum of axon lengths found in different situations.

Terminal axons. The length and degree of branching of the terminal, varicose part of the adrenergic axon varies with species and organs and, in some cases, axons become varicose before entering the effector organ (Olson, 1969; Burnstock, 1970; Costa & Furness, 1972). Indirect estimations, which need to be regarded with caution (Geffen & Livett, 1971), indicate that the total length of the terminal ramifications is 10 cm in the rat iris, 20 and 34 cm in the hind leg of rat and cat respectively (Dahlström, 1966). The varicosities show considerable variation in size from 0·4 to 2 μm in diameter and from 0·5 to 3 μm in length. There are about 250 to 300 varicosities per millimetre (Norberg & Hamberger, 1964; Malmfors, 1965; Dählstrom & Häggendal, 1966a; Merrillees, 1968; Burnstock, 1970) but there appears to be considerable variation in different tissues. Intervaricosity diameters range from 0·1 to 0·5 μm (Merrillees, 1968). Varicosities are packed with mitochondria and vesicles of various kinds, while intervaricosities contain predominantly neurofibrils and/or neurofilaments. The last few varicosities of a fibre usually become free of their Schwann cell investment (Merrillees, 1968; Verity & Bevan, 1968). Plate 7 shows the varicose appearance of the terminal part of the adrenergic axon.

Physiological and histochemical evidence that NA is released from the varicosities of the terminal axon has been presented (Malmfors, 1965; Bennett & Burnstock, 1968; Furness, 1970a; see Section 5.2). It is of interest that it has been reported that small areas (about 0·1 μm diameter) of the membrane of varicosities in cholinergic nerves are thickened and associated with clusters of vesicles, suggesting, by analogy with other synapses that they may represent transmitter release sites (McMahan & Kuffler, 1971).

Varicosities show considerable variation in the number of adrenergic vesicles they contain. This does not appear to depend on their proximity to smooth muscle cells, and consecutive varicosities from the same neuron appear to have a comparable vesicle density (Fillenz, 1971). It has been calculated that each varicosity in the rat iris contains approximately 1000 to 2000 vesicles (Dahlström, Häggendal and Hökfelt, 1966); a figure of about 500 vesicles was obtained from counts in serial sections (Hökfelt, 1969). Variation in the electron-density of adrenergic vesicles is related semi-quantitatively to NA content (Hökfelt & Ljungdahl, 1972; Thureson-Klein, Klein & Yen, 1973) and may reflect different levels of utilization of transmitter stores.

3.1.2 *Transmitter storage*

Intraneuronal distribution of NA. NA is not evenly distributed throughout the neuron. Most of the transmitter is located in the varicosities of the terminal axons (Norberg & Hamberger, 1964; Malmfors, 1965; Dahlström, 1966; Iversen, 1967; Geffen & Livett, 1971). Indirect estimations indicate that the concentration of NA in the cell body and in the non-terminal axon is of the order of 10 to 100 μg g^{-1}, while in the terminal varicosities it is about 10 000 μg g^{-1} (Dahlström, 1966; Geffen & Livett, 1971). It has been estimated that each nerve cell body in the cat superior cervical ganglion contains about 4×10^{-13}g NA. Estimates of the NA content of single varicosities in different tissues are in good agreement: 4 to 6 $\times$ 10^{-15}g for rat iris and vas deferens (Dahlström, Häggendal & Hökfelt, 1966); 10^{-14}g for rabbit pulmonary artery (calculated from Bevan, Chesher & Su, 1969) and 3.4×10^{-14}g for guinea-pig uterine artery (Bell & Vogt, 1971).

NA in adrenergic neurons is partly bound in subcellular particles, which can be isolated by ultra-centrifugation techniques and demonstrated as granular (or dense core) vesicles by electron microscopy, and partly stored in a extravesicular compartment. The proportion of bound and unbound NA in adrenergic neurons is difficult to establish because the techniques used to isolate the vesicles produce a loss of NA from the vesicles. The error introduced is variable depending on the isolation procedures used (von Euler, 1966). A further source of error is contamination by non-neuronal NA from chromaffin cells in sympathetic ganglia and in various other tissues studied. It has been proposed that only about 20 per cent of NA in the cell body is bound in a vesicular compartment, the remainder being stored in the cytoplasm in an unknown form. In non-terminal axons, a higher percentage of NA is bound to vesicles (50 to 80 per cent) and in the terminal varicosities it has been assumed that most of the NA is stored in vesicles (Geffen & Livett, 1971; De Potter, Chubb & De Schaepdryver, 1972). Extravesicular storage of NA is suggested by the localization of ^{3}H–NA even in those bovine splenic nerve axons in which no vesicles could be seen (Stjärne, Roth, Bloom & Giarman, 1970). Tranzer (1972, 1973) and Hökfelt (1973) presented structural evidence for extravesicular storage of NA in rat adrenergic neurons. At present it is not possible to correlate this extravesicular NA with the pool of 'transport' NA as described in biochemical experiments (Von Euler, 1966).

Electronmicroscopic identification of NA-containing vesicles. With electronmicroscopy the NA-containing vesicles appear, under suitable fixation, as both small and large vesicles with a dense core (termed 'dense core vesicles' or 'granular vesicles') (De Robertis and Pellegrino de Iraldi, 1961a; Richardson, 1964; Grillo, 1966; Burnstock and Robinson, 1967; Tranzer, Thoenen, Snipes & Richards, 1969; Jaim-Etcheverry & Zieher, 1971a; Hökfelt, 1971; Geffen & Livett, 1971; Bloom, 1972; Furness, Campbell, Gillard, Malmfors, Cobb & Burnstock, 1970; Bisby & Fillenz, 1971; De Potter, Chubb & De Schaepdryver, 1972; Smith, 1972b).

While there is general agreement between various authors on the size of the small granular vesicles, descriptions of the size and proportion of large granular vesicles differ, depending on the species and on the organ (Table 1). In the majority of adrenergic nerve terminals in adult rat and guinea-pig (e.g. vas deferens, heart, iris, pineal gland and mesentery), *'small granular vesicles'* (30 to 60 nm) are predominant, while there is a

Table 1

Percentages and average sizes of small and large granular vesicles in adrenergic neurons based on electron micrographs of serial sections

Tissue	s.g.v.	l.g.v.	Reference
Rat iris	96% (49·5 nm)	4% (94·7 nm)	Hökfelt, 1969
Rat vas deferens	96% (50 nm)	4% (85 nm)	Bisby & Fillenz, 1971
	94·7% (50 nm)	5·3% (100 nm)	Tranzer, 1973
Rat mesenteric nerves	~90% (50 nm)	~10% (100 nm)	Tranzer, 1973
Cat spleen	75·6% (50 nm)	24·4% (85 nm)	Bisby & Fillenz, 1971

small percentage (1 to 5 per cent) of *large granular vesicles* (60 to 120 nm); in addition there are a variable number of *agranular* (*'empty'*) *vesicles* (30 to 60 nm) (Hökfelt, 1968, 1969, 1973; Burnstock, 1970; Fillenz, 1971; Bloom 1972; Tranzer, 1973), the number of the latter depending on both technical and physiological factors (see further below).

In the cell bodies of the adrenergic neurons which supply the heart, iris and pineal gland of the rat, the majority of granular vesicles are small (30 to 50 nm) with a small percentage (2–5 per cent) of large granular vesicles (about 100 nm) (Hökfelt & Dahlström, 1971; Tamarind & Quiliiam, 1971; Eränkö, 1972a). Clusters of small granular vesicles are also found in the non-terminal axons supplying these organs (Hökfelt, 1969; Fillenz, 1971; Hökfelt & Dahlström, 1971a), and the mesentery and vas deferens of the rat (Tranzer, 1973). Although the proportion of small to large granular visicles in these non-terminal axons is the same as in their terminals, the overall number of vesicles is at least one hundred times less (Tranzer, 1973).

In the adrenergic nerve terminals in the cat spleen, the size of the 'large granular vesicles' is about 70–80 nm, and their percentage is higher (24 per cent) (Bisby & Fillenz, 1971). In the cell bodies and non-terminal axons supplying the spleen of cat, dog, sheep and calf, the majority of the granular vesicles are of the large type (Kapeller & Mayor, 1967, 1969a; Geffen & Ostberg, 1969; Bisby & Fillenz, 1971; Klein & Thureson-Klein, 1971; Lagercrantz, 1971). Tranzer (1973) showed that in bovine splenic nerves five minutes post mortem, about half of the granular vesicles are small but after forty-five minutes, most of the granular vesicles are large. In the intestine of several species there is evidence for two types of adrenergic profiles, one with predominant small granular vesicles (30 to 60 nm) and the other with almost exclusively large granular vesicles (70 to 80 nm) (see Furness & Costa, 1974).

Only a few workers have applied rigorous and extensive countings of vesicles with statistical analysis of results (Table 1 & Figure 4 a-d). Further investigation is needed to determine whether these differences in the proportions and size of large granular vesicles in adrenergic neurons in different locations is a sufficient basis for classification into different types.

It is important to consider the problem of *preservation of NA storage sites* in electronmicroscope sections, since some confusion has arisen because the appearance of adrenergic vesicles varies with different preparative procedures; furthermore, susceptibility to these procedures is not uniform in different organs and species. In general, permanganate fixation gives the best preservation of the granule of the small adrenergic vesicles, followed by glutaraldehyde-dichromate-OsO$_4$, glutaraldehyde-OsO$_4$ and formaldehyde-OsO$_4$ (see Bloom, 1972; Hökfelt & Ljungdahl, 1972). Permanganate fixation causes the loss of NA from the granules, but the density of the granule gives a good idea of the content of NA at the moment of fixation (Hökfelt, 1971; Jaim-Etcheverry & Zieher, 1971a). Glutaraldehyde reacts with NA forming an insoluble polymer which becomes electron dense after OsO$_4$. A high degree of specificity for NA is achieved by potassium dichromate (Tranzer, Thoenen, Snipes & Richards, 1969; Woods, 1969; Jaim-Etcheverry & Zieher, 1971).

An invaluable tool for localizing NA in adrenergic nerves is 5-hydroxy dopamine (5-OHDA), which is taken up by adrenergic nerves and storage vesicles and reacts with various fixatives to give a dense precipitate which serves as a *specific marker for adrenergic vesicles* (Tranzer, Thoenen, Snipes & Richards, 1969). Other amines can also act as markers for adrenergic vesicles, e.g. α-methylnoradrenaline, 6-hydroxy dopamine and 6-hydroxy tryptamine (Bennett, Burnstock, Cobb & Malmfors, 1970; Furness, Campbell, Gillard, Malmfors, Cobb & Burnstock, 1970; Thoenen & Tranzer, 1971).

Those vesicles in adrenergic terminals which appear empty (agranular vesicles) become granular following various procedures such as loading with NA (Tranzer & Thoenen, 1967a) or with 5-OHDA (Tranzer & Thoenen, 1968), by using a triple fixation with aldehydes potassium dichromate and OsO$_4$ (Tranzer & Snipes, 1968; Woods, 1969), or by incubating the tissue prior to fixation in a Ringer solution containing Mg^{2+} ions (Iwayama & Furness, 1971).

It has been shown that the dense core of the large granular vesicles consists, in addition to an amine, of an osmiophilic substance resistant to amine depleting drugs (see Jaim-Etcheverry & Zieher, 1971a; Hökfelt, 1971). Attention has also been drawn to the composition of the electron-transparent halo between the granule and the vesicle membrane which forms about three times the volume of the granular core and can be differentially strained with zinc iodide (Pellegrino de Iraldi & Suburo, 1971).

Changes in size of granules corresponding to the functional state of the nerves supplying the pineal gland and intestine have been claimed (Tafuri, 1964; Bondareff, 1965; Wurtman & Axelrod, 1968; Tranzer, Thoenen, Snipes & Richards, 1969; Machado, 1971), and it has been demonstrated that prolonged nerve stimulation reduces the proportion of small granular vesicles (Van Orden, Bloom, Barrnett & Giaman, 1966; Kupferman,

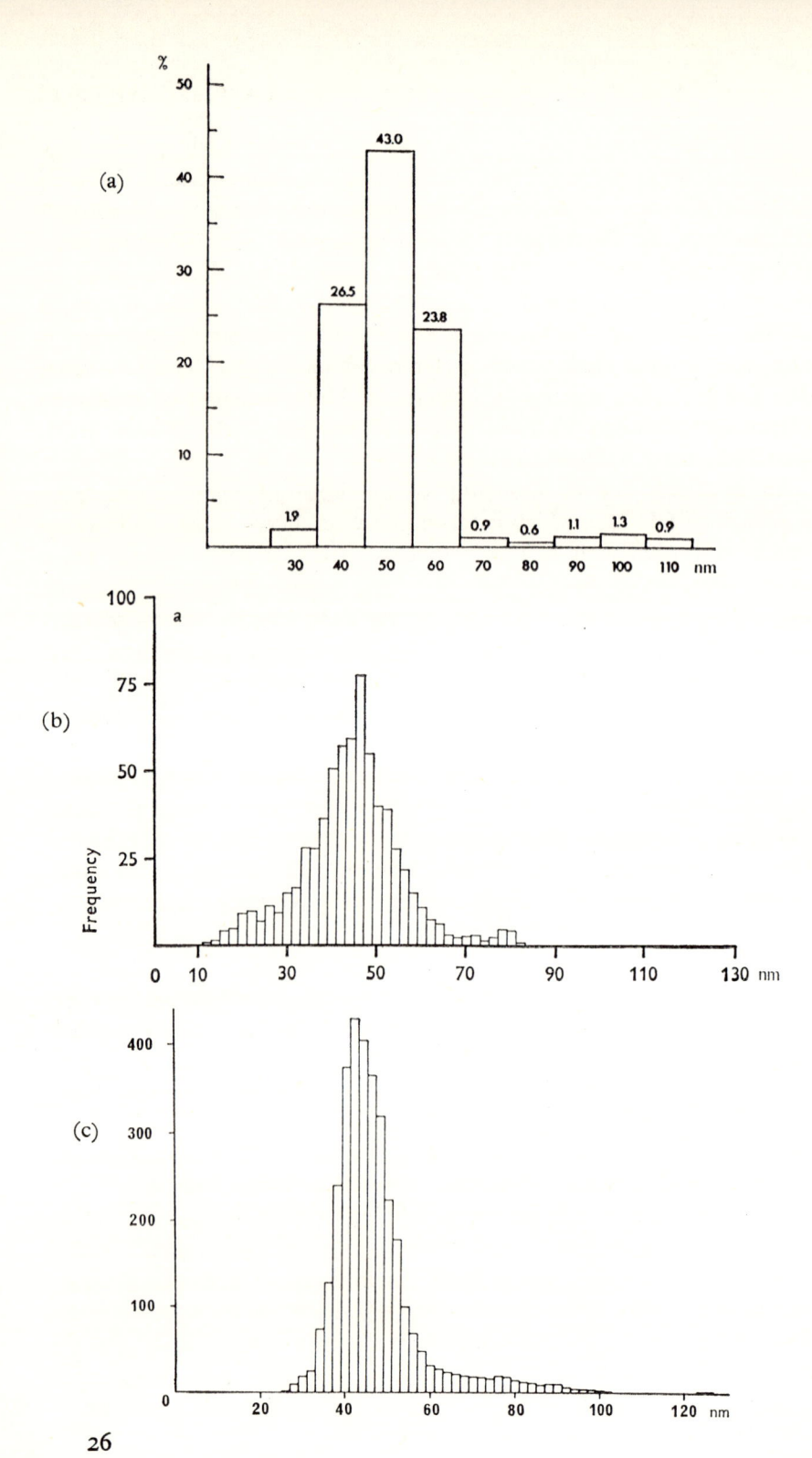

(d)

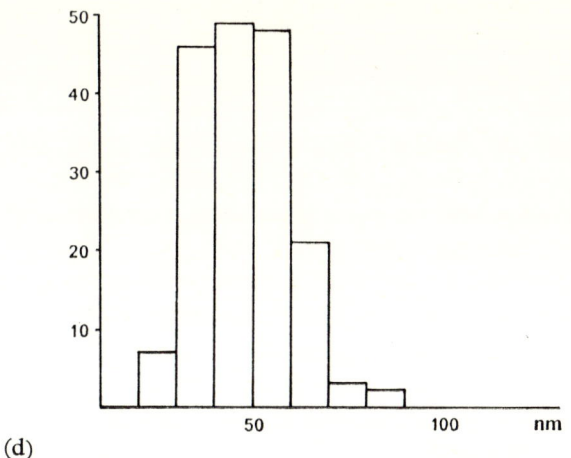

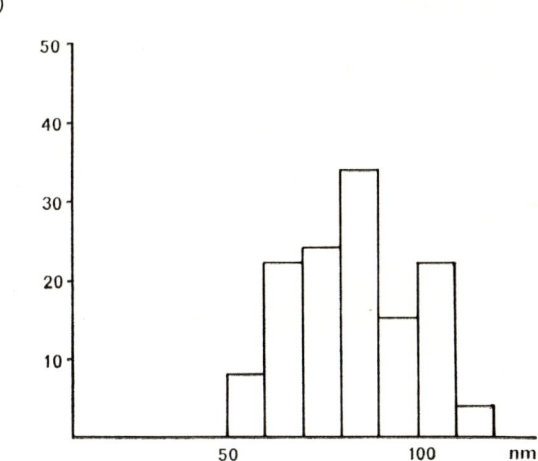

Figure 4 a-d Distribution of diameters of adrenergic vesicles in adrenergic terminal axons.
(a) distribution of diameters of granular vesicles in adrenergic varicosities in the rat iris. Ordinate: percentage of vesicles; Abscissa: diameters in nanometres. Reproduced with permission from Hökfelt, 1969.
(b) distribution of diameters of granular vesicles in adrenergic varicosities in cat spleen. Reproduced with permission from Geffen and Ostberg, 1969.
(c) distribution of diameters of granular vesicles in adrenergic varicosities in mouse vas deferens. Ordinate; number of vesicles. Abscissa, diameter in nanometres. Courtesy of Anna Ostberg.
(d) distribution of diameters of granular vesicles in adrenergic axons in periarterial adrenergic axons (top) and in intramuscular axons (bottom) in mouse pylorus. Ordinates: number of vesicles; Abscissae: diameter in nanometres. Reproduced with permission from Dermietzel, 1971.

Gillis & Roth, 1970; Bisby, Cripps & Dearnaley 1971; Coté, Palaic & Panisset, 1970).

Chemical composition of isolated vesicles. Since von Euler and Hillarp (1956) developed the method of ultracentrifugation of NA-containing particles from bovine splenic nerve (non-terminal axons), there have been many studies of the chemical composition and properties of this particulate fraction (see Stjärne, 1966; Potter, 1966; Roth, Stjärne, Levine & Giarman, 1968; Smith, De Potter & De Schaepdryver, 1969; Lagercrantz, 1971; Smith, 1972b; De Potter, Chubb & De Schaepdryver, 1972) and comparisons made with the composition of adrenal chromaffin granules (Stjärne, 1966; Douglas, 1968; Banks & Helle, 1971; Kirshner & Kirshner, 1971; Lagercrantz, 1971; Smith, 1971a; von Euler, 1972).

Most studies of the composition of adrenergic nerve vesicles have been made from extracts of non-terminal adrenergic splenic nerve trunks (Banks & Helle, 1971; Lagercrantz, 1971; von Euler, 1972; De Potter, Chubb & De Schaepdryver, 1972) and these granular vesicles (which are predominantly large) differ considerably from the vesicles, predominantly small, found in the terminal varicosities. Fewer reliable results on the chemical composition of terminal vesicles are available due largely to the greater technical problems involved (see Potter, 1966; Austin, Livett & Chubb, 1967; Bisby & Fillenz, 1971; Smith, 1972a). However, a comparison of the composition of large and small granular vesicles in terminal varicosities has recently been reported (see Bisby & Fillenz, 1971; De Potter, Chubb & De Schaepdryver, 1972; Bisby, Fillenz & Smith, 1973).

NA in isolated granules is associated with ATP and Ca^{2+} to form a complex. It has been claimed that the ratio NA:ATP in this complex is 4:1 as in bovine chromaffin granules (Shore, 1962; De Prada, Berneis & Pletscher, 1971). However, it has been suggested recently that, after taking into account mitochondrial contamination, the ratio is 7–12:1 (De Potter, Chubb & De Schaepdryver, 1972), a ratio closer to the more recent figures for the adrenal medulla of rat and rabbit (O'Brien, Da Prada & Pletscher, 1972). It is interesting that if NA, Ca^{2+} and ATP are mixed *in vitro* they form aggregates with physico-chemical properties remarkably similar to adrenergic vesicles (Berneis, Pletscher & Da Prada, 1970). Using biochemical analysis of isolated vesicles and immunohistochemical methods, the specific proteins chromogranin A and dopamine-β-hydroxylase (DβH), found originally associated with the adrenomedullary vesicles (Banks & Helle 1971; Blaschko, 1972), have also been demonstrated in association with adrenergic vesicles (Geffen & Livett, 1971; De Potter, 1971). Chromogranin A is present in adrenergic vesicles in both soluble (10 to 20 per cent) and insoluble form. It is not known whether the NA/ATP-Ca^{2+} complex is bound to the soluble or to the insoluble fraction (see Geffen & Livett, 1971; Banks & Helle, 1971). Small amounts of phospholipids and cholesterol were found in adrenergic vesicles (Lagercrantz, 1971). The finding of a cytochrome b561 in the adrenergic vesicles led to the suggestion of the

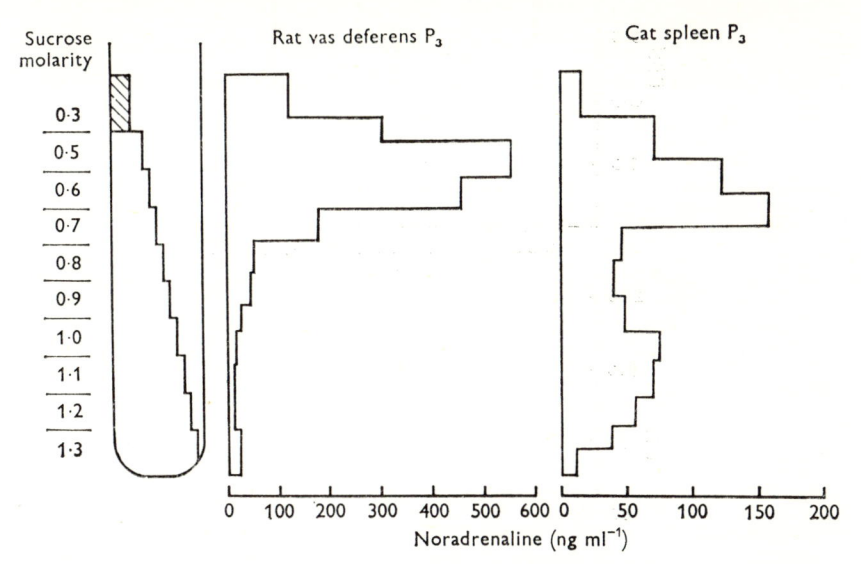

Figure 5 Noradrenaline distribution in sucrose gradients from rat vas deferens and cat spleen. In both organs NA is present in the 'light' fraction. The larger amount of NA is the 'heavy' fraction of the cat spleen compared to the vas deferens probably reflects the different percentages of large granular vesicles in these organs (see Table 1). Reproduced with permission from Bisby and Fillenz, 1971.

presence of an electron transport chain in the vesicles possibly related to DβH (see Lagercrantz, 1971). Another enzyme, an Mg^{2+}–dependent adenosine-3′-phosphatase, has been detected in these vesicles (Lagercrantz, 1971).

In studies made on adrenergic nerves, using sucrose gradient centrifugation, two types of NA storage particles have been isolated: a *heavy* particle and a *light* particle (Potter, 1966; Stjärne, 1966; Austin, Livett & Chubb, 1967; Bisby & Fillenz, 1971; Lagercrantz, 1971; De Potter, Chubb & De Schaepdryver, 1972) which have been equated with *large* and *small* granular vesicles respectively (Bisby & Fillenz, 1971; Klein & Thureson-Klein, 1971; Lagercrantz, 1971; Smith, 1972a; Figure 5). Both *heavy* particles (large granular vesicles) and *light* particles (small granular vesicles) in the vas deferens contain DβH (De Potter, Chubb & De Schaepdryver, 1972; Bisby & Fillenz, 1971; Bisby, Fillenz & Smith, 1973; Figure 6).

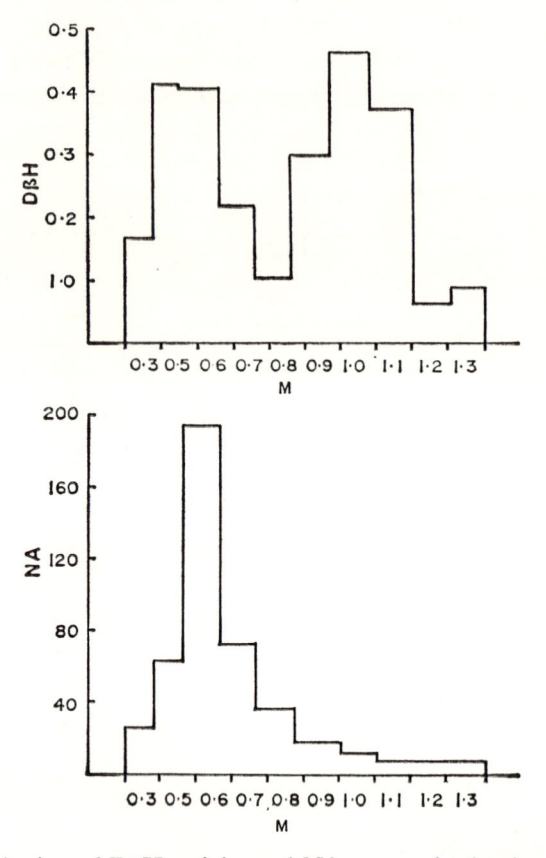

Figure 6 Distribution of DβH activity and NA content in density gradients prepared from microsomal pellets of rat vas deferens. The presence of the two peaks of DβH corresponding to the 'light' and 'heavy' fractions indicates that, in the rat vas deferens, both small and large granular vesicles contain DβH. Ordinates: DβH activity in units ml^{-1} and NA content in ng ml^{-1}. Abscissae: molarity of sucrose in the density gradient.
Reproduced with permission from Bisby, Fillenz and Smith, 1973.

However, in the spleen the presence of DβH in the small granular vesicles is dubious (see De Potter, Chubb & De Schaepdryver, 1972; Figure 7). Furthermore, the small granular vesicles of the spleen contain less NA than the small granular vesicles of the vas deferens (Bisby & Fillenz, 1971), although they contain about ten times more NA than the large granular vesicles of the same nerves (De Potter, Chubb & De Schaepdryver, 1972).

The properties of the isolated adrenergic vesicles have been studied extensively (see Stjärne, 1966; Iversen, 1967; Lagercrantz, 1971; von Euler, 1972). Adrenergic vesicles take up and store NA in a bound form. DA is also taken up and converted to NA by DβH but not stored. Bound NA is continuously interchanged with unbound NA in an equilibrium

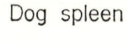

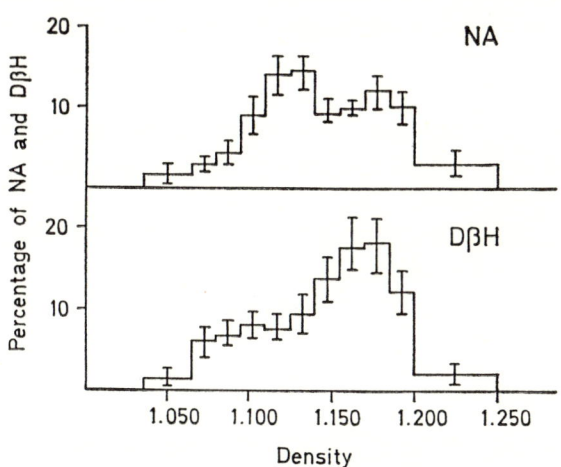

Figure 7 Distribution of DβH activity and NA content in a centrifugation gradient from a particulate fraction from dog spleen. The absence of a DβH peak in the region of 'light' fraction, suggests that small granular vesicles in spleen contain little DβH and that DβH is contained in the large granular vesicles. Reproduced with permission from De Potter, Chubb and De Schaepdryver, 1972.

depending on the properties of the binding complex more than on an active process of uptake and release. The high level of NA stored in the vesicles is a function of the balance between bound and unbound NA and on the synthesis and utilization of NA. The physiological implications of the vesicle storage of NA will be discussed in relation to the process of adrenergic neurotransmission in Sections 5.6 and 5.7.

3.1.3 *Axoplasmic transport of NA and related enzymes*

Following the presentation of the theory of axoplasmic flow by Weiss and Hiscoe (1948), extensive studies, beginning with the investigation of Dahlström and Fuxe (1964), have been made of axoplasmic flow in mammalian adrenergic nerves by observing the accumulation of organelles, transmitter substances and associated enzymes following nerve constriction (see Barondes, 1969; see reviews Geffen & Livett, 1971; Dahlström, 1971). In studies on non-adrenergic nerves two components of axoplasmic flow have been demonstrated with radio-labelled protein: a *fast flow* of particulate material at 100 to 500 mm d^{-1} (or 4 to 20 mm h^{-1}) and a *slow flow* of soluble axonal protein at 0·5 to 25 mm d^{-1} (or 0·02 to 1 mm h^{-1}). The wide range of both fast and slow flow rates is likely to reflect differences in flow rates between neurons of different species as well as different neuron systems within the same animal, but may also result from differences in measuring techniques (see Dahlström, 1971).

Accumulation of granular vesicles and mitochondria has been demonstrated proximal to a ligation in mammalian adrenergic nerves *in vivo* and *in vitro* (Kapeller & Mayor, 1967, 1969a; Banks, Mangnal & Mayor, 1969; Geffen & Ostberg, 1969; Hökfelt & Dahlström, 1971; Banks & Mayor, 1972). The rate of accumulation of NA (Härkönen, 1964; Dahlström, 1965; Dahlström & Häggendal, 1966b; Dahlström, 1971; Livett, Geffen & Austin, 1968), which is closely related to the accumulation of granular vesicles (Banks, Mangnal & Mayor, 1969; Kappeller & Mayor, 1969a; Lever, Spriggs & Graham, 1968), is similar to that of DβH (Laduron & Belpaire, 1968a, b; Geffen, Hunter & Rush, 1969) and chromagranin A (Geffen, Hunter & Rush, 1969), and corresponds to the fast component of axoplasmic flow.

Non-granular enzymes such as tyrosine hydroxylase (T-OH) (Laduron & Belpaire, 1968a), dopa-decarboxylase (Dahlström & Jonason, 1968; Laduron, 1970) and monamine oxidase (MAO) (Dahlström, Jonason & Norberg, 1969) accumulate at a much slower rate, which corresponds to the slow flow of axoplasm and mitochondria (Weiss & Pillai, 1965; Banks, Mangnal & Mayor, 1969). The results of Wooten & Coyle (1973) suggest that T-OH is associated with fast axonal flow and dopa-decarboxylase with an intermediate rate of flow (see Table 2).

There is no evidence at present for retrograde transport of NA in intact peripheral adrenergic nerve fibres (Geffen, Hunter & Rush, 1969). However, retrograde flow of DA in unligated central dopaminergic neurons in the nigroneostriatal system has been reported (Ungerstedt, Butcher, Butcher, Andén & Fuxe, 1969). Mitochondria and associated enzymes, including MAO (see McLean & Burnstock, 1972), appear to move in both directions, at least after ligation, at a rate of transport of about 0·6 mm h^{-1} (Banks, Mangnal & Mayor, 1969).

Studies by Dahlström and her collaborators (see Dahlström, 1971) suggest that as adrenergic vesicles are transported down the axon, there is an increase of their NA content. A similar increasing gradient of ACh towards the distal parts of cholinergic nerves has also been reported (Evans & Saunders, 1967).

Nerve impulse activity (at least in short-term experiments) does not appear to significantly alter the rate of NA transport (Geffen & Rush, 1968;

Table 2

Absolute rates of transport for enzymes present in adrenergic axons in rat sciatic nerve (from Wooten and Coyle, 1973)

DβH	138–185	mm d^{-1}
T-OH	106–167	mm d^{-1}
Dopa-decarboxylase	36–86	mm d^{-1}
MAO	3	mm d^{-1}
COMT	3	mm d^{-1}

Dahlström, 1971). The rate of transport of acetylcholinesterase (AChE) along cholinergic nerves was also unaffected by electrical stimulation (Jankowska, Lubinska & Niemierko, 1969). However, reflex increase in nerve impulses caused by reserpine induces synthesis of enzymes in the cell body (see Section 4.1) and increase the number of vesicles, which are then transported to the terminals (Dahlström & Haggendäl, 1969).

There is considerable interest in the mechanism of axonal transport (see Dahlström, 1971; Mayor, Banks, Tomlinson & Grigas, 1971). Local injections of vinblastine and colchicine in peripheral nerves block the transport of NA granules (Dahlström, 1968; Hökfelt & Dahlström, 1971). Since these drugs are known to disrupt microtubules, it was suggested that the microtubules in adrenergic axons are involved in the fast transport of amine storage granules (see Schmitt, 1968; Dahlström, 1971; Mayor, Banks, Tomlinson & Grigas, 1971; Banks & Mayor, 1972). It should be noted that axonal flow in adrenergic nerves has been examined only in non-terminal axons; no studies of axonal flow have been carried out on terminal varicose axons.

3.1.4 *Origin, life-span and fate of adrenergic vesicles*

Granular vesicles are synthesised in the cell body and are transported to the nerve terminals by axonal flow (Dahlström, 1971) Banks & Mayor, 1972;

Matthews, 1972). There is also some evidence that granular vesicles can be formed in the nerve terminals. There are indications from electron micrographs of adrenergic terminals (e.g. Pellegrino de Iraldi & de Robertis, 1968; Machado, 1971; Hökfelt 1973a) and particularly those of neurons grown in culture (Holtzmann, 1971; Chamley, Mark, Campbell & Burnstock, 1972a), that granular vesicles may arise by dilatation of regions of smooth endoplasmic reticulum in which a dense matrix accumulates. Other suggestions are that small granular vesicles are formed by budding from large granular vesicles, by the division of large granular vesicles into small ones (Fillenz, 1971), or by fission of the empty membranes of the large vesicles during retrieval of the membrane after exocytosis (Smith, 1971b) in a similar manner to the way neurosecretory granules disintegrate into small vesicles following exocytosis (Normann, 1965).

Adrenergic vesicles, which are present in large numbers in terminal varicosities, are able to synthesize NA from DA (see Geffen & Livett, 1971). The amount of NA that reaches the terminals in one day by axonal flow from the cell body has been calculated to represent less than 1 per cent of the peripheral store of NA (Geffen, Livett & Rush, 1970). This suggests that the main function of adrenergic vesicles originating in the cell body is to supply the terminals with the machinery capable of local synthesis and storage of NA.

The life span of an adrenergic vesicle, calculated in terms of the time of recovery of NA levels in the terminals after irreversible damage of the vesicles by reserpine, is of the order of three to four weeks. The life span of NA in the terminals, estimated by monitoring levels of ^{3}H-NA, is of the order of 1 to 2 days (see Banks, Mayor, Mitchell & Tomlinson, 1971; Dahlström, 1971; Geffen & Livett, 1971). These observations led to the conclusion that each vesicle is used several times and that NA alone (without the other vesicular components) is released from the vesicles (Folkow & Häggendäl, 1970; Smith & Winkler, 1972). This would be in conflict with the hypothesis of release of NA from adrenergic terminals by a process of exocytosis (see Section 5.3). However, calculation of the life span of the adrenergic vesicles, by using the enzyme DβH as a vesicular marker and by taking into consideration the tenfold higher concentration of NA in the terminal vesicles than in the non-terminal vesicles, suggests that the life span of vesicles is of the same order as that of vesicular NA, namely about forty-eight hours (De Potter & Chubb, 1971; De Potter, Chubb & De Schaepdryver, 1972); these figures are consistent with the exocytosis hypothesis. The ultimate fate of the vesicles after use is not known.

3.2 **Other catecholamine-containing cells**

Several types of cells are able to store catecholamines, including mast cells (Hopwood, 1971; Coupland, 1972) and some endocrine cells (Falck & Owman, 1968; Cegrell, 1970; Håkanson, 1970; Tjälve, 1971), but the most widespread non-neuronal tissue which contains CA is chromaffin tissue (Hopwood, 1971; Coupland, 1971). Although chromaffin cells vary considerably in distribution, morphology and properties, they all share a common embryological origin with sympathetic neurons, namely neural crest tissue (Coupland, 1965) (see Section 6.2). They are able to synthesize and store CA in subcellular vesicles which have many similarities with adrenergic vesicles (see Section 3.1).

It is now recognized that the term 'chromaffin' based on chrome staining is no longer strictly applicable, since cells such as mast and enterochromaffin cells also stain with chrome salts (Boyd, 1960; Hopwood, 1971; Coupland, 1972) and since in some species (e.g. guinea-pig and rat) staining of chromaffin cells is weak or even undetectable (Grigor'eva, 1962; Lempinen, 1964; Jacobowitz, 1967). Therefore chromaffinity cannot be used as a distinguishing feature for 'chromaffin' tissue (Boyd, 1960; Coupland, 1965; Jacobowitz, 1967; Costa, Eränkö & Eränkö, 1974a). The lack of chrome reaction of extra-adrenal cells in adult rat (Eränkö & Harkönen, 1965) led to the introduction of the term 'small intensely fluorescent' (or SIF) cells (Norberg, Ritzén & Ungerstedt, 1966) and this term has been applied to extra-adrenal chromaffin cells which are intensely fluorescent in all species. Nevertheless, these are not a newly discovered type of cell and it seems preferable to retain the term 'chromaffin tissue' for the time being (Jacobowitz, 1967; Costa & Furness, 1973).

In mammals, part of the chromaffin tissue is associated with steroid secreting tissue to form the adrenal glands in which the chromaffin tissue occupies the inner part (medulla). A significant number of chromaffin cells not associated with the adrenal gland are scattered in various organs, usually in association with autonomic nerves, and from what has been termed extra-adrenal chromaffin tissue. When clustered to form small chromaffin bodies in association with sympathetic nerves and ganglia, this tissue has been called 'paraganglionic' (Kohn, 1903; Boyd, 1960; Coupland, 1965).

3.2.1 *Adrenal chromaffin cells*

Two types of chromaffin cells have been distinguished in the adrenal medulla (Eränkö, 1960; Elfvin, 1967; Coupland, 1965, 1972). One type stores NA in large vesicles with a dense homogenous core; the other cell type stores A in larger vesicles with paler and larger core (Coupland, 1972). The distribution and proportion of these cell types in the adrenal medulla depends to some extent on cortical hormones and shows wide species variation (Wurtman & Axelrod, 1968; Coupland, 1972; Wurtman, Pohorecky & Baliga, 1972). Adrenal chromaffin cells receive innervation from cholinergic neurons whose origin is in the spinal cord and have therefore been equated with postganglionic adrenergic neurons (Coupland, 1965). The conditions for differential release of NA and A from the adrenal medulla and their roles in sympathetic responses are complex and deserve more extensive review.

3.2.2 *Extra-adrenal chromaffin cells*

Extra-adrenal chromaffin cells are widely distributed in the body (Köhn, 1903; Boyd, 1960; Grigor'eva, 1962; Coupland, 1965; Jacobowitz, 1970; Kobayashi, 1971; Bock, 1970, 1973; Costa & Furness, 1973). Many are concentrated in the pre-aortic region to form the organ of Zuckerkandl. Smaller paraganglia are scattered throughout the peritoneal region. Chromaffin cells, either in small groups or singly, are also found in association with sympathetic ganglia (Plate 8), vagus nerves, nerve plexuses, heart and blood vessels. Chromaffin cells form the major component of the carotid and pulmonary bodies (Biscoe, 1971; Howe & Neil, 1972). Clusters of these cells (paraganglionic) have also been reported in more peripheral locations such as visceral organs, popliteal cavity, orbital cavity and the middle ear. The paraganglia are highly vascularized. When chromaffin cells are found in even smaller groups than in paraganglia, they are characteristically closely associated with fenestrated capillaries (Plate 8).

Although there is some variation in structure, extra-adrenal chromaffin cells are characterized by high CA content and by the presence of large granular vesicles (Coupland, 1972; Costa, Eränkö & Eränkö, 1974a) which are similar to the NA-containing vesicles in adrenal chromaffin cells (Plate 8). Processes of these cells, which may be varicose, have been observed. They are associated with neurons, blood vessels and brown fat, but usually they join nerve trunks and plexuses and it is difficult to follow them to their terminations (Costa & Furness, 1973; Plate 8). In the superior cervical ganglion of the rat, chromaffin cells have been shown to receive preganglionic innervation and to establish synaptic contacts with postganglionic sympathetic neurons (Siegrist, Dolivo, Dunant, Foroglou-Kerameus, Ribaupierre & Rouiller, 1968; Matthews & Raisman, 1969; Williams & Palay, 1969; Yokota, 1973; see Section 2.2 and Figures 2 and 3).

There is ultrastructural, histochemical and biochemical evidence for substantial variation in the type of monoamine found in extra-adrenal chromaffin cells. For example, some cells in the abdominal cavity contain NA and A (Costa & Furness, 1973). DA-containing cells have been identified in the carotid body, superior cervical ganglion and in the heart of some species (see Biscoe, 1971; Ehinger, Falck, Persson & Sporrong, 1968).

Thus, extra-adrenal chromaffin cells show characteristics of endocrine cells and have been shown to release CA under a variety of conditions, including exposure to some hormones (Muscholl & Vogt, 1964; Brundin, 1966; Hervonen, 1971). Moreover, it has been suggested that they may function as interneurons in the superior cervical ganglion (Siegrist, Dolivo, Dunant, Foroglou-Karameus, Ribaupierre & Rouiller, 1968; Williams & Palay, 1969; Matthews & Raisman, 1969). In the carotid body, there is now evidence that the so-called 'chemoreceptor cells' are typical CA-containing chromaffin cells which are innervated by cholinergic fibres, and probably modulate chemoreceptor activity by releasing CA (Biscoe, 1971; Howe & Neil, 1972).

Extra-adrenal chromaffin cells are particularly resistant to CA depletion by α-methyl-p-tyrosine, reserpine, guanethidine, 6-OHDA and immunosympathectomy; this feature has been related to low turnover of CA in these cells (Norberg, Ritzén & Ungerstedt, 1966; Van Orden, Burke, Geyer & Lodoen, 1970; Burnstock, Evans, Gannon, Heath & James, 1971; Eränkö & Eränkö, 1972b).

3.3 Summary

Most of the nerve cell bodies in sympathetic ganglia are adrenergic and contain typical noradrenaline storage vesicles. Several types of neurons can be distinguished on the basis of size, distribution of their processes, their relationship with other neurons and fluorescence histochemistry, but no correlation between morphological types and function has been recognized.

Non-terminal adrenergic axons extend to the effector organ where they branch and form a varicose terminal network which, together with the terminal varicose fibres of other neurons, form the 'ground plexus'. Intervaricosities contain mainly neurotubules and/or neurofilaments. Varicosities contain most of the granular vesicles and are the sites of release of transmitter. The concentration of noradrenaline in the varicosities is very much higher than in the cell body and non-terminal axons. Most of the noradrenaline in the terminals is bound in granular vesicles, while in the cell body a large proportion of noradrenaline is unbound.

Small granular vesicles are predominant in the varicosities, but a small proportion of large granular vesicles is also present. The variable number of agranular vesicles represent 'empty' small granular vesicles. The size and proportion of large granular vesicles in neurons supplying the spleen differ from those seen in all other adrenergic neurons examined. This may form the basis of classification of adrenergic neurons into more than one type.

Noradrenaline in vesicles is bound with ATP, Ca^{2+} and the proteins chromagranin A and dopamine-β-hydroxylase. Dopamine-β-hydroxylase is present in both small and large granular vesicles in the vas deferens, but may not be present in small granular vesicles in the spleen. Vesicular storage allows a large amount of transmitter to be stored in adrenergic neurons in an inactive form. Adrenergic vesicles are mainly synthesized in the cell body and transported by fast axonal flow to the terminals, but some vesicles may be formed in the terminals. Noradrenaline synthesis mainly occurs in the terminal varicosities. Other components of the metabolic pathway for noradrenaline, namely monoamine oxidase and dopa-decarboxylase, are transported by slow axonal flow to the terminals, while tyrosine hydroxylase appears to be transported by fast axonal flow.

Adrenal and extra-adrenal chromaffin cells, which share a common embryological origin with adrenergic neurons, are able to synthesize, store and release catecholamine (noradrenaline, adrenaline or dopamine) and may play an important role together with adrenergic neurons in sympathetic function.

4 BIOSYNTHESIS AND METABOLIC DEGRADATION OF NA

4.1 Biosynthesis

NA is synthesized from tyrosine in adrenergic neurons, in adrenal and extra-adrenal chromaffin cells. Early evidence for this sympathetic pathway came from studies on the adrenal medulla and only more recently the same pathway has been found in adrenergic neurons (see Goodall & Kirshner, 1958; von Euler, 1972).

4.1.1 *Biosynthetic pathway and enzymes*

The pathway of NA synthesis illustrated in Figure 8 was originally suggested by Blaschko (1939) and by Holtz (1939) following the discovery of dopa-decarboxylase (Holtz, Heise & Lüdtke, 1938) and has been confirmed experimentally by many authors (see Blaschko, 1954, 1972a; Udenfriend, 1966a; Weiner, 1970; von Euler, 1972). The hydroxylation of tyrosine to dopa is catalysed by tyrosine hydroxylase (T-OH), an enzyme which is localized in adrenergic neurons (Nagatsu, Levitt & Udenfriend, 1964; Potter, Cooper, Willman & Wolfe, 1965; Udenfriend, 1966a, b; Sedvall & Kopin, 1967b; Thoenen, Mueller & Axelrod, 1970). The subsequent decarboxylation of dopa to dopamine is catalysed by a widely distributed enzyme, the aromatic amino acid decarboxylase (Lovenberger, Weissbach & Udenfriend, 1962; Sourkes, 1966), while the final stage involves the β-hydroxylation of dopamine to NA, by DβH (Kaufmann & Friedman, 1965; Kaufmann, 1966). The further conversion of NA to A in the adrenaline-containing cells of adrenal and extra-adrenal chromaffin tissue is catalysed by the enzyme phenylethanolamine-N-methyl transferase (PNMT) (Axelrod, 1962; Gennser & Studnitz, 1969; Stjärne, 1972a).

4.1.2 *Site of NA synthesis*

Synthesis of NA occurs in all parts of the adrenergic neuron namely in the cell body, in the non-terminal and terminal axon (Goodall & Kirshner, 1958; Musacchio & Goldstein, 1963; Spector, Sjoerdsma, Zaltzman-Nirenberg & Udenfriend, 1963; Iversen, 1967), the three enzymes involved in the synthesis of NA, namely T-OH, dopa-decarboxylase and DβH, being distributed throughout the entire adrenergic neuron (Goodall &

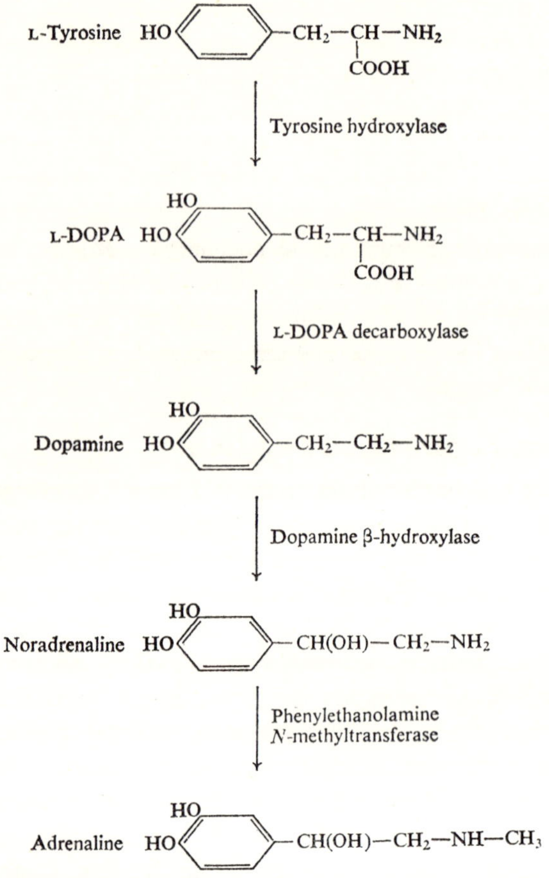

ADRENERGIC NEURONS

Figure 8 The intermediate and the enzymes involved in the synthesis of NA and A. Reproduced with permission from Blaschko, 1973.

Kirshner, 1958; Potter, Cooper, Willman & Wolfe, 1965; Dahlström & Jonason, 1968; Laduron & Belpaire, 1968a, b; Roth, Stjärne, Levine & Giarman, 1968; Smith, De Potter & de Schaepdryver, 1969; Laduron, 1970; Black, Hendry & Iversen, 1971b; Jarrott & Iversen, 1971; Livett, Geffen & Rush, 1971; de Potter, Chubb & De Schaepdryver, 1972).

Tyrosine hydroxylation and dopa decarboxylation occur in the axoplasm while β-hydroxylation of DA to NA occurs in the adrenergic vesicles. DβH is bound to the adrenergic vesicles in axons (Weiner, 1970; von Euler, 1972), although in the cell bodies a significant amount of this enzyme appears to be unbound (De Potter, Chubb & de Schaepdryver, 1972).

It has been suggested that there are two types of T-OH with differing subcellular distributions (Ikeda, Fahien & Udenfriend, 1966; Joh, Kapit & Goldstein, 1969). The findings of Coyle & Wooten (1972) and Wooten & Coyle, (1973) (see Table 2) suggest that T-OH travels down the adrenergic axons at a speed similar to that of the NA vesicle-bound DβH suggesting that at least part of the enzyme is vesicle-bound. The subcellular localization of these enzymes is difficult to determine because of the technical procedures of their isolation and because cytoplasmic enzymes can be absorbed on to membranes (Fonnum, 1967).

4.1.3 *Regulation of NA synthesis in adrenergic neurons*
Although there is both diurnal and seasonal variation in NA content of tissues in some species (Axelrod, 1971; von Euler, 1971), the levels of NA in adrenergic neurons are remarkably constant under physiological conditions. These are maintained by a balance between rates of synthesis and utilization (see Udenfriend, 1968; Weiner, 1970). The rate of NA synthesis in adrenergic nerves can be increased about five times by nerve activity (Gordon, Reid, Sjoerdsma & Udenfriend, 1966; Gordon, Spector, Sjoerdsma & Udenfriend, 1966; Roth, Stjärne & von Euler, 1967a, b; Austin, Livett & Chubb, 1967; Weiner & Rabadjija, 1968a, b; Axelrod, 1972). In contrast, the synthesis and turnover of NA in hibernating animals is strongly reduced (Draskóczy & Lyman, 1967).

The overall rate at which catecholamines in tissue stores are used and replaced by newly synthesized amines is defined as the rate of turnover, which is a useful parameter in comparative studies. For example, the 'half-time' for catecholamine turnover (i.e. the time taken to replace half of the total catecholamine store with newly synthesized material) varies from about 1·5 h for dopamine-containing neurons in the CNS, 2 to 12 h for sympathetic neurons, and about 300 h for the adrenal medulla (see Iversen & Callingham 1970). For most adrenergic neurons the turnover rate corresponds to the replacement of 6 to 29 per cent of the total store of catecholamine per hour (Blakeley, Brown, Dearnaley & Harrison, 1968).

The regulation of the biosynthesis of NA appears to be achieved by two different mechanisms: one is *fast* in onset and the other is *slow* (Weiner, 1970; Axelrod, 1972, see Cotten 1972). The fast mechanism occurs in the terminals and is based on a rapid feedback inhibition by NA of its own synthesis. The slow mechanism occurs mainly in the cell body and is based on enzyme induction of T-OH and DβH by preganglionic nerve activity.

Fast regulatory mechanism. The hydroxylation of tyrosine is the slowest of the three steps in the biosynthesis of NA and therefore T-OH is the rate-limiting enzyme (Udenfriend, 1966a; Spector, Sjoerdsma, Saltman-Nirenberg & Udenfriend, 1963; Levitt, Spector, Sjoerdsma & Udenfriend, 1965). The enzyme activity is regulated by the end product NA, which acts as a fast feed-back inhibitor on its synthesis (Nagutsu, Levitt & Udenfriend, 1964; Spector, Gordon, Sjoerdsma & Udenfriend, 1967; Weiner, 1970). Only a small pool of NA located in a separate compartment seems to be involved in the regulation of T-OH (Alouisi & Weiner, 1966; Weiner, 1966; Weiner, 1970; Weiner, Cloutier, Bjur & Pfeffer, 1972). Thus, a reduction of the level of NA in this compartment would reduce the tonic inhibition of T-OH rapidly leading to an increase of NA to a level which would restore the inhibition. Conversely any increase of NA in this compartment above basal levels would lead to a further inhibition of NA synthesis. For example, angiotensin, a drug which increases NA release from adrenergic endings (see Section 5.6) increases the formation of NA from tyrosine in isolated heart and vas deferens (Boadle-Biber, Hughes & Roth, 1969). Increase of sympathetic activity *in vivo* and *in vitro*, which increases the release of NA from adrenergic terminals (Brown & Gillespie, 1957; Folkow, Häggendal & Lisander, 1967), is associated with an increase in the rate of NA synthesis occurring at the level of tyrosine hydroxylation (Oliverio & Stjärne, 1965; Alouisi & Weiner, 1966; Gordon, Reid, Sjoerdsma & Udenfriend, 1966; Gordon, Spector, Sjoerdsma & Udenfriend, 1966; Austin, Livett & Chubb, 1967; Roth Stjärne & von Euler, 1967a, b; Sedvall & Kopin, 1967a, b; Weiner, 1970; Thoa, Johnson, Kopin & Weiner, 1971). It has been suggested that, since Na^+ or Ca^{2+} may affect T-OH activity (Gutman & Segal, 1972), the influx of these cations during nerve stimulation rather than the end product regulation is responsible for the increase in NA synthesis (Cloutier & Weiner, 1973).

Increase of tyrosine conversion to dopa is fast in onset in the terminals, but there is no associated change in the cell bodies or in the axons (Roth, Stjärne & von Euler, 1967b; Weiner & Rabadjija, 1968a, b; Thoenen, 1970; Bhatnagar & Moore, 1971a, 1972).

In contrast, reduction of activity of adrenergic neurons following decentralization (Hertting, Potter & Axelrod, 1962; Fisher & Snyder, 1965; Musacchio & Weise, 1965; Sedvall, Weise & Kopin, 1968), and following increase of cytoplasmic NA (Alouisi & Weiner, 1966; Spector, Gordon, Sjoerdsma & Udenfriend, 1967; Udenfriend, 1968; Weiner, 1970; Dairman & Udenfriend, 1971; Spector, Tarver & Berkowitz, 1972; Weiner

& Bjur, 1972) is associated with decreased T-OH activity and NA synthesis from tyrosine. This regulation of NA synthesis in adrenergic neurons is a rapid and highly sensitive mechanism of adaptation to variable requirements, which may play a physiological role during short lasting fluctuations of sympathetic activity.

Slow regulatory mechanisms. In contrast to the fast regulation in nerve terminals, increased sympathetic activity *in vivo* and *in vitro* produces an

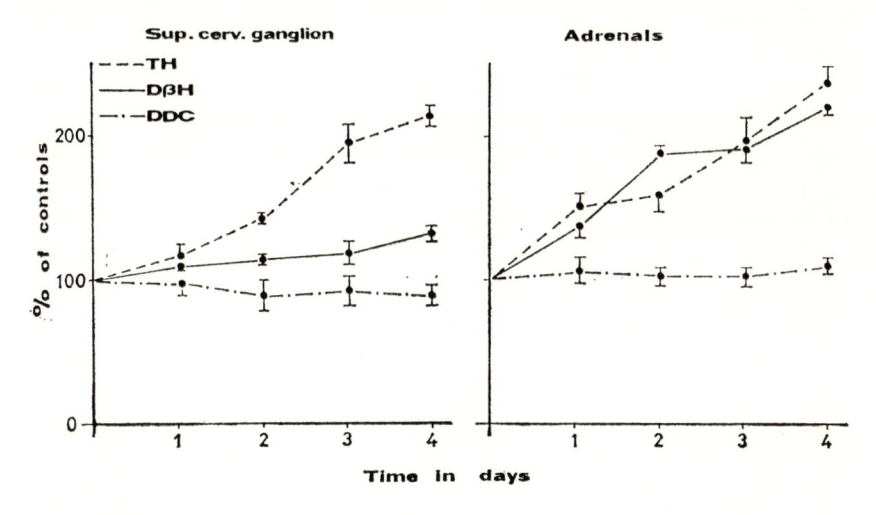

Figure 9. Trans-synaptic induction of synthetic enzymes for NA.
Effect of cold exposure at 4 °C which stimulates sympathetic activity, on tyrosine hydroxylase (T-OH), dopamine-β-hydroxylase (DβH) and dopa decarboxylase (DDC) activity in superior cervical ganglia and adrenals of the rat. In the rat superior cervical ganglion increased sympathetic activity leads to an increase of tyrosine hydroxylase activity by a process of induction (synthesis of new enzyme). Reproduced with permission from Thoenen, 1972.

increase in T-OH activity in the adrenergic cell bodies only after several hours and reaches a peak after several days (Axelrod, Mueller & Thoenen, 1970; Thoenen, 1970; Thoenen, Mueller & Axelrod, 1970; Thoenen, Kettler, Burkard & Saner, 1971; Bhatnagar & Moore, 1972; Figure 9). This slow increase is due to induction of the enzyme which is due to increased activity in preganglionic nerves, the signal being the depolarization of the membrane of the adrenergic neuron (Dairman, Gordon, Spector, Sjoerdsma & Udenfriend, 1968; Mueller, Thoenen & Axelrod, 1969a, c;

Thoenen, Mueller & Axelrod, 1969; Axelrod, Mueller & Thoenen, 1970; Brimijoin & Molinoff, 1971; Bhatnagar & Moore, 1971a, b; 1972; Weiner 1970). The time required to accomplish the single steps of T-OH induction, i.e. the duration of increased preganglionic activity is about two hours (Otten, Paravicini, Oesch & Thoenen, 1973). Appearance of increased T-OH in the terminals occurs only after a delay of one or two days (Thoenen, Mueller & Axelrod, 1970). These authors suggested that the increase is due to local induction and not to an axonal transport of enzyme induced in the cell body, although this hypothesis may need to be reconsidered following the finding of a fast axonal flow of T-OH (Coyle & Wooten, 1972; Wooten & Coyle, 1973; Table 2). Dopa-decarboxylase activity is not altered by increased sympathetic activity (Weiner & Rabadjija, 1968a; Black, Hendry & Iversen, 1971a). DβH shows an initial increase in activity in the cell bodies and a slight decrease in activity in the terminals; the level of DβH increases in the terminals only after several hours (Molinoff, Brimijoin, Weinshilboum & Axelrod, 1970). The increase in the cell bodies continues for several days after return of sympathetic activity to basal levels (Molinoff, Brimijoin, Weinshilboum & Axelrod, 1970; Thoenen, Angeletti, Levi-Montalcini & Kettler, 1971; Thoenen, 1970). This suggests that there is a mechanism to replace the DβH in the adrenergic terminals which is released with NA upon nerve stimulation (see Smith & Winkler, 1972). This is apparently achieved by transneuronal induction of the enzyme in the cell body and subsequent axonal transport to the terminals (Silberstein, Brimijoin, Molinoff & Lemberger, 1972; Axelrod, 1972). The trans-synaptic induction of T-OH and DβH in adrenergic cell bodies is accompanied by an increase in the content of NA as shown by increase in the intensity of specific fluorescence in the cell bodies following cold exposure (Costa & Eränkö, 1973; Figure 10).

The slow mechanism of regulation of the synthesis of NA has also been found in the adrenal medulla, where a similar mechanism also controls the level of PNMT (Axelrod, Mueller & Thoenen, 1970).

The metabolic degradation of the rate-limiting step enzyme T-OH is slow, with a half life of at least eight days (Mueller, Thoenen & Axelrod, 1969b) compared with a half life for DβH of 12 to 15 hours (Thoenen, Kettler, Burkard & Sanar, 1971; Axelrod, 1972). This suggests that if sympathetic activity returns to the original level, the enzyme too would slowly return to its original level. If the physiological sympathetic activity is interrupted by decentralization of the sympathetic ganglia, the level of enzymes in the adrenergic cell bodies slowly decreases, the decay being

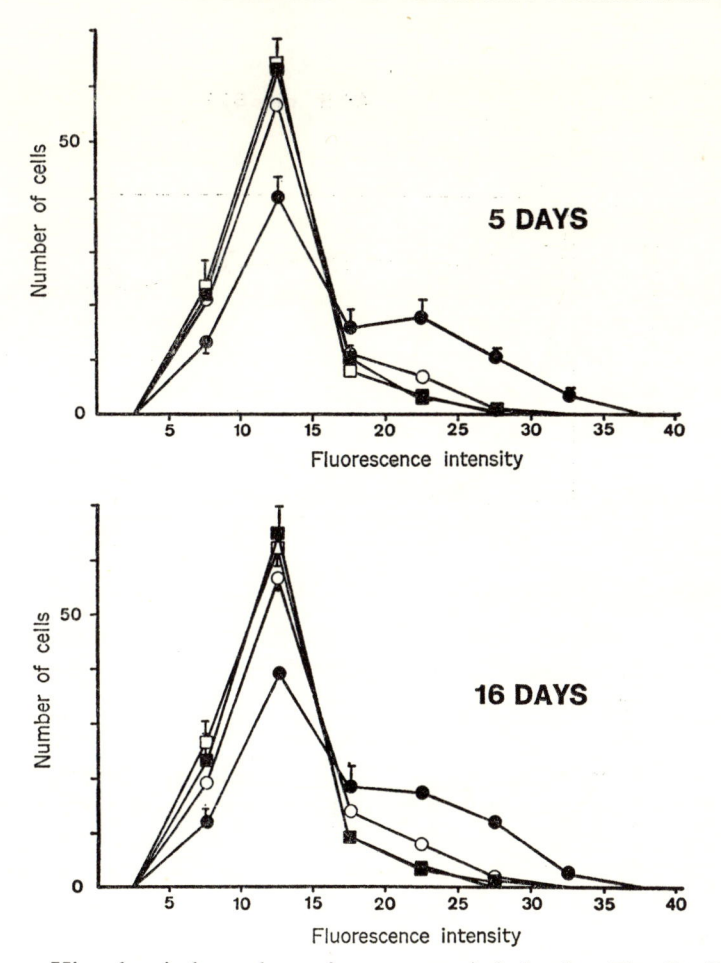

Figure 10. Histochemical correlates of trans-synaptic induction. The distribution of fluorescence intensities of nerve cell bodies in the rat superior cervical ganglion. In intact ganglia (filled circles) of rats kept at 5 °C for 5 or 16 days, the percentage of cell bodies showing intense fluorescence is significantly greater than in contralateral decentralized ganglia (filled squares), in intact (open circles) or decentralized (open squares) ganglia of rats kept at room temperature. Increased sympathetic activity produces a nerve-mediated increase of CA in the adrenergic cell bodies. Reproduced with permission from Costa and Eränkö, 1973.

dependent on the level of activity before decentralization (Hendry, Iversen & Black, 1973; Costa & Eränkö, 1973; Figure 11).

This slow, long term adaptation of the amount of rate-limiting step enzyme in response to increased nerve activity may play a role in development and in long lasting increases of sympathetic activity such as occurs during physical training, exposure to stressful environment, in pathological

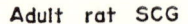

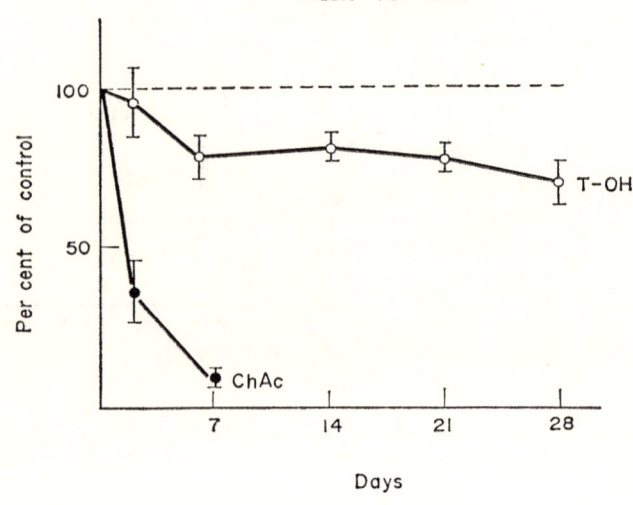

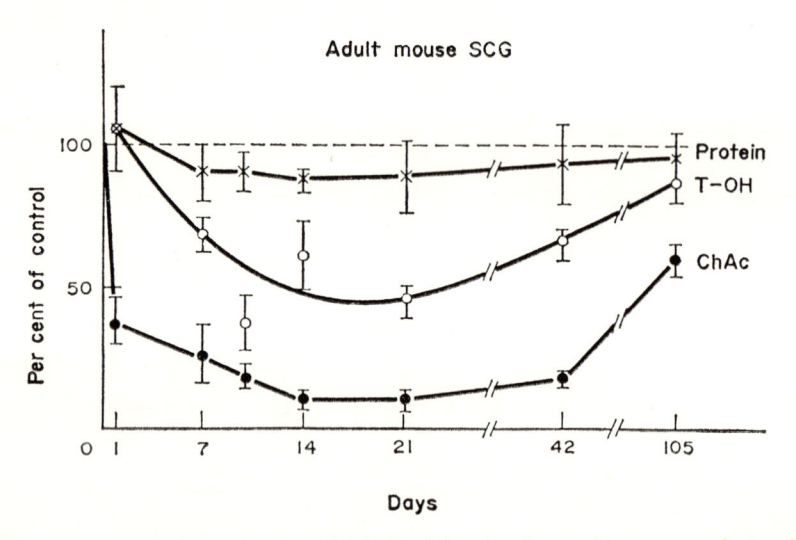

Figure 11a, b. Role of preganglionic activity in the maintenance of tyrosine hydroxylase activity in adrenergic neurons.
(a) Changes in tyrosine hydroxylase (T-OH) activity, protein and choline acetyl transferase (ChAc) activity in adult mouse superior cervical ganglia (SCG) following surgical decentralization.
(b) Same as for (a) but in adult rat superior cervical ganglion.
  Since T-OH activity present in adrenergic neurons is in part a reflection of their synaptic input, the larger fall in T-OH activity in the mouse suggests that the level of preganglionic activity is higher in mouse than in rat.
Reproduced with permission from Hendry, Iversen and Black, 1973.

situations, and in long lasting reduction of sympathetic activity during hibernation (Draskóczy & Lyman, 1967).

Physiological role of regulatory mechanisms. The rate of NA synthesis probably undergoes rapid variation in every adrenergic neuron. The range of rates is dependent on the total amount of T-OH in the neuron, whilst the amount of T-OH probably depends on the overall level of preganglionic activity. Adrenergic neurons, which are tonically active (see Chapter 2) would have a higher amount of T-OH and a higher NA turnover than adrenergic neurons which are only periodically active or silent. For instance the adrenergic neurons supplying the vas deferens, which are inactive most of the time (see Swedin, 1971) have a lower rate of NA synthesis than the adrenergic neurons which supply the submaxillary gland or the heart (Corrodi & Malmfors, 1966; Swedin, 1970). The resultant reflex increase of T-OH after reserpine treatment is more marked in adrenergic nerves to the vasculature than in those to the heart (Tarver, Berkowitz & Spector, 1971), a finding which is consistent with the different activity and reflex activation of the cardioaccelerator nerves and the vasomotor nerves (Folkow & Neil, 1971). Differences in activity, and therefore in NA synthetic enzyme levels and NA turnover, may be more important distinguishing features for adrenergic neurons than the length of their axons, i.e. 'short' and 'long' neurons (see Section 3.1).

4.2 Metabolic degradation of neuronal NA

Most NA in the body is excreted as deaminated or *o*-methylated products. It is therefore apparent that the two major routes of metabolism of NA involve the enzymes monoamine oxidase (MAO) and catechol-*o*-methyl transferase (COMT) (Axelrod, 1966; Kopin, 1972; Sharman, 1972).

4.2.1 *Localization and properties of MAO*

The term 'monoamine oxidase' designates a group of enzymes which catalyse the oxidative deamination of monoamines (Gorkin, 1966; see Blaschko, 1972b, for historical background).

MAO is localized within the double outer membrane of the mitochondria of various cell types including neurons (Snyder, Fischer & Axelrod, 1965; Tipton, 1967; Schnaitman, Erwin & Greenwaalt, 1967; Jarrott & Iversen, 1968; Giacobini & Kerpel-Fronius, 1970; Furness & Costa, 1971b; Jarrott, 1971a). There is both pharmacological and histochemical evidence for intracellular localization of MAO in all regions of adrenergic neurons

(Jarrott, 1971a; Klingman & Klingman, 1972b). Furthermore, MAO decreases in tissues following surgical or immunochemical adrenergic denervation (Burn & Robinson, 1954; Stromblad, 1956; Snyder, Fischer & Axelrod, 1965; Almgren, Andén, Jonason, Norberg & Olson, 1966; Iversen, Glowinski & Axelrod, 1966; Jarrott & Langer, 1971; Jarrott, 1971a; Jarrott & Iversen, 1971; Klingman & Klingman, 1972a). Extraneuronal MAO has also been demonstrated in both smooth and cardiac muscle cells supplied by adrenergic nerves (Su & Bevan, 1971; Jarrott, 1971a; Jarrott & Langer, 1971; de la Lande & Johnson, 1972; Burnstock, McLean & Wright, 1971; Burnstock, McCulloch, Story & Wright, 1972). The levels of MAO appear to be particularly high in vascular smooth muscle compared to COMT, the reverse being the case for most visceral muscles (Su & Bevan, 1971; de la Lande & Johnson, 1972; Burnstock, McCulloch, Story & Wright, 1972).

Multiple forms of MAO, which have different substrate specificity, inhibitor and heat sensitivity and electrophoretic mobility, have been described in a variety of tissues as well as in adrenergic neurons (Jarrott, 1971a; Goridis & Neff, 1971b; Tipton, 1973). A specific form of MAO, very sensitive to clorgyline and capable of oxidizing 5HT *in vivo*, has been found in adrenergic neurons (Goridis & Neff, 1971a, b; Coquill, Goridis, Mach & Neff, 1973) which however is also associated with non-adrenergic enteric neurons (Furness & Costa, 1971b). An increase in MAO activity has been reported in sympathetic ganglia following decentralization (Giacobini, Karjalainen, Kerpel-Fornius & Ritzén, 1970), which suggests that MAO activity may be influenced by nerves. MAO activity also appears to be influenced by hormones (Neff & Goridis, 1972).

4.2.2 *Localization and properties of COMT*
COMT is an enzyme widely distributed in many tissues (Axelrod, Albers & Clemente, 1959; Jarrott, 1971b). In adrenergically innervated tissues most of the enzyme is extraneuronal, but part is intraneuronal (Jarrott, 1971b; Jarrott & Langer, 1971; Klingman & Klingman, 1972a, b). The significantly high content of COMT in adrenergically innervated smooth muscle compared with little or no COMT present in non-innervated smooth muscle (Burnstock, McCulloch, Story & Wright, 1972) suggests that adrenergic nerves may control COMT levels.

4.2.3 *Metabolic routes of neuronal NA inactivation*
The NA that is released by nerve activity is mainly inactivated by *o*-

methylation, partly in the effector tissue and partly in the liver and kidney (Brown & Gillespie, 1957; Kopin, 1964; Axelrod, 1966; Kalsner & Nickerson, 1969b; Su & Bevan, 1971; Tarlov & Langer, 1971; Langer, Stefano & Enero, 1972). It is not possible to give reliable figures of the relative amounts of NA which follow the different metabolic pathways, since metabolic degradation of NA is affected by adrenergic activity, the relative activity and levels of the two enzymes, the efficiency of neuronal and extraneuronal uptake of NA and the geometry of the neuro-effector junction. For example, more unaltered NA than deaminated NA is recovered in the effluent from

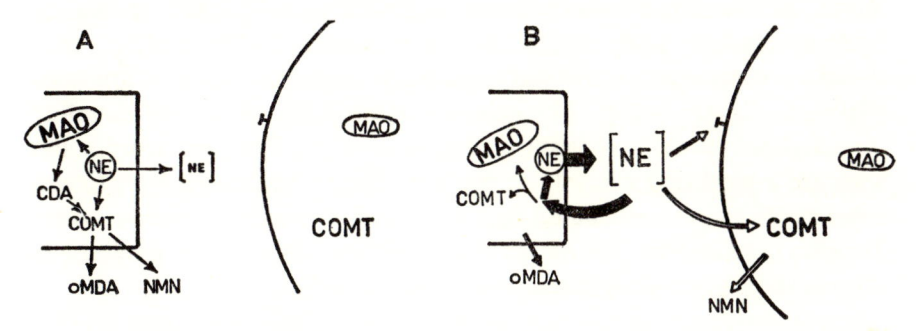

Figure 12. Schematic representation of the metabolic pathway of noradrenaline in in spontaneous (A) and in stimulation-induced (B) release. MAO: monoamine-oxidase; COMT: catechol-*o*-methyl transferase; NE: norepinephrine (noradrenaline); CDA: catechol-deaminated products; NMN: normetanephrine; OMDA: *o*-methylated deaminated metabolites.
A. Spontaneous release. Noradrenaline (NE) retained in granules leaks into the cytoplasm of terminal axons. Both NMN and OMDA products are formed in the cytoplasm before noradrenaline reaches the synaptic gap.
B. Stimulation-induced release. Nerve stimulation releases unchanged noradrenaline which is metabolized to NMN in postsynaptic sites. Released noradrenaline which is taken up by nerves is deaminated and *o*-methylated in adrenergic terminal axons to OMDA.
Reproduced with permission from Langer, Stefano and Enero, 1972.

rat and guinea-pig heart following nerve stimulation. Inhibition of MAO produces a shift towards *o*-methylation and vice versa. Impairment of neuronal uptake significantly decreases the amount of deaminated products of NA released by nerve activity, but has little effect on spontaneous release of deaminated products of NA. Conversely impairment of extraneuronal uptake reduces *o*-methylation (Langer, 1970; Tarlov & Langer, 1971; Su & Bevan, 1971). *o*-Methylation and deamination following extraneuronal uptake appears to be important in tissues where the nerves are remote from the muscle, such as blood vessels (Trendelenburg, 1972; de la Lande & Johnson, 1972; de la Lande, Jellett, Lazner, Parker & Waterson, 1974).

Intraneuronal MAO is concerned with the breakdown of cytoplasmic NA and is therefore involved in the regulation of the amount of NA incorporated into vesicular stores (see Chapter 5). Metabolism of NA through MAO in adrenergic nerves leads to the production of a deaminated glycol, 3,4-dihydroxyphenylglycol (DOPEG) (Graefe, Stefano & Langer, 1973; Dubocovich & Langer, 1973), which is capable of inhibiting T-OH activity and so participating in the regulation of T-OH activity (Rubio & Langer, 1973). Intraneuronal o-methylation of DOPEG by COMT to 3,4-dihydroxymandelic acid, which does not inhibit T-OH activity, may therefore contribute to the fast regulatory mechanism of NA synthesis (Rubio & Langer, 1973). The deaminated products of NA are therefore o-methylated in the cytoplasm or diffuse passively from the neuron, and then are o-methylated largely in the liver before being excreted (Enero, Langer, Rothlin & Stefano, 1972; Langer, Stefano & Enero, 1972). Figure 12, taken from Langer, Stefano & Enero, (1972), shows a schematic representation of the metabolic fate of neuronal NA.

4.3 Summary

Noradrenaline is synthesized throughout the adrenergic neuron, but to the greatest extent in the terminals from tyrosine in a biosynthetic pathway which involves the enzymes tyrosine hydroxylase, dopa-decarboxylase and dopamine-β-hydroxylase. The first two steps of noradrenaline synthesis occur in the cytoplasm, whilst β-hydroxylation of dopamine occurs in the granular vesicles.

Regulation of noradrenaline biosynthesis is achieved by both fast and slow mechanisms. The fast mechanism occurs in the terminals and is based on a rapid feedback inhibition by noradrenaline of its own synthesis. The slow mechanism occurs mainly in the cell body and is based on enzyme induction of tyrosine hydroxylase and dopamine-β-hydroxylase by preganglionic nerve activity. Rapid fluctuations in physiological activity of the adrenergic neuron cause fast variations in noradrenaline synthesis in a range which depends on the total amount of synthetic enzymes available. This is, in turn, determined by the long term changes in physiological activity.

Noradrenaline is metabolized within the neuron mainly by monoamine oxidase, and in extraneuronal tissues mainly by catechol-o-methyl-transferase.

5 ADRENERGIC NEUROEFFECTOR TRANSMISSION

Readers are referred to a number of reviews which approach this topic from widely different points of view (see for example, Burnstock & Holman, 1963, 1966; Bennett & Burnstock, 1968; von Euler, 1969; Holman, 1969, 1970; Speden, 1970; Axelsson, 1971; Geffen & Livett, 1971; de Potter, Chubb & De Schaepdryver, 1972; Bennett, 1973a, d; Bevan & Su, 1973). While a number of different tissues are supplied by adrenergic nerves, in this account the emphasis will be on neuromuscular transmission.

5.1 Relation between adrenergic terminal axons and muscle effectors

5.1.1 *Varicose terminal axons*

In the vicinity of the effector tissue, adrenergic axons become varicose (see Burnstock, 1970; and Figure 1) and branches intermingle with other axons to form the 'autonomic ground plexus' (Hillarp, 1946) (see also Section 3.1 and Plate 9). The extent of the branching and therefore the area of effector tissue affected by individual adrenergic neurons appears to vary with the tissue (see Norberg & Hamberger, 1964; Merrillees, 1968). Branching of single adrenergic axons has been demonstrated by fluorescence histochemistry in the iris (Malmfors & Sachs, 1965a; Olson & Malmfors, 1970) and in the intestinal wall (Furness & Costa, 1974). Adrenergic axons combined in bundles are enveloped by Schwann cells. Within the effector tissue they partially lose the Schwann cell envelope and the last few varicosities are usually completely naked (Richardson, 1964; Simpson & Devine, 1966; Merrillees, 1968; Lever, Spriggs & Graham, 1968; Verity & Bevan, 1968).

The density of adrenergic innervation (in terms of the number of axon profiles per muscle cell in cross section) varies considerably in different organs (Norberg & Hamberger, 1964; Norberg, 1967; Yamauchi & Burnstock, 1969; Burnstock, 1970). For example, it is very high in the vas deferens (Plate 9), nictitating membrane and spincteric parts of the gastrointestinal tract; it is low in the ureter, uterus and non-sphincteric part of the gastrointestinal tract. The arrangement of the adrenergic plexus varies between tissues. For example, the vas deferens is supplied by an homogenous network of adrenergic terminal axons throughout the muscle, while in blood vessels the adrenergic nerves are confined to the adventitial-medial border and rarely penetrate the smooth muscle coat (Plate 9). The implications of the different patterns of innervation on adrenergic transmission will be discussed in Section 5.6.

5.1.2 *Junctional cleft*

There is considerable variation in the relationship of individual adrenergic nerve varicosities with smooth muscle cells in different organs (see Taxi, 1965; Burnstock, 1970; Verity, 1971; Burnstock & Iwayama, 1971). Neuromuscular distances in the regions of closest apposition are in the range 10 to 30 nm in the vas deferens (Richardson, 1962; Merrillees, Burnstock & Holman, 1963; Lane & Rhodin, 1964; Yamauchi & Burnstock, 1969; Furness & Iwayama, 1972), nictitating membrane (Evans & Evans, 1964; Van Orden, Bensch, Langer & Trendelenburg, 1967; Esterhuizen, Graham, Lever & Spriggs, 1968), sphincter pupillae (Evans & Evans, 1964; Richardson, 1964; Uehara & Burnstock, 1972), rat parotid gland (Hand, 1972) and atrioventricular and sinoatrial nodes in the heart (see Yamauchi, 1969). Adrenergic nerve profiles in the vas deferens have been observed to penetrate, and perhaps terminate, deep inside smooth muscle cells with neuromuscular distances as small as 10 nm (Watanabe, 1969; Furness & Iwayama, 1971, 1972). Junctions with neuromuscular distances of less than 30 nm will be termed 'close junctions' (Plate 10).

In contrast, the closest apposition between adrenergic nerves and smooth muscle cells in blood vessels is about 50 to 80 nm (see Somlyo & Somlyo, 1968, Speden, 1970; Burnstock, Gannon & Iwayama, 1970; Graham & Keatinge, 1972; Burnstock, 1975. However, considerable variation in the minimum neuromuscular distance has been reported, from 50 to 60 nm in bronchial and auricular arteries of the rat, 70 to 100 nm in pancreatic, renal, parathyroid, mesenteric, uterine and cerebral arteries and arterioles, to about 400 nm in the large pulmonary artery (Lever & Esterhuizen, 1961; Zelander, Ekholm & Edlund, 1962; Lever, Graham, Irvine & Chick, 1965; Simpson & Devine, 1966; Verity & Bevan, 1968; Lever, Spriggs & Graham, 1968; Bell, 1969a; Iwayama, Furness & Burnstock, 1970 and Plate 10).

The longitudinal muscle coat of the intestine is in most cases devoid of direct adrenergic innervation, most adrenergic terminal axons being confined to Auerbach's plexus which lies adjacent to it (Norberg, 1964; Jacobowitz, 1965; Read & Burnstock, 1969b; Costa & Gabella, 1971).

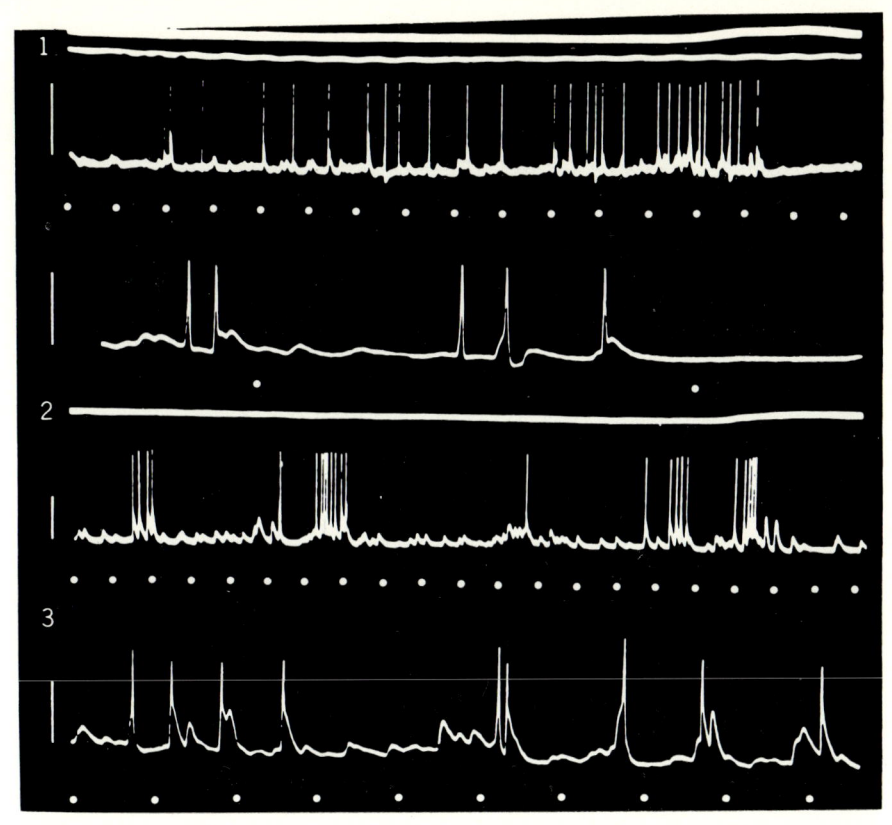

Plate 1. Spontaneous activity of adrenergic neurons. Intracellular potentials recorded from three tonically active neurons (1, 2 and 3) of the rabbit's superior cervical ganglion. Neuron 1: the upper record shows: upper trace, respiration; middle trace, blood pressure; lower trace, intracellular potentials. Below this trace, intracellular potentials of the same neuron shown with a greater time scale. Neuron 2: upper trace, respiration; lower trace, intracellular potentials. Neuron 3: intracellular potentials. Calibration 50 mV. Time mark on each record 0·5 s. *Reproduced with permission from Skok, 1973.*

Plate 2. Adrenergic neurons. *Top*: Freeze-dried section of the superior cervical ganglion of the rat prepared by the fluorescence histochemical method for CA. The adrenergic cell bodies show different intensities of fluorescence which probably reflect different NA levels. *Courtesy of Barbara Evans.* *Middle:* Whole mount preparation of pelvic plexus of the rabbit showing (white arrow) an isolated adrenergic cell body with several processes. The black arrow points to one of the extra-adrenal chromaffin cells with a long process. Fluorescence histochemical method for CA. *Bottom:* Electron micrograph of a nerve cell body of the superior cervical ganglion of the rat. The nerve cell and the axon profiles (*top right*) are surrounded by Schwann cell processes. Glutaraldehyde-Osmium fixation. *Courtesy of J. Heath.*

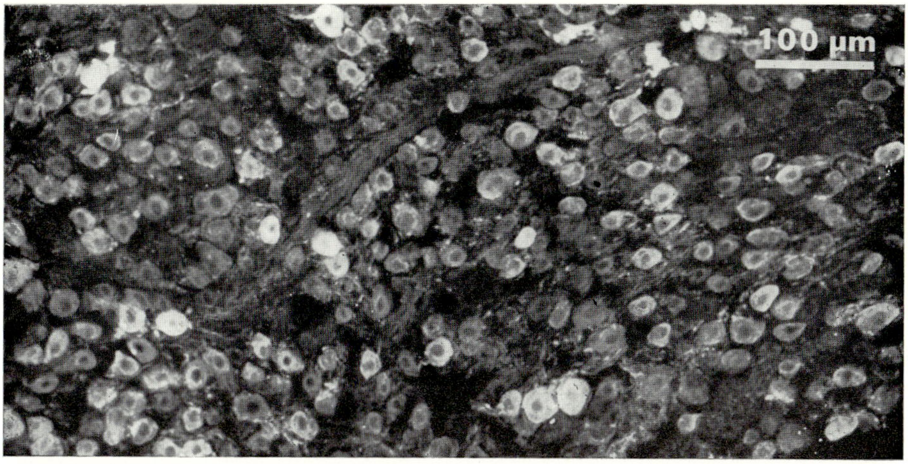
100 µm

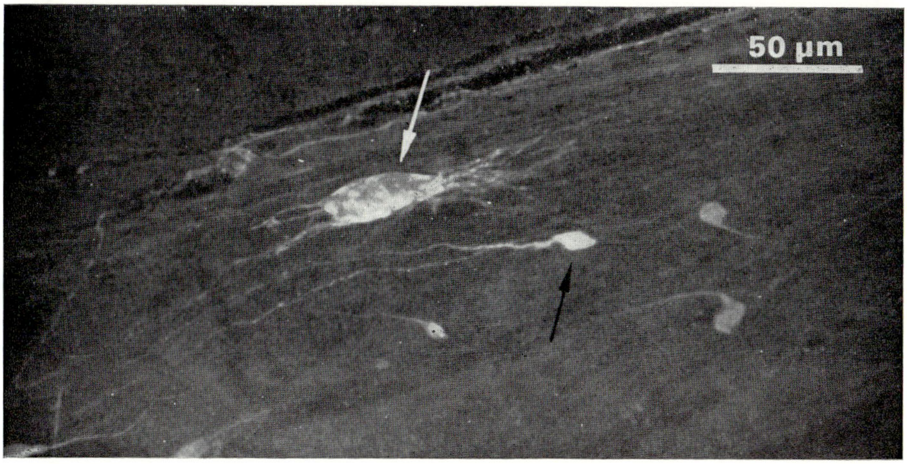
50 µm

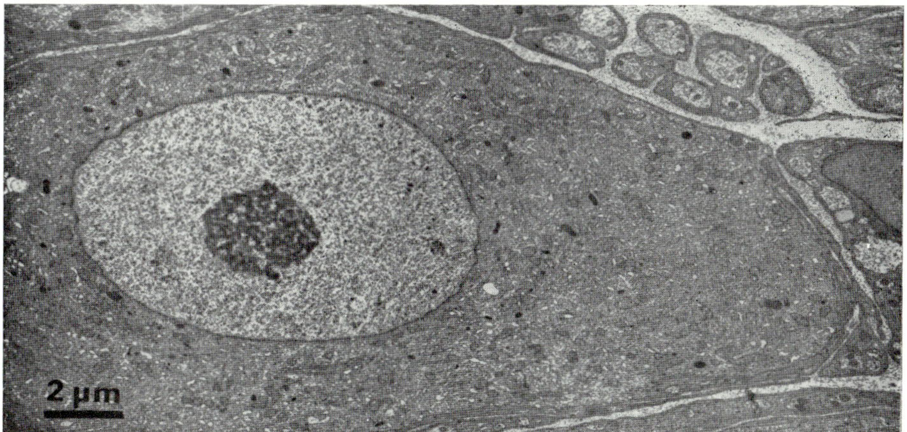
2 µm

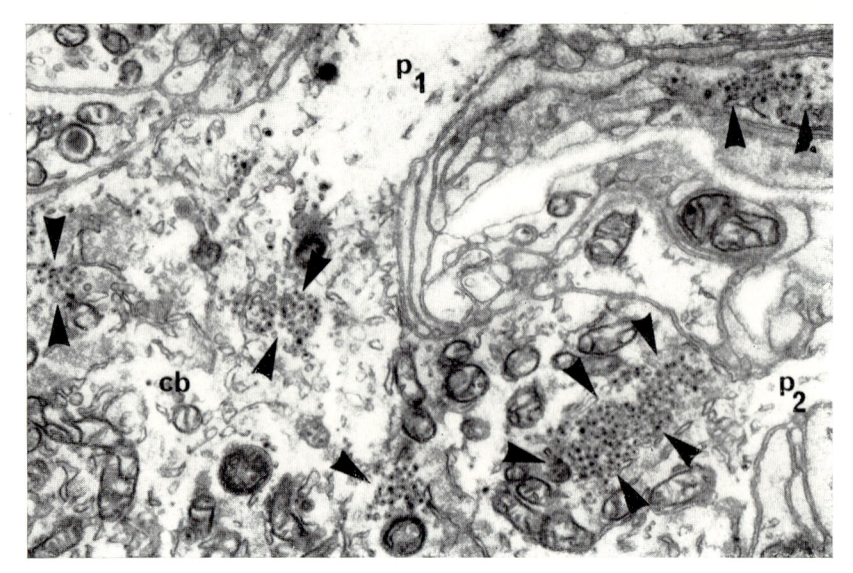

Plate 3. Superior cervical ganglion of the rat adrenergic cell body (cb) with processes (p₁, p₂ and p₃) in which clusters of small granular vesicles (*arrows*) are shown. ×36 000. *Reproduced with permission from Hökfelt, 1969.*

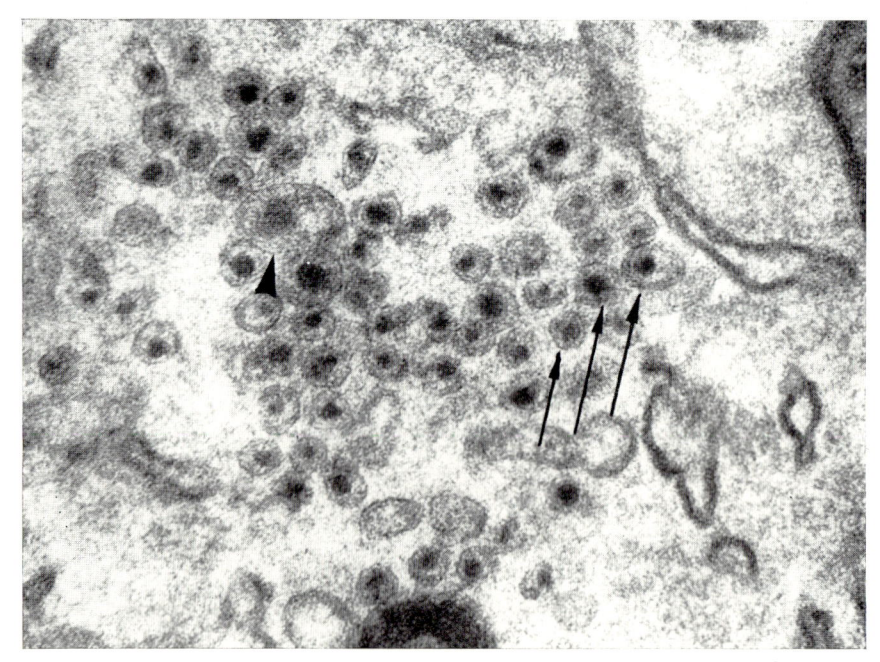

Plate 4. Small granular vesicles (*small arrows*) and a large granular vesicle (*large arrow*) in the cell body of an adrenergic neuron of the superior cervical ganglion of the rat. ×200 000. *Reproduced with permission from Hökfelt, 1969.*

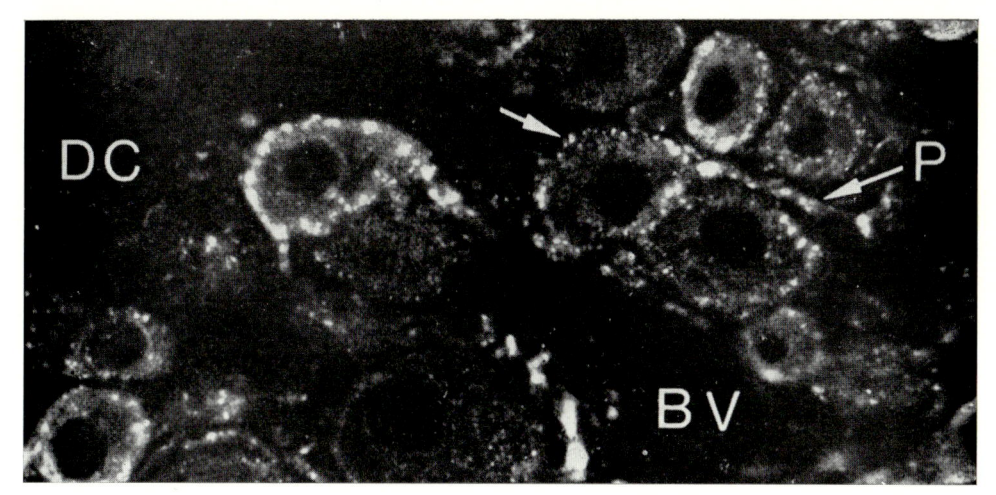

Plate 5. Adrenergic cell bodies in the superior cervical ganglion of the rat. Fluorescence histochemical technique for CA. Fluorescence intensity varies between the individual cells, one of which (DC) shows no fluorescence. Noradrenaline fluorescence appears localized in granular form (*arrow*) each granule probably corresponding to a cluster of small granular vesicles (see Plates 3 and 4). BV: blood vessel; P: neuronal process. ×525. *Reproduced with permission from Eränkö, 1972a.*

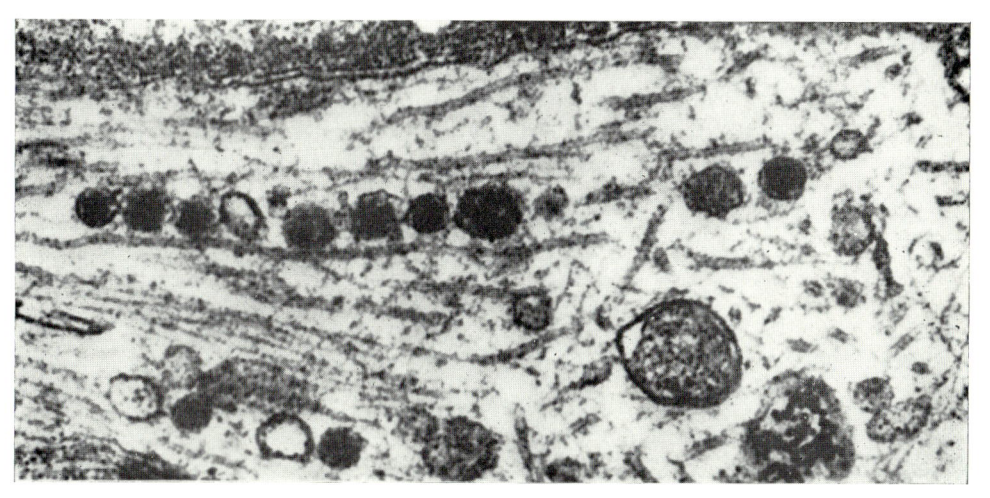

Plate 6. Adrenergic vesicles arranged parallel to microtubules in an axon of the bovine splenic nerve. Glutaraldehyde and Osmium tetroxide. ×56 000. *Reproduced with permission from Thureson-Klein, Klein and Yen, 1973.*

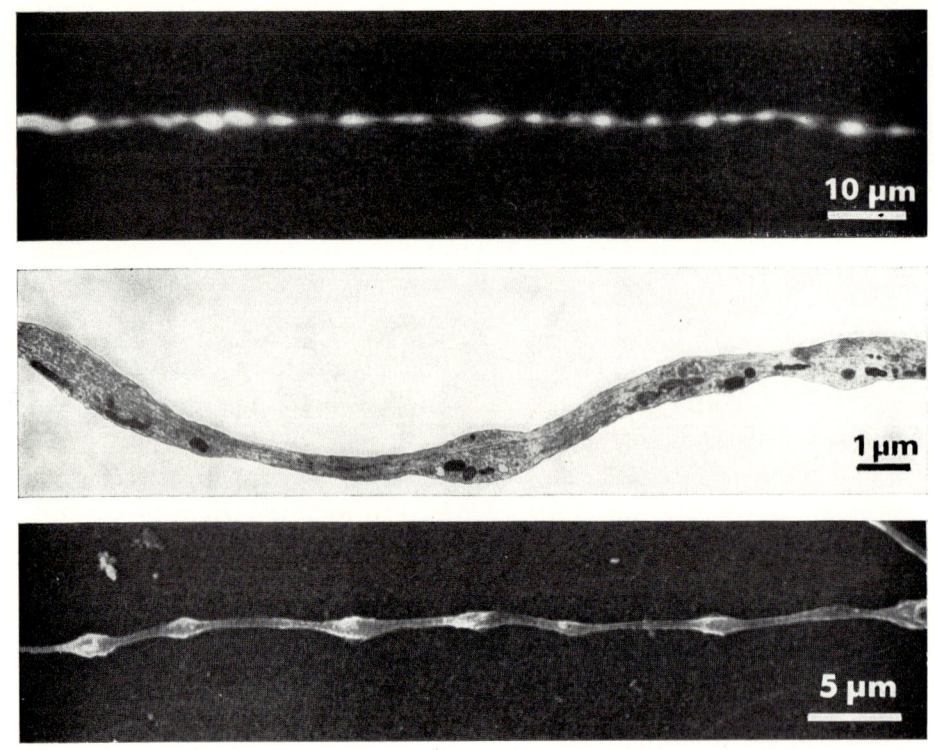

Plate 7. The varicose nature of the adrenergic axon terminal. *Top:* a single adrenergic axon in the guinea-pig mesentery. Fluorescence histochemical method for CA on a whole mount preparation. *Middle:* longitudinal section through a single sympathetic axon growing from a sympathetic ganglion in tissue culture. *Reproduced with permission from Chamley, Mark, Campbell & Burnstock, 1972. Bottom:* scanning electron microscopic view of a single sympathetic axon in a culture of new born guinea-pig sympathetic chain. *Courtesy of Julie Chamley.*

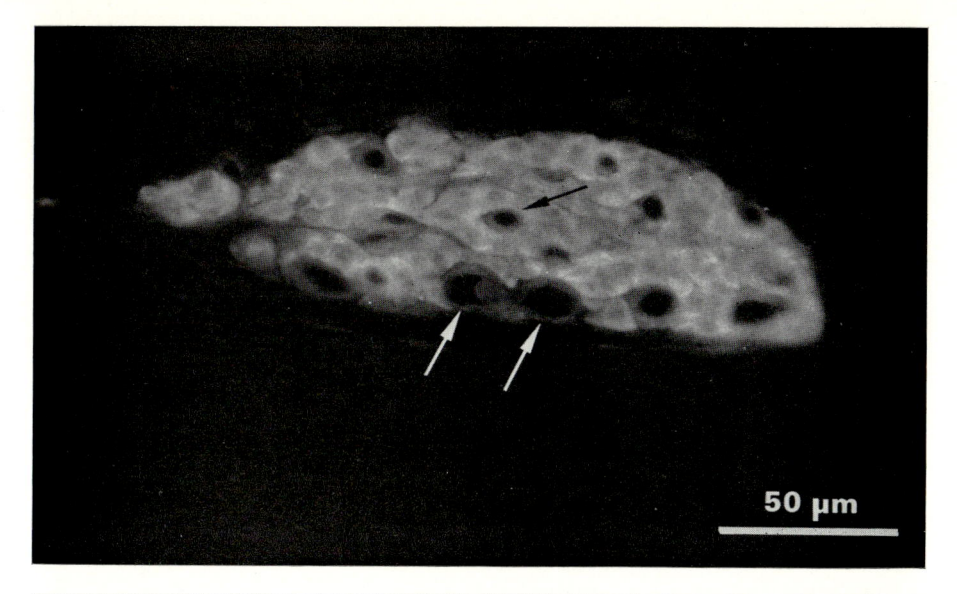

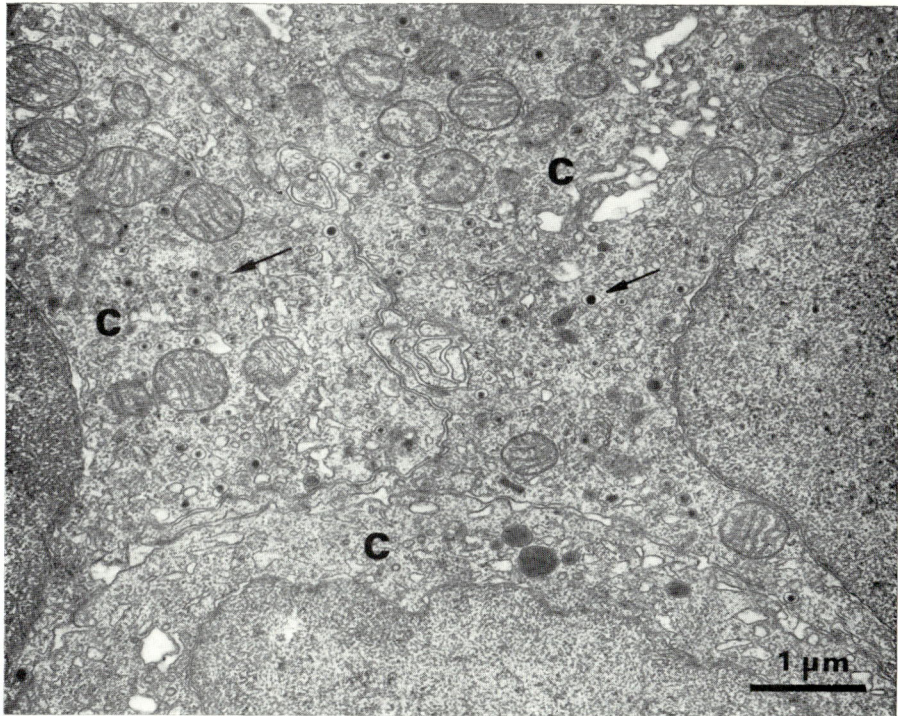

Plate 8. Extra-adrenal chromaffin cells. *Top:* section through a small paraganglion associated with the inferior mesenteric ganglion of the guinea-pig. Clusters of intensely fluorescent cells surround small vessels (*arrows*). The nuclei of the small cells can be recognized as round less fluorescent areas. Fluorescence histochemical method for CA. *Bottom:* electron microscopic view of three extra-adrenal chromaffin cells in the abdominal cavity of a young guinea-pig. A high number of mitochondria and large granular vesicles (*arrows*) characteristic of these cells are present. *Courtesy of J. Heath.*

Plate 9. Fluorescence histochemical demonstration of adrenergic nerve fibres. *Top left:* the adrenergic ground plexus in the accessory muscle of the anal sphincter of the rabbit. Whole mount preparation. Small bundles of fluorescent fibres form a dense network. *Top right:* whole mount preparations from the submucosa of the rabbit rectum showing the branching of adrenergic terminal axons. *Bottom:* comparison between the adrenergic innervation of the densely innervated vas deferens of the guinea-pig *(left)* and the rabbit ear artery *(right)* in which the adrenergic fibres are confined to the adventitial-medial border. The inner elastic membrane shows a non-specific fluorescence (autofluorescence).

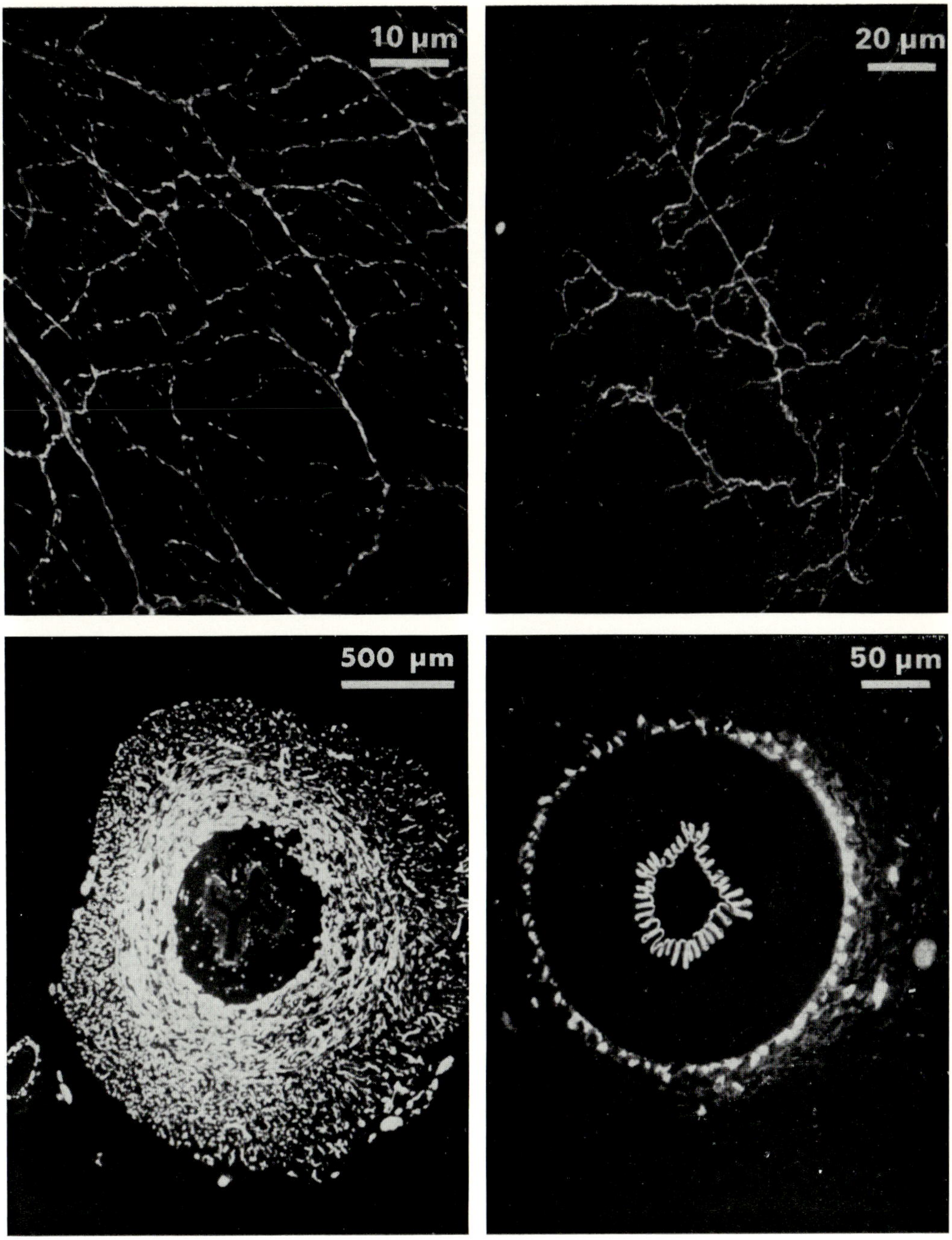
10 µm
20 µm
500 µm
50 µm

Plate 10. Electron microscopy of adrenergic fibres. *Top:* adrenergic axon profile (A) approaching to within 100 nm of the surface of a smooth muscle cell in the media of the sheep carotid artery. Basement membrane lies between the axon and the muscle cell, and many vesicles, some of which are granulated are seen in the axon profile. *Reproduced with permission from Burnstock, Gannon & Iwayama, 1970.* Bottom: close junctions between adrenergic axon terminals and muscle cells of the rat vas deferens (*left*) and of mouse vas deferens (*right*). The axons are in grooves in the surfaces of the muscle cells, and between the axon and muscle cell membranes there is no basement membrane. The rat vas deferens has been incubated in Kreb's solution containing 6-OHDA, while the mouse was injected with 5-OHDA prior to sacrifice. Most of the small vesicles have dense cores. *Reproduced with permission from Furness and Iwayama, 1971.*

A

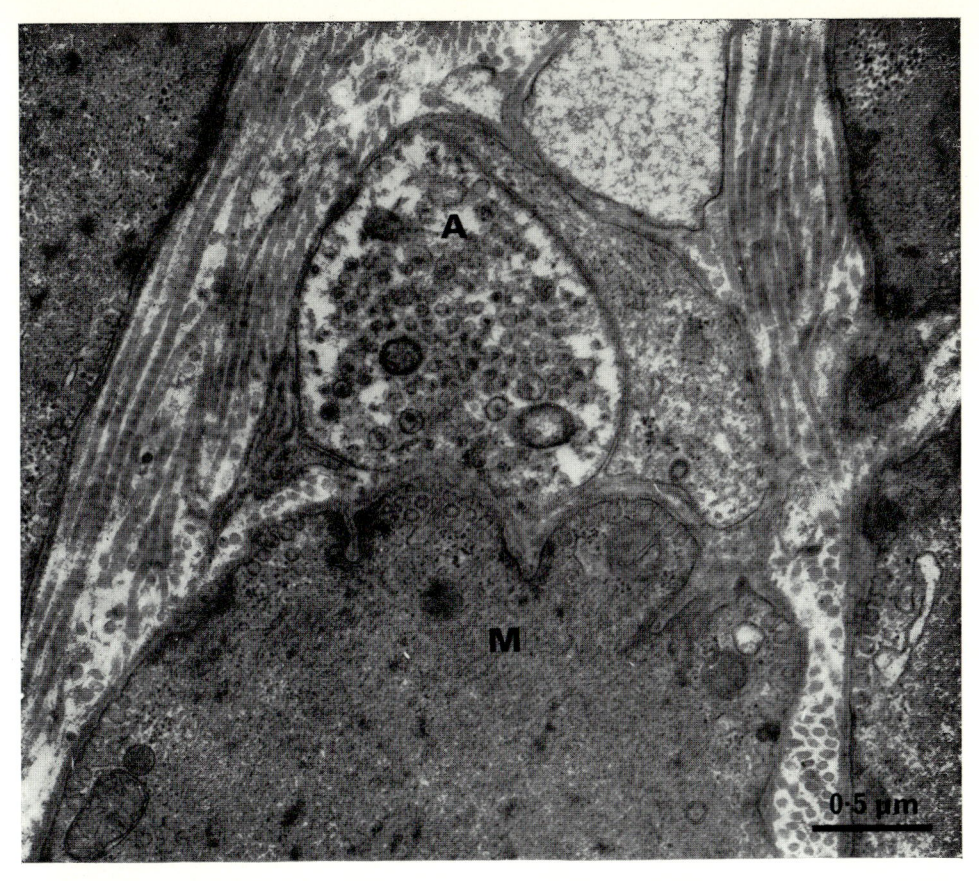

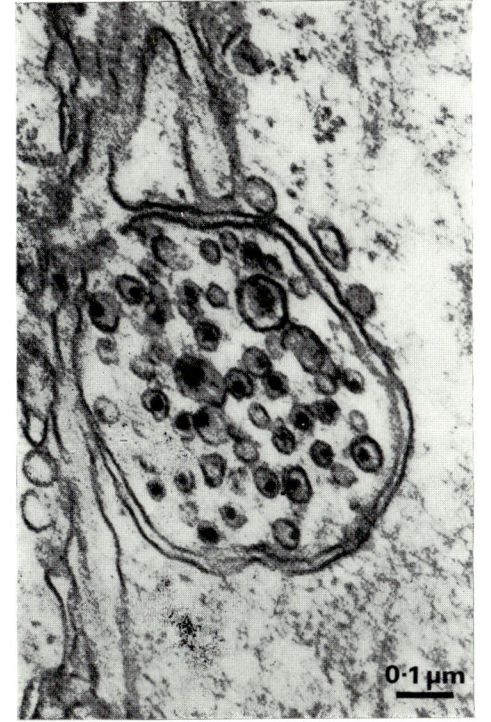

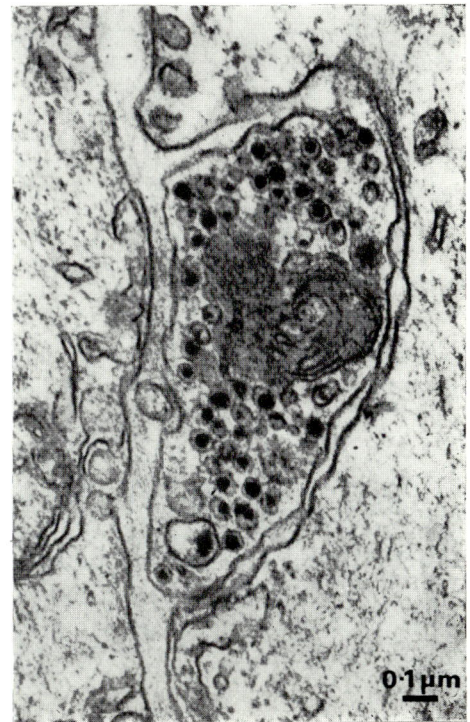

M

0·5 µm

0·1 µm

0·1 µm

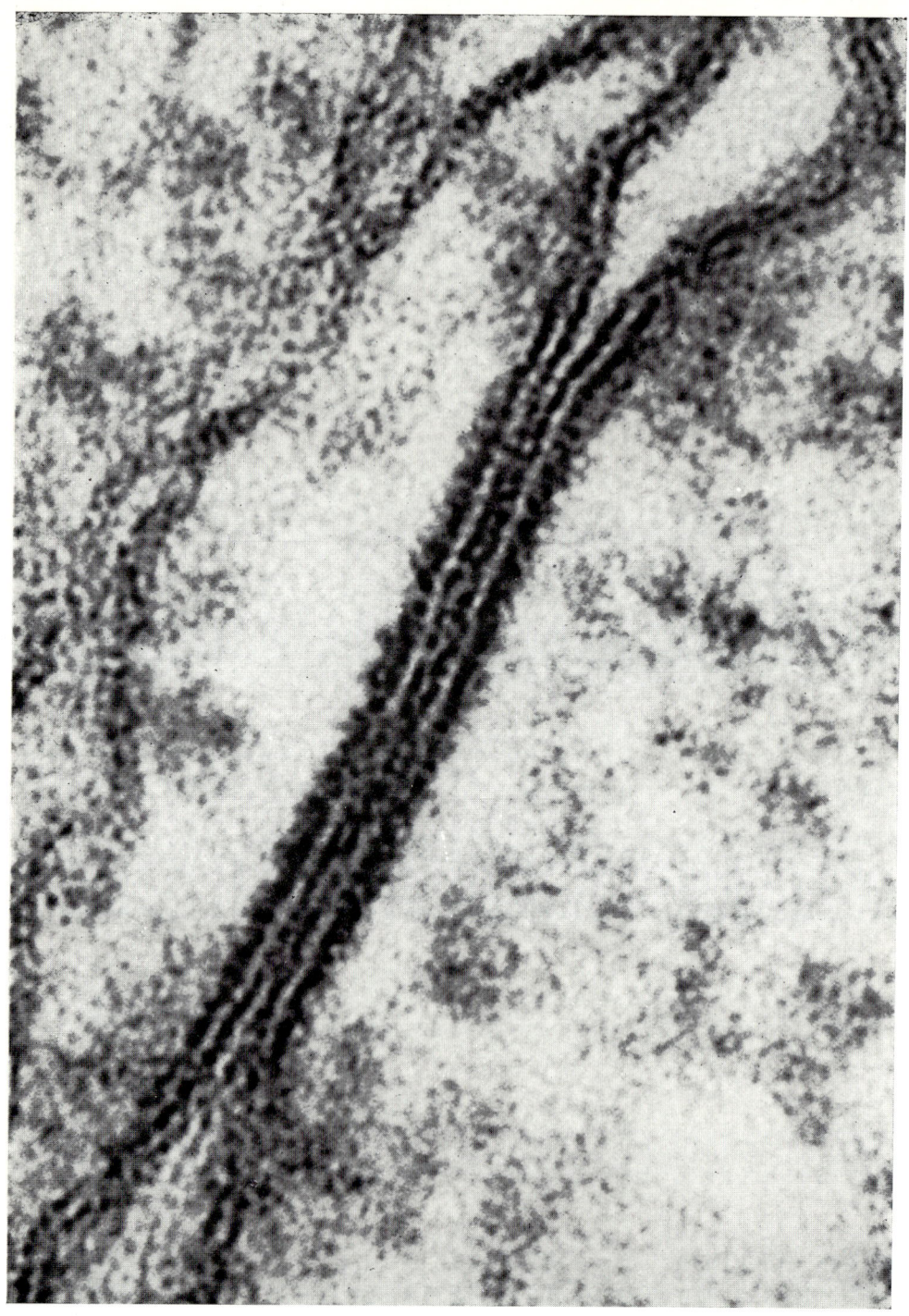

Plate 11. Gap junction between two smooth muscle cells grown in tissue culture (S_1; S_2). *Reproduced with permission from Campbell, Uehara, Mark and Burnstock, 1971.*

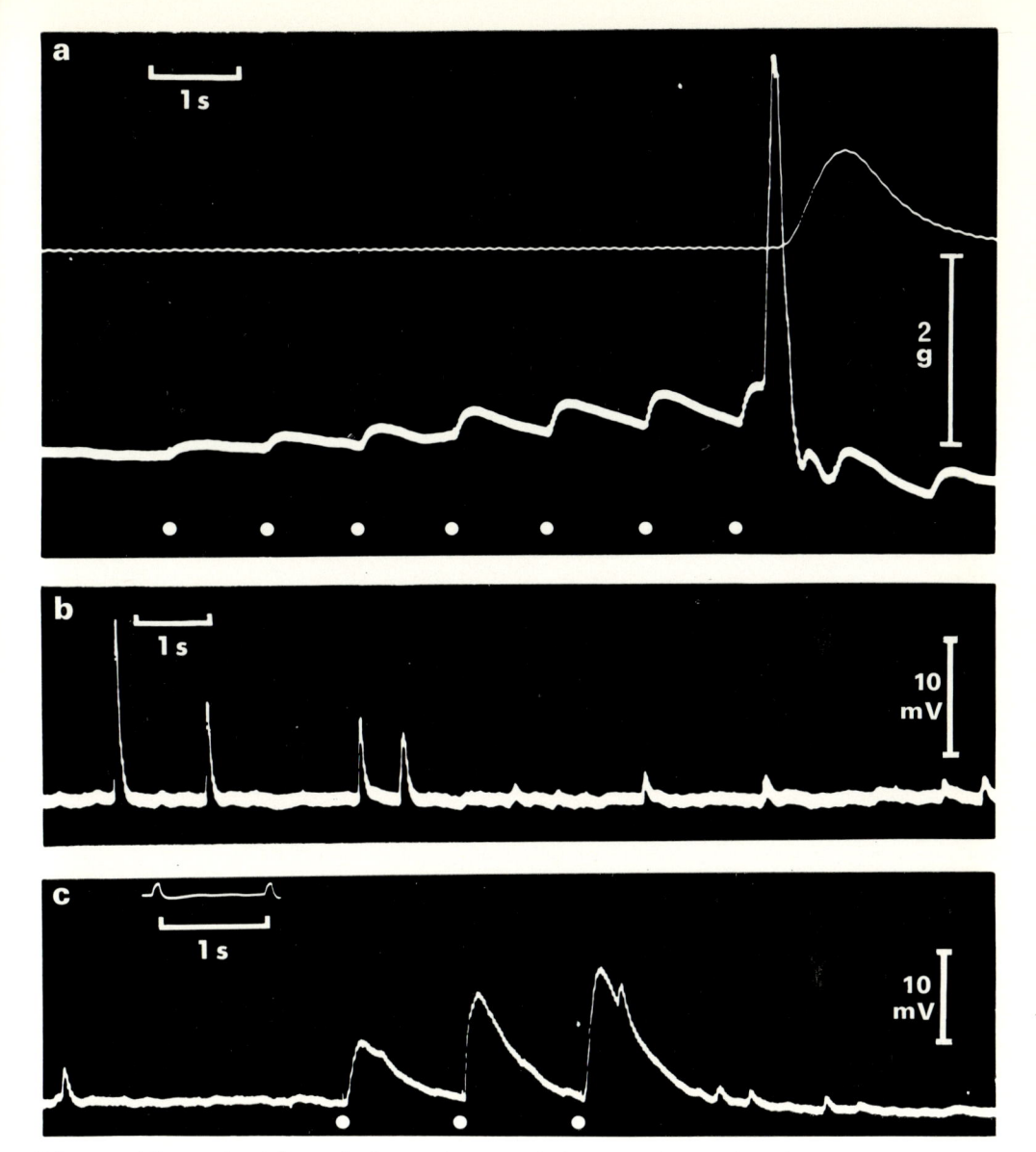

Plate 12. Electrophysiology of adrenergic transmission at excitatory junctions.

(a) Excitatory junction potentials in response to repetitive stimulation of adrenergic nerves (white dots) in the guinea-pig vas deferens. The upper trace records the tension, the lower trace the electrical activity of the muscle recorded extracellularly by the sucrose gap method. Note both summation and facilitation of successive junction potentials. At a critical depolarization threshold an action potential is initiated which results in contraction.

(b) Microelectrode recording of spontaneous junction potentials in a single smooth muscle cell of the guinea-pig vas deferens.

(c) Comparison of the time course of excitatory junction potentials in smooth muscle cells of the guinea-pig vas deferens in response to nerve stimulation (*white dots*) and spontaneous junction potentials. Note also post stimulation enhancement of spontaneous discharge. *Reproduced with permission from Burnstock and Holman, 1962a.*

Plate 13. Electrophysiology of adrenergic transmission at inhibitory junctions.

(*a, b* and *c*). Hyperpolarizations at different frequencies of stimulation, 4, 10 and 20 Hz in *a, b* and *c*, respectively, of the perivascular inhibitory adrenergic nerves recorded in single spontaneously active smooth muscle cells of the guinea-pig taenia coli. Nerve stimulated for the period indicated by the arrow. *Reproduced with permission from Bennett, Burnstock & Holman, 1966a.*

(*d*). Inhibitory junction potentials in response to stimulation of intrinsic inhibitory (purinergic) neurons. The taenia coli was stimulated transmurally at 1 Hz. Every pulse produced a large hyperpolarization.

Note that adrenergic nerves to the intestinal muscle appear to be much less effective than the intrinsic inhibitory (purinergic) nerves. *Reproduced with permission from Bennett, Burnstock and Holman, 1966b.*

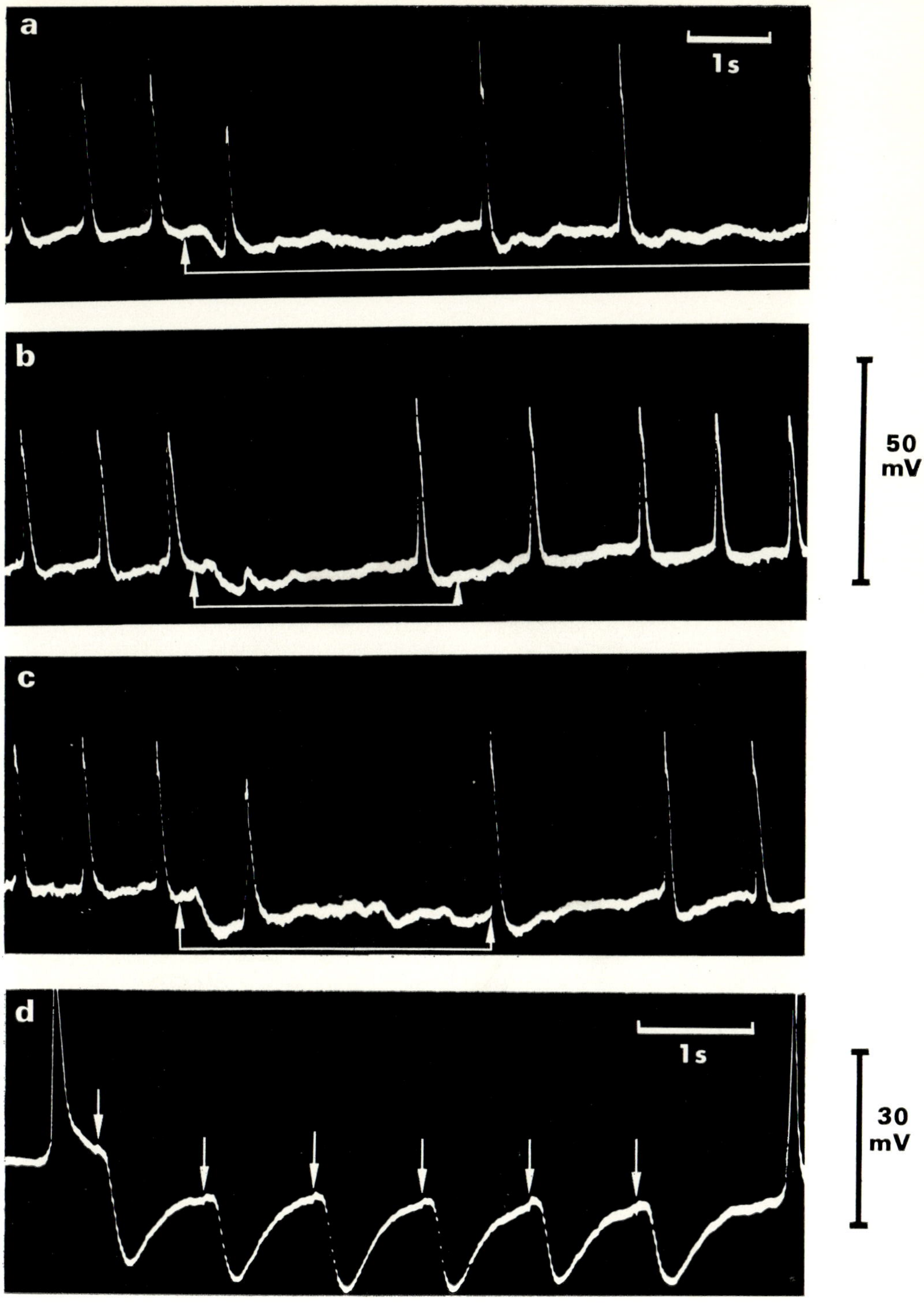
a
1s
b
50
mV
c
d
1s
30
mV

This situation is comparable to the innervation of the medial smooth muscle coat in blood vessels by the adventitial-medial plexus (Burnstock, Gannon & Iwayama, 1970). No close apposition of bordering adrenergic nerves and smooth muscle of the longitudinal coat of the intestine has been observed (Gabella, 1972). In the taenia coli of the guinea-pig, which is adrenergically innervated (Bennett & Rogers, 1967; Costa & Gabella, 1971), the closest approach of nerve varicosities to muscle cells appears to be of the order of 100 nm (Bennett & Rogers, 1967). In contrast to the longitudinal coat (with the exception of the taenia), varicose adrenergic nerve fibres have been observed within the circular muscle coat of some regions of the gastro-intestinal tract (Read & Burnstock, 1969b; Costa & Gabella, 1971). Close apposition (15 to 30 nm) of nerve and muscle membrane has been reported in the circular muscle coat in different species, including 'multiaxonal junctions' where groups of up to seven axon profiles lie in close apposition with single muscle cells (Brettschneider, 1962; Rogers & Burnstock, 1966b; Thaemert, 1966; Nagasawa & Mito, 1967). However, there is no clear evidence that these nerve profiles are adrenergic.

5.1.3 *Post-junctional specialization*

Until recently, no well-established postsynaptic structures have been established at autonomic neuroeffector junctions (see Burnstock & Iwayama, 1971).

Interruption of the basement membrane of the effector cell occurs where there are close neuroeffector junctions (Burnstock, 1970; Furness & Iwayama, 1971; Hand, 1970, 1972; Verity, 1971; Uehara & Burnstock, 1972). Various postjunctional features have been described in the regions of close apposition of muscle and nerve varicosities, although these structures are not consistently present at all junctions (see Uehara & Burnstock, 1972; Gabella, 1973). Postjunctional structures of three main kinds have been reported:

(a) aggregations of caveolae intracellulares in the smooth muscle cell membrane apposed to nerve varicosities. These have been described in arterioles of the pancreas (Lever, Graham, Irvine & Chick, 1965) and in the muscle coat of the intestine (Lane & Rhodin, 1964; Burnstock & Iwayama, 1971);

(b) areas of increased electron density of the postjunctional membrane (Taxi, 1965; Nagasawa & Mito, 1967) or desmosome-like structures (Ivanov, 1971) at the close-contact area between axons and muscle cells. However, these appear to be a rare feature and may represent mechanical attachment points between axon and muscle cell, not necessarily confined to neuromuscular regions (Ivanov, 1971);

(c) subsurface cisternae have been observed in the smooth muscle where it is closely apposed (less than 20 nm) to nerve terminals in the vas deferens (Richardson, 1962; Merrillees, Burnstock & Holman, 1963; Burnstock & Iwayama, 1971), in parotid gland (Hand, 1972) and sphincter pupillae (Uehara & Burnstock, 1972). High resolution electronmicrographs of the guinea-pig sphincter pupillae show that there is a highly organized cytoplasmic zone between the cell membrane and the subsurface cisternae where an electron opaque layer, 4 to 6 nm thick, lies between the muscle membrane and the distal membrane of the cisterna. The functional significance of this postsynaptic structure is not known.

5.1.4 *Muscle effector bundle and nexuses*

There is now clear evidence that in most tissues the muscle effector is not represented by the single smooth muscle cell, but rather by a bundle of cells in electrical continuity with each other (see Bennett & Burnstock, 1968; Burnstock & Iwayama, 1971). Sites of electrotonic coupling are represented morphologically by areas of close apposition between the plasma membranes of adjacent cells called 'nexuses' (Dewey & Barr, 1962). High resolution electronmicrographs have shown that the membranes at these nexal junctions consist of 'gap junctions' (Plate 11) rather than 'tight junctions' (Uehara & Burnstock, 1971; Campbell, Uehara, Mark & Burnstock, 1971). The number and arrangement of nexuses in muscle effector bundles in different organs and their relation to the density of adrenergic innervation has not yet been analysed. In some tissues (e.g. large arteries) it is possible that there is little electrical continuity between muscle cells (Holman, 1969; Speden, 1970).

5.2 Release of NA

There is strong evidence that nerve impulses release only the NA which is bound and stored in vesicles and not extravesicular NA (see Smith & Winkler, 1972). NA is released 'en passage' from all the varicosities of the adrenergic axon (Hillarp, 1959; Burnstock & Holman, 1963; Malmfors, 1965; Bennett & Burnstock, 1968; Furness, 1970a.)

Physiological or electrical stimulation of adrenergic neurons at low frequencies results in release of NS from terminals, although the NA content remains constant (Brown & Gillespie, 1957; Hertting & Axelrod, 1961; Davies, 1963; Dearnaley & Geffen, 1966; Iversen, 1967), the fast mechanism of biosynthesis of NA being activated by nerve stimulation (see Section 4.1). At higher 'supraphysiological' frequences of stimulation, transmission is impaired and is abolished when the total NA store is reduced by only 2 to 3 per cent (Folkow, Häggendal & Lisander, 1967). This indicates that only a small fraction of total NA store is available for release (Iversen, 1967) and it has been found that this fraction consists largely of newly synthetized NA (Kopin, Breese, Krauss & Weise, 1968). Failure of reserpine to block adrenergic transmission, even following depletion of 99 per cent of the NA store, is consistent with the concept of a small compartment of readily available NA (Lee, 1967; Wakade & Krusz, 1972, see Section 5.7).

Most workers have reported that the amount of NA released per impulse increases with frequency of stimulation (Hughes, 1972), at least at physiological frequencies of 2-16 Hz (Blakely & Brown, 1966; Folkow, Häggendal & Lisander, 1967; but see Farnebo & Malmfors, 1971), while at higher frequencies, the amount of NA per impulse decreases (Brown & Gillespie, 1957; Zimmerman & Whitmore, 1967; Folkow, Häggendal & Lisander, 1967; Häggendal, Johansson, Jonason & Ljung, 1970).

NA is released spontaneously in the absence of nerve activity in packets apparently in a similar manner to the spontaneous release of ACh at the skeletal neuromuscular junction (see *Section 5.5*). Nerve impulses release packets of NA in progressively increasing amounts (facilitation) until there is a sufficient amount of transmitter to evoke the effector response (Burnstock, Holman & Kuriyama, 1964; Burnstock & Holman, 1966).

5.2.1 *Mechanism of release*

Several questions arise concerning the mechanism of both spontaneous and nerve-induced release of NA from adrenergic nerves.

The possibility that NA is released from adrenergic nerves by a process of exocytosis similarly to the release of CA from the adrenal medulla has been extensively reviewed (see Smith & Winkler, 1972; De Potter, 1973).

There is strong evidence that the mechanism of secretion of catecholamines from the adrenal medulla is a process of exocytosis in which the adrenal granular vesicle discharges its entire soluble contents into the extracellular space, leaving the vesicle membrane behind in the cytoplasm (de Robertis, 1964; Kirshner & Kirshner, 1971; Smith & Winkler, 1972). The main evidence for exocytosis of CA vesicles from the adrenal medulla is that during splanchnic nerve stimulation the appearance of CA in the effluent from adrenal glands is accompanied by all the soluble components of the storage granules, but not the cytoplasmic proteins such as lactate dehydrogenase or PNMT (Schneider, Smith & Winkler, 1967). Chromogranin A, DβH and ATP are released in approximately the same proportions as those in which they are found in isolated vesicles (see Douglas, 1968; Kirshner & Kirshner, 1971; Geffen & Livett, 1971). Further support for exocytosis comes from subcellular fractionation studies, which have shown that there is no decrease in the lipid (and therefore granule membrane) content of the vesicle fractions from CA-depleted glands (Poisner, Trifaró & Douglas, 1967; but see Winkler, 1971) or in membrane-bound DβH (see Smith & Winkler, 1972). There has also been electron microscopic demonstration of granule-free vesicles following stimulation of the adrenal medulla (Malamed, Poisner, Trifaro & Douglas, 1968) and of granule-containing invaginations of the chromaffin cell membranes reminiscent of the exocytosis process (Grynszpan-Winograd, 1971).

Release of neurotransmitter at the skeletal neuromuscular junction is probably also carried out by the process of exocytosis as first discussed by Katz (1962). The demonstration of synaptic vesicles (Robertson, 1956), and miniature end plate potentials (del Castillo & Katz, 1954) led to the hypothesis that vesicles represented the morphological basis of quantal packets of neurotransmitter (del Castillo & Katz, 1954; Katz, 1969, 1971). Proteins and possibly ATP are released together with ACh during stimulation of cholinergic fibres (Musich & Hubbart, 1972; Silinsky & Hubbard, 1973). There are convincing visual demonstrations with the freeze-etching technique of synaptic vesicles fusing with the presynaptic membrane and opening into the synaptic cleft (Nickel & Potter, 1970, 1971; Pfenniger, Akert, Moore & Sandri, 1971).

The hypothesis that NA might also be released from adrenergic nerve terminals by exocytosis is becoming widely accepted (see Smith & Winkler, 1972; De Potter, 1973), although the roles of the different types of granular vesicles (see Section 3.1) in this process are not clearly established. The evidence for exocytosis of NA in adrenergic nerves is based on measurements showing the release both *in vitro* and *in vivo* of the vesicular proteins, DβH and chromogranin A together with NA while proteins such as dopa-decarboxylase located outside the vesicles are not released (see De Potter, Chubb & De Schaepdryver, 1972; Smith & Winkler, 1972).

ATP may be released together with proteins and NA from adrenergic nerves. Stimulation of adrenergic nerves is followed by release of small amounts of tritium from taeniacoli incubated in ^{3}H-adenosine, which is blocked together with NA by guanethidine (Su, Bevan & Burnstock, 1971; Paddle & Burnstock, 1974).

Electronmicroscopic pictures showing fusion of adrenergic vesicles with axon membranes support the concept of release of NA by exocytosis (Fillenz, 1971). A problem which is still debated concerns the relative amounts of NA and protein released from nerve terminals. Weinshilboum, Thoa, Johnson, Kopin and Axelrod (1971) found that the ratio of the NA to DβH released was the same as that of NA to the soluble DβH in the isolated vesicle fraction of vas deferens. On the other hand, several workers using the spleen found that the ratio of DβH to NA released was lower than that found in the soluble lysate of the adrenergic vesicles (De Potter, Chubb & De Schaepdryver, 1972). It is possible that these different results could be explained by the differences in the adrenergic vesicle population of the two tissue studies (Section 3.1). The similar turnover of NA and DβH at the adrenergic terminals supports the exocytosis hypothesis. DβH plasma level is a more reliable marker for adrenergic nerve activity than NA, since DβA found in the blood stream appears to originate almost exclusively from adrenergic nerves and, unlike NA, it is not rapidly inactivated (Axlerod, 1972; Rush & Geffen, 1972).

Good correlation between adrenergic activity and serum level of DβH has been found in experimental animals (see Axelrod, 1972), but in humans the enzymic assay for serum DβH gives a wide normal range of activity making detection of pathological deviations difficult (Goldstein, Fuxe & Hökfelt, 1972; Nagatsu & Udenfriend, 1972). A sensitive and specific radioimmunoassay for DβH that is not dependent on its labile enzymatic activity has been developed (Rush & Geffen, 1972) and applied to human pathological situations (Geffen, Rush, Louis & Doyle, 1973a, b).

The finding that drugs capable of disrupting microtubules, such as colchicine, vinblastine and cytochalasin B prevent the release of NA and DβH from nerves following stimulation, lead to the suggestion that exocytosis involves a contractile process (Axelrod, 1972).

A cation-exchange hypothesis for the mechanism of NA release from adrenergic nerves has been proposed recently which is based on the presence of sulphomucopolysaccharides (SMPS)-protein complexes in adrenergic vesicles as well as in adrenal medulla and cholinergic vesicles (Uvnäs, 1973). According to this hypothesis, a number of transmitter vesicles become attached to the varicosity membrane as a result of the impulse-induced depolarization of the nerve membrane. The membrane of these vesicles and of the varicosities fuse to form transient loci of increased cation permeability. This transient increase in cation permeability, similar to that occurring in the nerve membrane, occurs at the sites of membrane fusion. Cations enter and release amines from the SMPS-protein complex during depolarization periods. The size of the quantal package released at a nerve terminal does not depend on the quantity of transmitter in a vesicle, but rather on the ion conductance of the fused vesicle-nerve terminal membrane. Reuptake of NA is also accounted for by this hypothesis (Uvnäs, 1973).

5.2.2 *Role of calcium in NA release*

Nerve mediated release of NA from adrenergic varicosities, as for the release of CA from adrenal medullary cells (see Douglas, 1968; Viveros, Arqueros & Kirshner, 1969; Viveros, Arqueros, Connet & Kirshner, 1969) is dependent on the presence of calcium ions in the extracellular fluid (Smith & Winkler, 1972). Output of NA and DβH from perfused rabbit heart or cat spleen and colon following adrenergic nerve stimulation is strongly reduced in calcium-free media (Huković & Muscholl, 1962; Boullin, 1967; Kirpekar & Misu, 1967; Smith & Winkler, 1972). Conversely increase of Ca^{2+} leads to an increase in NA release during stimulation (Stjärne, 1973a, b). It seems likely, by analogy with the skeletal neuromuscular junction (Katz, 1969), that when an impulse reaches the nerve terminals, calcium enters the terminals and provides an essential link in 'stimulus secretion coupling' (see Douglas, 1968) for DβH and NA to be released from the nerve terminal into the synaptic cleft. It is noteworthy that tyramine, which does not need CA^{2+} to release NA from nerves, is incapable of releasing DβH (see Chubb, De Potter & De Schaepdryver, 1972). It has been suggested that release of NA is due to the outward movement of potassium during membrane depolarization which modulates the calcium entry into the adrenergic terminal (Kirpekar, Pratt, Puig & Wakade, 1972; Smith & Winkler, 1972).

It has also been suggested that calcium activates contractile microfilaments in the neuronal membrane to produce an opening which allows the discharge of NA as well as the macromolecules released by exocytosis (Axelrod, 1972).

5.2.3 *Physiological control of NA release*

The amount of NA which is released from adrenergic nerves mainly depends on the firing rate of the action potentials which invade the terminals (see Haefely, Hürlimann & Thoenen, 1965; Iversen, 1967). However, in recent years various local mechanisms for control of NA release involving substances such as DA and NA, prostaglandins, angiotensin and ACh have been proposed (see Figure 3).

NA and DA released from adrenergic nerves act on adrenergic receptors of the α-type that are particularly sensitive to DA, located on the membrane of the adrenergic terminals; this leads to inhibition of further release of NA. Evidence for this hypothesis is based on the following arguments: α-adrenoreceptors are located on adrenergic terminals (Farnebo & Hamberger, 1971; Kirpekar & Puig, 1971; Langer, Adler, Enero & Stefano, 1971; Enero, Langer, Rothlin & Stefano, 1972; Starke, 1972a, b); DA is released together with NA (Austin, Livett & Chubb, 1967; Musacchio, Fischer & Kopin, 1966; Collins & West, 1968a; Snider, Almgren & Carlsson, 1973) and is very effective on presynaptic α-adrenoceptors, the effect being antagonized by apomorphine and haloperidol (Snider, Almgren & Carlsson, 1973; Langer, 1973a); stimulation of the presynaptic receptors leads to a decrease of NA release from nerves (Farnebo & Hamberger, 1971; Farnebo & Malmfors, 1971; McCulloch, Rand & Story, 1972; Starke, 1972a, b; Starke, Wagner & Schümann, 1972; Stjärne, 1972b; Werner, Starke & Schümann, 1972; Armstrong & Boura, 1973; Bralet & Rochette, 1973; Kirpekar, Furchgott, Wakade & Pratt, 1973); blockade of the presynaptic receptors by phenoxybenzamine (PBA) or phentolamine enhances the release of NA from adrenergic nerves (Cripps, Dearnaley & Howe, 1972; Enero, Langer, Rothlin & Stefano, 1972; Bennett, 1973b; McCulloch, Rand & Story, 1972; Hughes, 1973; Starke, 1972a, b; Langer, 1973a) and of DβH (De Potter, Chubb, Put & De Schaepdryver, 1971; Johnson, Thoa, Weinshilboum, Axelrod & Kopin, 1971) in the same proportions with DβH as found in intraneuronal vesicles (De Potter, Chubb, Put & De Schaepdryver, 1971); the effects of PBA and phentolamine are not related to their ability to inhibit both neuronal and extraneuronal NA uptake (Cripps & Dearnaley, 1971; Farnebo & Hamberger, 1971; Langer, Adler, Enero & Stefano, 1971; Starke, Montel & Schümann, 1971; Enero, Langer, Rothlin & Stefano, 1972; Langer, 1973a); nor are these effects due to the blockade of postsynaptic α-receptors, since the increase of NA release by PBA also occurs in tissues where there are only β-adrenoceptors (Langer, Adler, Enero & Stefano, 1971; McCulloch, Rand & Story, 1972; Starke, 1972a, b; Langer, 1973b).

The amount of NA and DA released from adrenergic nerves must reach a threshold concentration outside the nerve terminals in order to activate feed-back inhibition of their release (Enero & Langer, 1973). In some tissues the amount of NA released spontaneously reaches this threshold concentration, because PBA enhances the amount of NA released spontaneously (Hughes, 1972). As soon as adrenergic neurons fire action potentials, the feed-back inhibitory mechanism is activated, as indicated by a marked decline of NA released after the first pulse, a decline which is prevented by PBA (McCulloch, Rand & Story, 1972; Langer, 1973a).

Another mechanism of control of NA release involves prostaglandins of the E types (PGEs) (Wennmalm, 1971; Smith, 1972a; Johnson, 1973; Hedqvist, 1973a; Horton, 1973). Stimulation of sympathetic nerve is followed by an increase of formation and release of PGEs in the effector tissue (Swedin, 1971; Ferreira, Moncada & Vane, 1973; Junstad & Wennmalm, 1973). PGEs have been shown to inhibit both the amount of NA released by nerve stimulation (see Hedqvist, 1970, 1973a) and the effect of NA on postsynaptic receptors (Weiner & Kaley, 1969; Kadowitz, 1972; Ferreira, Moncada & Vane, 1973). Conversely inhibition of PGE synthesis leads to an increase of NA release and to a potentiation of the adrenergic response (Freholm & Hedqvist, 1973; Ferreira, Moncada & Vane, 1973). PGEs inhibit NA release, probably by depressing the affinity of the secretory mechanism for Ca^{2+} (Stjärne, 1973b).

These two mechanisms, involving presynaptic α-adrenoceptors and PGEs, are independent (Fredholm & Hedqvist, 1973; Starke & Montel, 1973; Hedqvist, 1973b) and both are more effective with low frequency adrenergic nerve activity (Stjärne, 1973c).

Another group of prostaglandins of the F type (PGFs) may also have a modulatory effect on adrenergic transmission, but in this case by a mechanism of facilitation of NA release during nerve activity (see Kadowitz, Sweet & Brody, 1972; Powell & Brody, 1973). PGFs have been found in skeletal muscle and in sympathetic nerves in the dog (Karim, Hillier & Devlin, 1968); they are released in tissues during sympathetic stimulation (Ferreira & Vane, 1967) and enhance sympathetic neurotransmission in some vascular beds (Kadowitz, Sweet & Brody, 1972; Powell & Brody, 1973).

Angiotensin has been found to potentiate adrenergic responses by increasing NA release from nerves during stimulation (Benelli, Della Bella & Gandini, 1964; Zimmerman & Gisslen, 1968; Kirah and Khairallah, 1969; Schümann, Starke & Werner, 1970; Hughes & Roth, 1971; Bell, 1972; Starke & Schümann, 1972). It has been suggested that this hormone may have a physiological role in the modulation of adrenergic transmission (Schümann, 1970; Bell, 1972; Smith, 1972a).

Another possible mechanism of control of NA release involves ACh which in low concentrations reduces NA release from adrenergic terminals by acting on muscarinic receptors (see Section 2.5). It has been suggested that this mechanism has physiological significance in the heart, where stimulation of the vagus reduces the amount of NA released by stimulation of the sympathetic nerves (Löffelholtz & Muscholl, 1970; Muscholl, 1973).

5.2.4 *Quantal hypothesis and quantitative estimation of NA release*

While there is electrophysiological evidence for release of packaged NA from adrenergic terminals (Burnstock & Holman, 1962a, 1966), it is still not possible to say whether the packages are quantal i.e. uniform in size (see Holman, 1970; Katz, 1971). The amplitude of spontaneous excitatory junction potentials (SEJP's) recorded in smooth muscle cells of the vas deferens show a skewed frequency distribution. Consecutive SEJP's fluctuate randomly in amplitude, although there is some hint of grouped discharges of 2 or 3 comparably sized SEJP's at intervals of ten minutes or more (Burnstock & Holman, 1962a). Stimulation of hypogastric nerves with graded voltages, produced some evidence for step-wise reduction in EJP amplitude (Furness, 1970a), but this indicates multiple innervation rather than evidence for quantal release. Interpretation is complicated by electrical coupling between neighbouring muscle cells and because transmitter is released from successive varicosities of one or more nerves at various distances from muscle cells. Thus, junction potentials recorded in single muscle cells cannot be used for the analysis of transmitter released from individual varicosities.

Attempts have been made to calculate the amount of NA released from a single varicosity in adrenergic nerve terminals. The authors listed in Table 3 were well aware that some of the assumptions necessary to the calculations were speculative and that the oversimplifications in their models could lead to misleading figures. Nevertheless there is good agreement between several authors concerning the fraction of the total NA store in a varicosity which is released per pulse.

In most of the experiments in Table 3, phenoxybenzamine was used to block the binding of NA to the receptors and also to block uptake of NA into nerves and muscles; in this way, it was assumed that a better approximation of the true release of NA was obtained. However, since phenoxybenzamine increases the amount of transmitter released per impulse (see

Table 3

Tissue	Fraction of total NA store released per pulse from a single varicosity	Authors
Cat spleen	10^{-4}	Kirpeker & Cervoni, 1963
Cat spleen	2×10^{-4}	Brown, 1965
Cat spleen	4×10^{-4}	Haefely, Hürlimann & Thoenen, 1965
Cat spleen	4.8×10^{-4}	Gillespie & Kirpekar, 1966
Cat spleen	4.6×10^{-4}	Kirpekar & Misu, 1967
Cat spleen	10^{-4}	Hedqvist & Stjärne, 1969
Cat spleen	3.6×10^{-4}	Blakeley, Brown & Geffen, 1969
Cat skeletal muscle	2×10^{-5}	Folkow, Häggendal & Lisander, 1967
Cat skeletal muscle	6×10^{-5}	Stjärne, Hedqvist & Bygdeman, 1969
Cat nictitating membrane	2.4×10^{-4}	Langer & Vogt, 1969
Rabbit pulmonary artery	5×10^{-5}	Bevan, Chesher & Su, 1969
Guinea-pig uterine artery	4.8×10^{-4}	Bell & Vogt, 1971
Rabbit vas deferens	6.6×10^{-5} to 15.9×10^{-5}	Hughes, 1972
Rabbit portal vein	5.6×10^{-5} to 16.2×10^{-5}	Hughes, 1972
Rat portal vein	10^{-5}	Häggendal, Johansson, Jonason & Ljung, 1970
Rabbit ear artery	2.5×10^{-4}	Vogt, 1973
Whole rabbit ear	1.4×10^{-5}	De la Lande, Paton & Wand, 1968
Rabbit heart	4.3×10^{-5}	Hukovic & Muscholl, 1962

This table is based on data from Häggendal, Johansson, Jonason & Ljung, 1970; Su & Bevan, 1971; Bell & Vogt, 1971; Hughes, 1972; Smith & Winkler, 1972; Vogt, 1973.

above) the values given in Table 3 for the fractional release may be too high by a factor of two to five. It is significant that lower figures for the number of NA molecules released per impulse were suggested by Folkow, Lisander & Häggendal (1967) who did not use phenoxybenzamine (see Smith & Winkler, 1972).

Smith and Winkler (1972) and De Potter, Chubb & De Schaepdryver (1972) suggested that the entire vesicle content of NA is released in each impulse, while Folkow and his collaborators (Folkow, Häggendal & Lisander, 1967; Folkow & Häggendal, 1970) argued that the NA content per vesicle is too high to represent the 'quantum' of transmitter. According to their calculations, only about three per cent of the vesicle NA content is discharged per nerve impulse. At present no definite answer can be given to these quantitative problems and more direct calculations are needed.

Following release of NA from the adrenergic varicosities, high transient

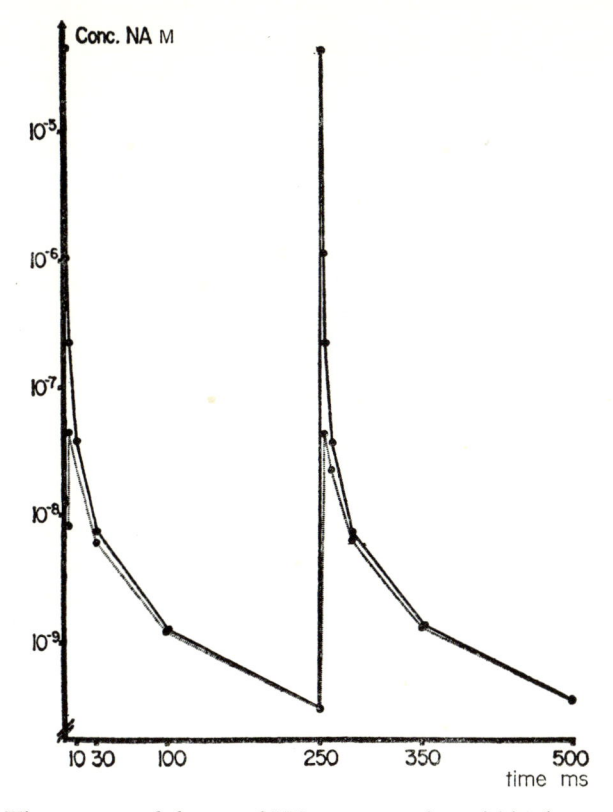

Figure 13. Time course of change of NA concentration within the vascular wall at 1000 nm (dashed line) and 100 nm (full line) from site of release calculated for spherical diffusion assuming instantaneous release of 5×10^{-20}g NA from one point source at 250 ms time intervals.
Reproduced with permission from Ljung, 1970.

concentration peaks occur near the muscle surface, close to the site of release, which results in high levels of initial receptor occupancy (Figure 13; Johansson, Johansson, Ljung & Stage, 1972). The peak concentration depends on the dimensions of the neuromuscular junction (Bevan & Su, 1971). These authors (Bevan & Su, 1973) calculated the concentration of NA during continuous stimulation at 10 Hz at the closest postjunctional smooth membrane to be: 6×10^{-9}M in the pulmonary artery with a neuromuscular distance of 4000 nm; 6×10^{-6}M in the ear artery with neuromuscular distance of 500 nm; 10^{-3} or 10^{-4}M in the portal vein with a neuromuscular junction of 150 nm. In Figure 13 the time course of changes in NA concentration at 100 and 1000 nm from the site of release has been calculated to peak at 10^{-5}M and 10^{-7}M respectively (Ljung, 1970). Bell and Vogt (1971) calculated a peak concentration of 4×10^{-4}M NA in the uterine artery. A concentration of 6×10^{-6}M of NA at the receptors during transmission was calculated for the cat hind leg and portal vein (Folkow, Häggendal & Lisander, 1967; Hughes, 1972); and in the rabbit ear artery, a concentration of 1·5 to $1·75 \times 10^{-7}$M NA was proposed (Waterson, 1973).

In organs where there are close neuromuscular junctions (20 nm) as in vas deferens, a concentration of 6×10^{-1}M would be expected at the receptors according to the corrected calculations of Furness (1974).

5.3 Postjunctional action of NA

5.3.1 *Adrenergic receptors*

The responses of tissues to NA are mediated by interaction with specific cell membrane constituents which have been called adrenergic 'receptors'. A detailed analysis of their classification, nature and distribution will not be made in this book and the reader is referred to recent reviews on the subject (Ahlquist, 1966; Axelsson, 1971; Furchgott, 1972; Triggle, 1972; Jenkinson, 1973). The original concept and subdivision of adrenoceptors into α and β introduced by Ahlquist (1948) and extended by Furchgott (1972) is based on the relative potencies of various adrenergic agonists and on the susceptibility of blockade by specific drugs.

The α-adrenoceptor is one which mediates a response pharmacologically characterized by: (a) a relative molar potency series in which adrenaline ⩾ noradrenaline > phenylephrine > isoprenaline; and (b) the susceptibility to specific blockade by phentolamine, dibenamine or phenoxybenzamine at relatively low concentrations.

The β-adrenoceptor is one which mediates a response pharmacologically characterized by: (a) a relative molar potency series (in the presence of a blocker of nerve amine uptake) which is either isoprenaline > adrenaline > noradrenaline > phenylephrine *or* isoprenaline > noradrenaline > adrenaline > phenylephrine, the two series of relative potencies depending on the tissue; and (b) a susceptibility to specific blockade by either propranolol or pronethalol at relatively low concentrations. Recently several subtypes of β-adrenoceptors have been differentiated (Lands, Arnold, McAuliff, Luduena & Brown, 1967; see Furchgott, 1972; Jenkinson, 1973).

Both α and β-adrenoceptors are often present in an effector tissue and mediate either antagonistic or synergistic actions; when the α- and β-effects are antagonistic, one of the effects predominates (Table 4). In non-sphincteric parts of the alimentary tract both α- and β-adrenoceptors mediate inhibition, while in sphincteric parts α-excitatory adrenoceptors predominate over β-inhibitory adrenoceptors (see Furness & Burnstock, 1975; Furness & Costa, 1974). In other smooth muscles, α-adrenoceptors mediate contraction, e.g. the nictitating membrane, spleen capsule, most blood vessels, vas deferens, retractor penis, while β-adrenoceptors mediate relaxation, e.g. uterus, coronary artery, bladder fundus (see Furchgott,

Table 4

Some NA effects mediated by α and β adrenoceptors (From Ahlquist, 1966; Bowman, Rand & West, 1971; Jenkinson, 1973)

Tissues and organs	Action of NA	Receptor
Heart	increase rate, force of contraction and ventricular excitability	β
Most blood vessels	vasoconstriction	α
Coronary arterials	vasodilatation	β
Tracheal and bronchial smooth muscle	relaxation	β
Non sphincteric parts of gastro-intestinal tract	relaxation	α and β
Sphincteric parts of gas-trointestinal tract	constriction	α
Uterine smooth muscle	contraction or relaxation depending on species and stages of oestrus cycle and pregnancy	α β
Bladder detrusor muscle	relaxation	β
Bladder trigone-sphincter muscle	contraction	α
Seminal vesicles	contraction	α
Retractor penis	contraction	α
Vas deferens	contraction	α
Pilomotor muscles	contraction	α
Nictitating membrane	contraction	α
Dilator pupillae	contraction (mydriasis)	α
Ciliary muscle	relaxation	β
Splenic capsule	contraction	α

1972). In the heart both sympathetic nerve stimulation and NA act on β-adrenoceptors to produce positive chronotropic and inotropic effects.

It is usually assumed that drugs applied directly to isolated smooth muscles act uniformly on the membrane surface of the cells, i.e. that the adrenoceptors are distributed homogeneously over the membrane. Nevertheless, by analogy with the skeletal neuromuscular junction, it seems likely that the sensitivity of smooth muscle is higher in the region of close apposition with the axon varicosities, the junctional region, than in the rest of the membrane, the non-junctional region, (Burnstock & Holman 1966; Burnstock & Bell 1974). In contrast, in organs where the transmitter reaches the smooth muscle membrane by diffusing from remotely placed nerve plexuses, such as in longitudinal muscle of the intestine and medial coat of large blood vessels, the sensitivity over the membrane may be more uniform; the action of exogenously applied NA closely mimics the effects of NA released from nerves in these tissues.

It is not known whether α- and β-adrenoceptors are located on the smooth muscle membrane of single cells or of different cells within the organ; iontophoretic application of small amounts of CA to smooth muscle with their narrow extracellular space has proved to be very difficult (see Bennett, 1973a) although iontophoretic application of ACh to cultivated smooth muscle cells has been reported (Purves, 1974). Iontophoretic application of α agonists to isolated guinea-pig liver parenchymal cells has been achieved (Green, Dale & Haylett, 1972).

5.3.2 *Chemical nature of adrenergic receptors*

While the concept of the 'receptor' has been an invaluable pharmacological tool, it has helped little in the understanding of the mechanism of interaction between NA and the effector tissue (Ahlquist, 1948; Triggle, 1965; De Robertis, 1973). Attempts have been made to determine the chemical nature of adrenoceptors. It has been suggested that the β-adrenoceptor is an integral part of the adenyl cyclase system (Robinson, Butcher & Sutherland, 1970), since most of the effects associated with β-adrenoceptors are associated with an increase in the intracellular level of cyclic AMP (see Axelsson, 1971; Furchgott, 1972; Triggle, 1972; Figure 14).

Speculations have also been made about the chemical nature of the α-adrenoceptor and it has been proposed that 'α-receptor effects are mediated by a decrease in the intracellular level of cyclic AMP' (Robison, Butcher & Sutherland, 1970). Attempts to isolate adrenergic receptors have been made. A proteolipid with a high affinity for binding NA has been isolated from bovine spleen capsule. An artificial bilayer lipid membrane with this proteolipid incorporated shows transient conductance changes in response to NA, which are blocked by phentolamine. It has been suggested that this proteolipid may play a receptor role in the spleen capsule (Ochoa, Fiszer, De Plazas & De Robertis, 1972). Belleau (1960) was the first to suggest that ATP might form an integral part of the adrenergic receptor surface, while Bloom & Goldman (1966) visualized the involvement of a CA–ATP complex and ATPase in α-adrenoceptors. In later theories linkage of CA with

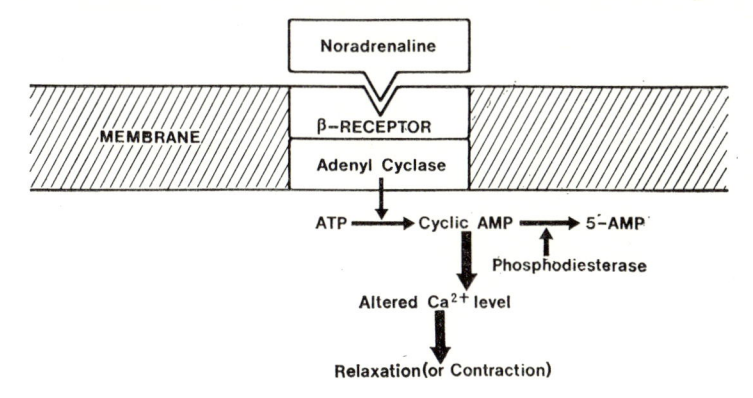

Figure 14. Schematic representation of the possible involvement of the adenyl cyclase system in NA actions mediated by β-adrenoceptors.

calcium has been proposed as a basis for α-adrenoceptor action (Belleau, 1967). In the theory of Honig and Stam (1967) it is proposed that the CA directly stimulates ATPase activity and thereby the actomyosin system.

5.3.3 *Effect of NA on membrane conductivity and calcium activity*

CA acts on those smooth muscles that are *inhibited* by sympathetic nerves by producing hyperpolarization of the muscle membrane, resulting in reduction or cessation of spike activity and consequent relaxation (Bülbring, 1957; Burnstock, 1958; Bülbring & Kuriyama, 1963). There has been considerable debate about the mechanism of action of CA on the postsynaptic muscle 'receptors' in producing this action (see Burnstock, Holman & Prosser, 1963; Bülbring & Tomita, 1969a, b, c; Setekleiv, 1970; Tomita, 1970; Kuriyama, 1970; Axelsson, 1971; Triggle, 1972). The relaxation of intestinal smooth muscle mediated by α-adrenoceptors involves a calcium-dependent increase in conductivity of the membrane to K^+ and Cl^-, while the relaxation mediated by β-adrenoceptors appears to involve a calcium-dependent stabilization of the membrane, through a process which requires energy and involves the formation of cyclic AMP (Jenkinson & Morton, 1967; Bülbring & Tomita, 1969c; Axelsson, 1971; Furness & Costa, 1974). An earlier hypothesis, namely that CA causes hyperpolarization by stimulating an electrogenic Na pump (Burnstock, 1958) is still being debated (see Setekleiv, 1970).

When CA is applied to those smooth muscles that are *excited* by sympathetic nerves, it produces depolarization and initiation or increase in frequency of spike activity, which is indistinguishable from the excitatory action of ACh on smooth muscles such as pig oesophageal muscularis mucosae (Burnstock, 1960) and dog retractor penis (Orlov, 1962). It has been suggested that in this situation, adrenaline produces depolarization by increasing the conductivity to Na^+, K^+ and probably other ions (Burnstock, 1960; Bolton, 1972) in a comparable way to ACh action at the skeletal motor end plate (see Katz, 1969).

A detailed analysis of the possible operational steps involved in the coupling between NA and the physiological effect of the effector cell would be out of place in this review. Nevertheless the coupling appears to involve calcium binding and potential changes are not always involved (see Section 5.5). The scheme below, summarizes the conclusions of Triggle (1972).

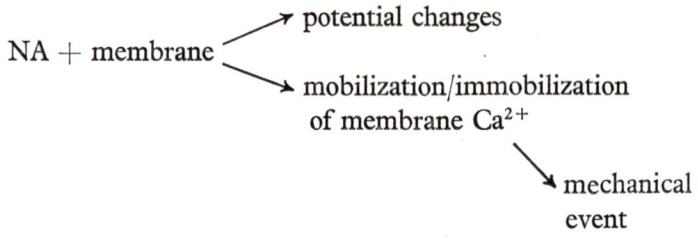

$$NA + membrane \nearrow potential\ changes \searrow mobilization/immobilization\ of\ membrane\ Ca^{2+} \searrow mechanical\ event$$

5.4 Inactivation of NA

Both the concentration of NA at the level of the receptors, and the termination of the effects of NA depend on at least five factors: re-uptake of NA into the nerve terminals; uptake of NA into non-neuronal cells; attachment of NA to 'receptor' sites on both pre and post synaptic membranes; metabolic degradation of NA by MAO and COMT; and diffusion of NA from the neuroeffector junction into the blood stream.

5.4.1 *Neuronal uptake of NA*

Intravenously injected radioactive-labelled NA was found to be accumulated avidly by adrenergic nerve terminals (Muscholl, 1961; Strömblad & Nickerson, 1961; Whitby, Axelrod & Weil-Malherbe, 1961). Neuronal uptake of NA (which has been termed 'uptake$_1$') occurs over the entire surface of the adrenergic neuron and the concentration of NA accumulated in the neuron rises to a level ten thousand times higher than that of the medium (see Iversen, 1967). The neuronal NA uptake system appears to be capable of clearing NA completely from the extraneuronal fluid space in milliseconds rather than in seconds (Iversen, 1971). Most NA taken up is stored in adrenergic vesicles (Crout, 1964; Iversen, 1967). The process is not specific for NA since other β-phenyl-ethylamines can be taken up by the same membrane mechanism (see Muscholl, 1972). The neuronal uptake mechanism for l-NA is temperature-dependent, requires the presence of sodium ions in the external medium, is a metabolically active process and obeys Michaelis-Menten kinetics (Iversen, 1967). In the absence of saturation, the rate of neuronal uptake is linearly related to the concentration of the amine in the medium. However, at a concentration of 10^{-7}g ml^{-1}, which gives 50 per cent of the maximal response in most tissues, there is a decline in the ability of the nerve fibres to remove NA from the region of the receptors (see Trendelenburg, 1972). It has been suggested that uptake of NA occurs only between nerve impulses (Bogdanski & Brodie, 1966; Palaic & Panisset, 1969; Häggendal & Malmfors, 1969). There is evidence that uptake is impaired by increased nerve activity (Trendelenburg, 1966; Gillis, Schneider, van Orden & Giarman, 1966; Häggendal & Malmfors, 1969; but see Palaic & Panisset, 1969; Bhagat & Zeidman, 1970; Yamamoto & Kirpekar, 1972; Kirshner, Schanberg & Ferris, 1972; Hughes, 1972). There is also evidence that membrane uptake is a function not only of the concentration of NA outside the neuron, but also of the intracytoplasmic concentration of NA. When MAO, the enzyme which controls the cytoplasmic level of NA, is inhibited, NA increases in the cytoplasm and a parallel decrease of net uptake of NA occurs (Trendelenburg, 1972).

The neuronal uptake mechanism of l-NA is inhibited by a wide range of sympathomimetic amines structurally related to NA, and by other drugs such as cocaine and imipramine derivatives (Iversen, 1971, 1973) which have become tools to study the role of uptake$_1$ in transmission (see Section 5.6).

5.4.2 *Extraneuronal uptake of NA*

Certain non-neuronal tissues also take up exogenous amines. The mechanism of uptake is different from the neuronal NA-uptake mechanism. This process was first detected in the rat heart when perfused by high concentrations of A and NA and was termed uptake$_2$ (Iversen, 1967). Accumulation of NA in smooth and cardiac muscle has been demonstrated with the fluorescence histochemical method (Avakian & Gillespie, 1968;

Draskóczy & Trendelenburg, 1970; Gillespie, Hamilton & Hosie, 1970; Gillespie & Muir, 1970; Eisenfeld, Axelrod & Krakoff, 1967; Burnstock, McLean & Wright, 1971; Burnstock, McCulloch, Story & Wright, 1972; O'Donnell & Saar, 1973). Most of the NA taken up by these muscles is readily metabolized by MAO and/or COMT in the cytoplasm (see Burnstock, McLean & Wright, 1971; Langer, Stefano & Enero, 1972). Thus, the accumulation of NA in these tissues underestimates the actual uptake, and following inhibition of MAO and COMT, accumulation in smooth muscle is detectable at 100 fold lower concentration of NA (Lightman & Iversen, 1969). Endothelial cells, fibroblasts and chondroblasts are also capable of taking up NA (Jacobowitz & Brus, 1971; O'Donnell & Saar, 1973).

Unlike uptake$_1$, uptake$_2$ is not stereochemically specific for l-NA and has a higher affinity for A than for NA; it has an even higher affinity for isoprenaline (Callingham & Burgen, 1966), although this CA does not appear to be accumulated by nerves (Iversen, 1971). Extraneuronal uptake is linear over a wide range of NA concentration and has a lower affinity than neuronal uptake, so that at low concentrations of extracellular NA, neuronal uptake is the more effective (Iversen, 1971). Extraneuronal uptake is inhibited by a number of drugs (e.g. normetanephrine, metanephrine, phenoxybenzamine and certain steroids) which are different from those found to be effective as neuronal uptake inhibitors (Iversen, 1971) (see Section 5.6). The process is also temperature-dependent and has little specificity in ionic requirement (Avakoin & Gillespie, 1968; Clark, Jones & Linley, 1969; Gillespie & Toward, 1973).

There are marked species differences in extent of accumulation of NA in smooth muscle cells. For example, NA accumulation in smooth muscle from guinea-pig is low compared to that of rat and mouse and rabbit, and greatest in mouse tissues (Gillespie & Muir, 1970; Jarrott, 1970; Bell & Vogt, 1971). In all species studied, arterial smooth muscle shows a greater ability to accumulate NA than visceral smooth muscle, especially the coronary arteries and arteries of the vas deferens. The mouse vas deferens and rabbit colon are exceptions (Gillespie, Hamilton & Hosie, 1970; Burnstock, McCulloch, Story & Wright, 1972). These marked differences in accumulation may not reflect differences in uptake efficiency, since other factors are involved.

A study of NA uptake into the non-sympathetically innervated smooth muscle of the human umbilical artery and chick amnion was carried out by Burnstock, McLean and Wright (1971). Low concentrations of NA are taken up into muscle and firmly bound intracellularly, particularly in the nucleus. Unlike uptake$_2$, this uptake is not blocked by phenoxybenzamine or normetanephrine and is unaffected by COMT inhibitors. Therefore, uptake$_2$ does not adequately account for all extraneuronal uptake.

5.4.3 *Metabolic degradation of NA*

Since both MAO and COMT are located intracellularly both in the neuron and in smooth muscle or other non-neuronal cells (see Section 4.2), their physiological role in the inactivation of NA released from nerves depends on the relative importance of the two uptake mechanisms described above. Most of the NA released by nerve stimulation is metabolized to normetanephrine in extraneuronal tissue. Released NA that is taken up by nerves is deaminated and *o*-methylated in adrenergic nerve endings (Langer, Stefano & Enero, 1972). MAO and COMT transform the NA, functionally inactivated by the uptake mechanisms, into metabolites which are almost biologically inactive (Langer, Stefano & Enero, 1972). The impairment of their activity therefore would only indirectly alter the effect of nerve-released NA on effector tissues. Figure 17 shows the main metabolic routes of NA released spontaneously and by nerve stimulation.

5.4.4 *Diffusion of NA from the neuromuscular junction into the blood stream*

Rosenblueth (1950) thought that most of the adrenergic transmitter was inactivated by diffusion into the blood stream. There appears to be little overflow of NA into the blood stream when adrenergic nerves are firing at frequencies up to 10 Hz, although some NA overflows at 30 Hz (Celander, 1954; Blakeley & Brown, 1966; Folkow, Häggendal & Lisander, 1967). However, even at rest some NA of nerve origin has been detected in human plasma (Engelman & Portnoy, 1970; see Bevan & Su, 1974).

After incubating rabbit aorta in ^{3}H-NA, Su and Bevan (1970a) found that 30 per cent of the labelled compounds recovered in the lumen was unchanged NA, and that the percentage increased to 50 per cent during nerve stimulation. A high percentage also escaped through the adventitia. The percentage of NA released from nerves, that escapes into the circulation can be markedly affected by changes in local blood flow and by the degree of contraction of the muscle (Axelrod, Kopin & Rosell, 1963; Carlsson, Folkow & Häggendal, 1964).

In vitro, the diffusion of NA away from receptors is estimated by measuring the NA collected in the organ bath (Hughes, 1972) or from the superfusate (Su & Bevan, 1970) or by the time course of termination of the response (Kalsner & Nickerson, 1968; Trendelenburg, 1974). Any condition which reduces the diffusion of NA away from the tissue will reduce the amount of NA collected and will prolongate the response. For example, it has been shown that by replacing the Ringer solution with mineral oil (Kalsner & Nickerson, 1968; Kalsner, 1971) or by avoiding the use of Ringer solution by suspending the tissue in humidified air (Gillis, 1971) little NA escapes from the tissue and the response is prolonged.

Another factor which influences the amount of NA that diffuses away from receptor sites is the spatial arrangement of the adrenergic nerve plexus. If there are many varicosities, the released NA will be taken up before it escapes from the plexus and therefore the net overflow of transmitter is low (Bevan, Bevan, Purdy, Robinson, Su & Waterson, 1972). This 'node crowding' may also explain the increase of NA overflow in the presence of cocaine from the rat vas deferens. The vas deferens is densely and homogeneously innervated and this increase in overflow is higher than the increase from the rat portal vein, which is innervated asymmetrically (Hughes, 1972). Physiological or experimental changes which influence the degree of diffusion of NA also influence the time course of the adrenergic responses. Little is known of the role of receptor sites in determining the time course of the response to NA released during nerve activity.

The interactions of all the inactivating mechanisms of NA and their relative role during adrenergic transmission will be discussed in Section 5.7.

5.5 Electrophysiology of adrenergic transmission

Burnstock and Holman (1960, 1961) were the first to apply electrophysiological methods to the study of transmission from adrenergic nerves to smooth muscle and this method has become a major tool for the examination of the detailed mechanisms involved in adrenergic function (see reviews by Burnstock & Holman, 1963, 1966; Bennett & Burnstock, 1968; Holman, 1969, 1970; Bennett, 1973a, d).

5.5.1 *Adrenergic excitatory transmission*

Burnstock and Holman (1960, 1961) reported transient depolarizations of the muscle membrane of impaled smooth muscle cells of the vas deferens in response to hypogastric nerve stimulation. Similar 'excitatory junction potentials' (EJP's) have been recorded in a variety of different tissues in response to adrenergic nerve stimulation since this time (see Table 5 and

Burnstock & Holman, 1966; Holman, 1969, 1970; Speden, 1970). When the depolarization produced by a single EJP or by a train of EJP's reaches a critical level, an action potential is initiated and contraction occurs (Plate 12).

In the absence of nerve stimulation, a random discharge of spontaneous depolarizations occurs (Plate 12). Since the occurrence of these potentials is reduced by reserpine or adrenoceptor blocking drugs (Burnstock & Holman, 1962b, 1964), they probably represent the spontaneous release of packets of NA from the nerves (see Burnstock & Holman, 1966). Spontaneous EJP's of similar time course have also been recorded in dog retractor penis (Orlov, 1962) and in guinea-pig seminal vesicles (Kajimoto, Kirpekar & Wakade, 1972).

EJP'S have been recorded in a few arterial smooth muscles, but their maximum amplitudes are considerably lower than those recorded in the vas deferens (Holman, 1969; Bell, 1969b). This difference may be due to the neuromuscular relationships in the two tissues; the minimum separation of nerve and muscle membranes is of the order of 20 nm in the guinea-pig vas deferens compared to about 80 nm in arteries (see Section 5.1). Despite the differences in EJP amplitude between these preparations, their time course is comparable (Table 5). No EJP's have been recorded in *large* arteries in response to adrenergic stimulation. However, a variety of electrical responses associated with contraction of the muscle have been reported; these include sustained depolarization, no change in membrane potential or even slight hyperpolarization (Su & Bevan, 1971; see Holman, 1969; Axelsson, 1971). The muscle coat of large arteries is separated in bundles by elastic lamellae and it is possible that the smooth muscle cells are not all electrically coupled (Bozler, 1948; Holman, 1969; Mekata, 1973; Bevan & Su, 1974). Since many of the muscle cells are not electrically coupled and since the adrenergic plexus is remotely placed from most of them, they are probably activated largely by NA reaching them by diffusion either from the adventitial adrenergic nerves or from the blood stream (see Section 5.6). Under certain conditions, NA may produce contraction of smooth muscle without detectable changes in the membrane potential (see Section 5.3, Piper, Griebel & Wende, 1971, Heusler, 1972; Bohr, 1973; Mekata, 1973). Some smooth muscle cells, as for example those in mouse vas deferens, show graded rather than all-or-none action potentials (Furness & Burnstock, 1969).

Facilitation of successive EJP's in response to repetitive nerve stimulation occurs and appears to be due to increased release of NA with each stimulus, rather than due to increased postsynaptic sensitivity (Burnstock,

Table 5

Temporal characteristics of adrenergic junction potentials, compared with those of transmission at the skeletal neuromuscular junction

Tissue	Stim Site*	Minimum latency (ms)	Rise time (ms)	Duration (ms)	Reference
Guinea-pig vas deferens	Preg.	20	15–20 (half rise time)	1000	Burnstock & Holman, 1961
Mouse vas deferens	Postg. (T/M)	6	40	500	Kuriyama, 1963
	Postg. (T/M)	10	10–20	100–250	Holman, 1970
Rat vas deferens	Postg. (T/M)	5	28–34	—	Furness, 1970a
Cat nictitating membrane†	Postg.	10	10–20	150	Holman, 1970
Rabbit ear and mesenteric arteries	Postg. (T/M)	20	50–80	500	Eccles & Magladery, 1937
Guinea-pig uterine artery	Postg. (T/M)	12	70–100	500–1000	Speden, 1967
Rat skeletal neuromuscular junction	—	20	—	900–1000	Bell, 1969b
		0·2	0·6	2–3	Boyd & Martin, 1956

* Preg. = preganglionic; Postg. = postganglionic; T/M = transmural; † = Extracellular recording

Holman & Kuriyama, 1964). Spontaneous release of NA is enhanced following nerve stimulation (Burnstock & Holman, 1966). These results suggest that the mechanism of storage and release of NA from sympathetic nerves is comparable to the release of ACh at the skeletal neuromuscular junction (Eccles, 1964; Katz, 1969). However, there are some notable differences. For example, the time course of the EJP is very long, some 100 times slower than the end plate potential, although the spontaneous EJP's are only about 10 times slower. There is a long (minimum 6 ms) variable delay time for the appearance of EJP's following nerve stimulation in different cells. Reduction of the number of nerve fibres stimulated leads to a general reduction in amplitude of EJP's in all cells rather than complete loss of EJP's in localized areas (Burnstock & Holman, 1961). Finally, the amplitude of EJP's in the majority of cells in the vas deferens is unaffected by depolarizing and by hyperpolarizing currents (Bennett & Merrillees, 1966; Holman, 1967; Furness, 1970a). These differences from the classical transmission characteristics known for the skeletal neuromuscular junction can be explained by the electrotonic coupling of activity between neighbouring cells within muscle effector bundles, and by 'en passage' release of transmitter from extensive terminal varicose nerve fibres (see Bennett & Burnstock, 1968; Burnstock & Iwayama, 1971).

The time course of the EJP depends on various factors. The rise time appears to depend on the spread of time taken for NA to reach the postjunctional receptors, which in turn depends on the number of varicosities which contribute and on the dimensions of the neuromuscular cleft. The amplitude of the EJP is a function of the maximum concentration of transmitter reached at the receptor site. The decay time probably depends on both the time taken for the transmitter to be inactivated and the passive electrical characteristics of the postjunctional membrane (Tomita, 1967a, b; Bennett & Burnstock, 1968; Burnstock & Holman, 1966; Bell, 1969b; Holman, 1970; Furness, 1970b; Bennett, 1973a, d).

Whether the long latency of the adrenergic EJP is due largely to presynaptic or to postsynaptic factors is not yet known (see Burnstock & Holman, 1966; Bennett & Burnstock, 1968; Holman, 1970). It seems unlikely to represent the time required for transmitter to diffuse from the nerve terminals to the effector sites (Furness, 1970a; Bell & Vogt, 1971). The demonstration of long delays following iontophoretic application of ACh to cultured smooth muscle cells might indicate that smooth muscle receptors are characterized by long activation times (Purves, 1974).

5.5.2 *Adrenergic Inhibitory Transmission*

In non-sphincteric intestinal smooth muscle, which is relaxed by adrenergic nerve stimulation, no inhibitory junction potentials in response to single stimuli have been recorded. Repetitive nerve stimulation at 5 Hz or greater is necessary to produce a small graded hyperpolarization of the membrane accompanied by reduction or cessation of spontaneous action potentials (Gillespie, 1962; Bennett, Burnstock & Holman, 1966a; Furness, 1969; Plate 13). The absence of discrete junction potentials may be explained in terms of the absence of close contacts between adrenergic nerve terminals and smooth muscle cells in the taenia coli and distal colon of the guinea-pig (see Section 5.1.) The transmitter reaches the muscle by diffusion from a distance of 100 nm or more, so that the slow initial phase of the hyperpolarization and the long decay time could be explained in these terms, especially since reuptake of NA into the nerves is ineffective at these distances (see Section 5.4). It is of interest that stimulation of non-adrenergic ('purinergic') inhibitory nerves to these tissues produces inhibitory junction potentials of up to 40 mV in response to single pulses (Bennett, Burnstock & Holman, 1966b; Burnstock, 1972; Plate 12).

5.6 Pharmacology of adrenergic transmission

Emphasis is placed here exclusively on those actions of drugs which interfere with adrenergic transmission. Drugs can inhibit, block or potentiate adrenergic transmission in a variety of ways by affecting NA synthesis, storage, release, uptake and inactivation, as well as the pre- and post-synaptic action of NA. Many drugs affecting adrenergic transmission have multiple sites of action, and impairment of one process may be balanced by enhancement of others in the adrenergic transmission mechanism and therefore may not lead to a detectable effect.

5.6.1 *Drugs which affect synthesis of NA*

α-Methyl-p-tyrosine, a potent inhibitor of T-OH (see Iversen, 1967), reduces the storage level of NA in adrenergic neurons and the rate of depletion increases with frequency of stimulation (Bhagat, 1967; Malmfors, 1969; Swedin, 1970; Almgren, 1971). This drug also impairs adrenergic transmission (Spector, Sjoerdsma & Udenfriend, 1965; Thoenen, Haefely, Gey & Hürlimann, 1966; Kopin, Breese, Krauss & Weise, 1968; see Kalsner, 1972), indicating that the synthesis of NA from tyrosine is essential in the maintenance of adrenergic transmission during prolonged activity (Alousi & Weiner, 1966; Bhagat, 1967; Kopin, Breese, Krauss & Weise, 1968; Weiner & Rabadjija, 1968b; Bhagat & Friedman, 1969; Kupferman, Gillis & Roth, 1970; Bhatnagar & Moore, 1971b). Adrenergic transmission is impaired by α-methyl-p-tyrosine even when NA stores are not completely depleted (Spector, 1966; Thoenen, Haefely, Gey & Hürlimann, 1966). Consequently, it has been suggested that newly synthesized NA is released preferentially from a small pool with rapid turnover of NA (Kopin, Breese, Krauss & Weise, 1968; Gewirtz & Kopin, 1970; Stjarne & Wennmalm, 1970; Bralet, Belay, Lallemant & Bralet, 1972). This hypothesis is in agreement with the view that there are at least two pools of NA in the adrenergic terminal; one is a large pool with a slow turnover rate of NA not readily available for nerve release and sensitive to reserpine depletion; the other is a small pool with a faster turnover of NA, which is readily available and more resistant to reserpine depletion (see Section 5.6; Trendelenburg, 1961; Kopin & Gordon, 1962; Potter & Axelrod, 1963; Chidsey & Harrison, 1963; Kirpekar & Furchgott, 1964; Sedvall & Thorson, 1965; Carlsson, 1966; Iversen, 1967; Kalsner, 1972).

Drugs which inhibit DβH, such as disulfiram and its reduced product diethyldithio carbamate, reduce NA stores and increase the DA content of tissues (Corrodi, Fuxe, Hamberger & Ljungdahl, 1970). Adrenergic transmission to the rabbit ileum, nictitating membrane and the spleen of the cat is impaired by these drugs (Thoenen, Haefely, Gey & Hürlimann, 1965).

5.6.2 *Drugs which affect storage of NA*

Drugs such as reserpine, prenylamine and tetrabenazine are potent NA depletors (see Shore, 1966). Depletion seems to be due to the blockade of the ATP/Mg^{2+}-dependent uptake and subsequent binding of NA into adrenergic vesicles (von Euler & Lishajko, 1963; Iversen, 1967). Any NA released from granular vesicles is inactivated by deamination in the cytoplasm by MAO (Shore, 1962; Kopin, 1964; Smith, 1966).

As a consequence of the irreversible damage to the granular vesicles produced in this way, the depletion of NA stores is not balanced by binding of new DA and NA. Therefore, it has been argued that the time course of the depletion is proportional to both the dose of reserpine and to the rate of NA utilization (Carlsson, 1966). Tissues, such as the heart, where the rate of turnover of NA is high as a result of continuous nerve activity, are less resistant to reserpine (Swedin, 1970). Consistent with this view is the observation that stimulation *in vitro* of adrenergic fibres which are relatively resistant to reserpine, increases the rate of NA depletion (Gillespie & McGrath, 1974). This suggests that differences in sensitivity to reserpine are due to differences in the level of nerve activity rather than to differences in axon length (see Section 3.1).

Depletion of NA stores by reserpine leads to impairment of adrenergic transmission (Bertler, Carlsson & Rosengren, 1956; Carlsson, Rosengren, Bertler & Nilsson, 1957; Muscholl & Vogt, 1958; Gillis & Yates, 1960; Fleming & Trendelenburg, 1961; Shore, 1962; Burnstock & Holman, 1962b; Andén & Henning, 1966; Haefely, Hürlimann & Thoenen, 1966; Day & Warren, 1968). However, transmission is rarely abolished and there are remarkable differences between the level of NA which remains in the tissue and degree of impairment of transmission in different organs. For example, in some organs a significant decrease in the response to nerve stimulation occurs only when the NA content falls below 50 per cent of normal (Lee, 1967). In the guinea-pig vas deferens, depletion of 99 per cent of the NA reduces the response only to 60 per cent of the maximal control response (Antonaccio, 1970; Wakade & Krusz, 1972). However, even when there is no apparent reduction of the response after reserpine treatment, transmission is not sustained (Enero & Langer, 1973). In addition there is poor correlation between the time course of recovery of adrenergic transmission and the recovery of the NA store. The content of NA in the cat nictitating membrane following a single dose of reserpine remains low for several days, while the response to nerve stimulation is restored in 48 h (Andén & Henning, 1966). An explanation is that recovery of transmission coincides with the arrival of new vesicles by axonal flow to replace the old vesicles irreversibly damaged by reserpine, and with the ability of the nerve to take up and retain NA (Häggendal & Dahlström, 1970).

Only a small fraction of the total store of NA appears to be involved in the maintenance of adrenergic transmission (Andén & Henning, 1966, 1968; Wakade & Krusz, 1972; Antonaccio & Smith, 1974). The ability of small amounts of endogenous NA to maintain transmission suggests that the 'safety factor' for adrenergic transmission is high. The level of this safety factor is probably related to the dimensions of the neuromuscular junction. For example, in the vas deferens, where neuromuscular junctions are about 20 nm wide, the peak concentration of NA at the receptors is of the order of 10^{-1} gml^{-1} (see Section 5.2). Therefore, even with a depletion of 99 per cent of the NA content, a peak concentration of about 10^{-3} gml^{-1} remains which is still effective on the vas deferens. In tissues with wide neuromuscular junctions, the peak concentration of NA at the receptors may be close to the threshold, so that any reduction of the amount of NA released due to depletion of the stores would result in marked impairment of transmission.

Reserpine has also been shown to block adrenergic transmission acutely (Nakazato & Ohga, 1972). A single large dose of reserpine injected intravenously selectively inhibits gastric relaxation to adrenergic stimulation in dogs. Acute blockade by reserpine of adrenergic transmission *in vitro* has been reported (Day & Owen, 1968; Day & Warren, 1968; von Euler, 1969; Kubo, Mishio & Misu, 1970; but see Wakade & Krusz, 1972), but this effect may not be specific since nonadrenergic nerves in the intestine are also blocked (Day & Warren, 1968).

Reserpine initially produces various short-lasting sympathomimetic effects by releasing NA from nerves (de Jongh & Van Proosdij-Hartzema, 1955, 1958; Maxwell, Plummer, Osborne & Ross, 1956; Domino & Rech, 1957; Krayer & Fuentes, 1958; Brimijoin & Trendelenburg, 1971).

5.6.3 *Drugs which affect release of NA*

Drugs which increase release of NA. Amphetamine and tyramine both release NA from adrenergic nerves. Amphetamine releases NA from an extra-vesicular compartment (Carlsson, Fuxe, Hamberger & Lindqvist, 1966) while tyramine releases vesicular NA, but not by exocytosis since DβH is not released (see Muscholl, 1966; Chubb, De Potter & De Schaedryver, 1972). Both amines potentiate adrenergic transmission (Ryall, 1961; Bhargava, Kar & Parmar, 1965). The increase in NA released by α-adrenoreceptor blocking drugs, angiotensin and prostaglandins of the F type has been considered previously (Section 5.2).

Drugs which prevent release of NA. Invaluable tools for studying adrenergic nerve function are the adrenergic neuron blocking drugs which prevent the release of NA and DA in response to stimulation of adrenergic nerves (Boura and Green, 1965; Collins & West, 1968; Kirpekar, Wakade, Dixon & Prat, 1969).

The most widely used of this group of drugs are guanethidine and bretylium. Their antagonism to adrenergic transmission is long lasting (Chang, Costa & Brodie, 1965) and is specifically reversed by amphetamine (Day & Rand, 1963; Rand & Wilson, 1967). A number of other effects of guanethidine and bretylium have been reported.

The long lasting toxic effects on adrenergic neurons by these drugs will be treated in Section 6.4; guanethidine has a depletory action on NA neuronal stores which does not appear to be related to its blocking action (Cass & Spriggs, 1961; Kuntzman, Costa, Gessa & Brodie, 1962; Chang, Costa & Brodie, 1965; Spriggs, 1966; Lundborg & Stitzel, 1968). Blockade of the neural NA uptake mechanism is another effect of this drug which causes a cocaine-like supersensitivity to NA (Maxwell, Plummer, Schneider, Poualski & Daniel, 1960; Abercrombie & Davies, 1963; Bentley, 1965). Anaesthetic action on nerve fibres has been reported for guanethidine and bretylium but at higher doses than those which effectively block adrenergic nerves (Rand & Wilson, 1967; Haeusler, Haefely & Hürlimann, 1969).

These drugs have also a ganglion blocking effect, acting on nicotinic receptors, which is easily reversed by washing (Maxwell, Plummer, Schneider, Poualski & Daniel, 1960; Kosterlitz & Lees, 1961; Gertner & Romano, 1961).

The mechanism by which guanethidine prevents NA release is not fully understood. It has been suggested that since the uptake of guanethidine into the nerves is a prerequisite for its effective action (Thoenen, Hürlimann & Haefely, 1966), it blocks the conduction at the level of preterminal axons by anaesthetic action (Campbell, 1970).

Other substances which share the adrenergic neuron blocking action include bethanedine, pempidine, mecamylamine and epsilon-aminocaproic acid and dimethyl phenyl piperazinium (DMPP) (Birmingham & Wilson, 1965; Burn & Rand, 1965; Andén, Henning & Obianwu, 1968).

The cocaine-like activity of some of these drugs may interfere with the efficiency of their blockade of transmission. The dose of bethanidine necessary to abolish adrenergic transmission to the heart is much lower than that required to inhibit transmission to the nictitating membrane. This difference has been interpreted as being due to the effect of a cocaine-like potentiation of the response; thus potentiation is more marked in densely innervated tissues (see Section 5.6; and Armstrong & Boura, 1970).

An indication that the efficacy of adrenergic neuron blocking agents in impairing adrenergic transmission is related to the neuromuscular distance, as is the efficacy of the NA depleting agents, comes from the work of Hughes (1972). Bretylium decreases the output of NA from both the rat portal vein and vas deferens to undetectable levels ($<$ 10 per cent) but, while it blocks transmission in the portal vein (neuromuscular distances of about 150 nm), it fails to abolish transmission in the vas deferens (neuromuscular distances of about 20 nm).

5.6.4 Drugs which affect uptake of NA

It has been known for a long time that cocaine increases the sensitivity of various tissues to CA (see Rosenblueth & Rioch, 1933; Cannon & Rosenblueth, 1949). It was only shown relatively recently that this effect is due to blockade of the neuronal uptake of NA (Macmillan, 1959; Trendelenburg, 1959, 1972), and because of this site of action, cocaine is said to induce 'prejunctional supersensitivity'. The degree of potentiation of the response to NA appears to be proportional to the density of the adrenergic innervation of the tissue (Gillespie & Rae, 1970; Trendelenburg, 1972; Bevan & Purdy, 1973). The response to NA of densely innervated smooth muscle such as the nictitating membrane, the iris and the vas deferens (see Section 3.1; and Plate 9) is greatly potentiated by cocaine (Haefely, Hürlimann & Thoenen, 1964; Basset, Cairncross, Hacket & Story, 1969; Wakade & Krusz, 1972). On the other hand, non-innervated smooth muscles such as those in umbilical vessels, show no potentiation of the NA response with cocaine (Trendelenburg, 1972). In blood vessels that have an asymmetrical adrenergic innervation (see Section 5.1, and Plate 9), marked potentiation by cocaine occurs when NA is applied from the adventitial side, but little potentiation when NA is applied from the intimal side (de la Lande, Frewin & Waterson, 1967). No clear potentiation has been reported for sparsely innervated smooth muscle such as the uterus and the non-sphincteric parts of the intestine (Burn & Tainter, 1931; Labate, 1941; Schofield, 1952; Roszkowski & Koelle, 1960; Stafford, 1963; Govier, Sugrue & Shore, 1969).

It has been suggested that cocaine also potentiates the response of NA by acting on the effector tissue, thus causing 'postjunctional supersensitivity' (Reiffestein, 1968; Kalsner & Nickerson, 1969a).

Rosenblueth & Rioch (1933) and Bacq & Frederiq (1934) first realized the implications of cocaine supersensitivity to exogenous CA for adrenergic transmission. They argued that if the transmitter liberated by adrenergic nerves was adrenaline or an analogue, then cocaine should potentiate the effect of adrenergic nerves; they showed this for the cat nictitating membrane. Since then, this result has been confirmed (Trendelenburg, 1959; Kirpekar & Cervoni, 1963; Haefely, Hürlimann & Thoenen, 1964) and similar results have been reported for adrenergic excitatory transmission to the heart (Hukovíc & Muscholl, 1962; Graham, Abboud & Eckstein, 1965; Moore, 1966; Gillis & Schneider, 1967; but see Koeker & Moran, 1971) some blood vessels (Bentley, 1965; Rogers Atkinson & Long, 1966; Su &

Bevan 1970b; Toda, 1971; Kadowitz, Sweet & Brody, 1971; Waterson, 1973) vas deferens (Huković, 1961; Bell, 1967; Hughes, 1972) and rabbit uterus (Varagic, 1956).

Little is known about effects of cocaine on adrenergic *inhibitory* transmission. Burn and Tainter (1931) did not find potentiation by cocaine of sympathetic nerve stimulation to the intestine. Govier, Sugrue & Shore (1969) also failed to show potentiation of inhibition in response to transmural stimulation of the intestine. It is now known that transmural stimulation activates mainly non-adrenergic inhibitory 'purinergic' nerves (see Burnstock, 1972) and therefore the results of the above authors may not be related to adrenergic nerves. Labate (1941) claimed to have observed potentiation by cocaine of uterine inhibition in response to sympathetic stimulation, but his published figure does not substantiate this claim.

Impramine and desmethylimipramine (DMI), which are potent antagonists of the neural uptake of CA (Glowinski & Axelrod, 1964; Iversen, 1967), have been reported to potentiate adrenergic responses (Sigg, Soffer & Gyermek, 1963; Kaumann, Basso & Aramedia, 1965; Basset, Cairncross, Hacket & Story, 1969; Glover & McCulloch, 1970), but the effect is masked by depression of CA actions on postsynaptic membranes (Turker & Khairallah, 1967; Basset, Cairncross, Hacket & Story, 1969; Scriabine, 1969; Glover & McCulloch, 1970; Hrdina & Ling, 1970; Toda, 1971; Westfall, 1973). Both cocaine and DMI increase the amount of NA which overflows from tissues during adrenergic stimulation (Cripps & Dearnaley, 1970; Langer, 1970; Su & Bevan, 1970a; Bell & Vogt, 1971; Starke, Montel & Schümann, 1971; Hughes, 1972; Starke & Schümann, 1972) but not the amount of DβH (De Potter, Chubb, Put & De Schaepdryver, 1971) which is released together with NA during nerve stimulation (Section 5.2).

The potentiation of adrenergic responses by cocaine is more marked at low frequencies than at high frequencies of nerve stimulation (Cairne, Kosterlitz & Taylor, 1961; Haefely, Hürlimann & Thoenen, 1964; Gillis & Schneider, 1967; Hughes, 1972).

The efficacy of cocaine in potentiating adrenergic responses depends on several factors including the neuromuscular distance, the extent of extraneuronal uptake, the sensitivity of the postsynaptic receptor to NA and the inhibition of further release of NA by the released NA through presynaptic inhibitory feedback as suggested by Enero, Langer, Rothlin & Stefano (1972) and Kirpekar, Furchgott, Wakade and Pratt (1973).

The neuromuscular geometry plays an important role and it is likely that cocaine is less effective the wider the neuromuscular distance since clear potentiation of the adrenergic response has been reported (see above) only for tissues with close neuromuscular junctions. On the basis of the effect of cocaine on the responses to exogenous NA, it has been proposed that significant effects of neural uptake on the concentration of NA at receptors is limited to 100 nm from the nerve membrane (Verity, 1971). However, in studying the effect of cocaine and NA released from nerves, it has to be kept in mind that both the amplitude and the duration of the response can be potentiated by cocaine.

The role played by extraneuronal uptake in adrenergic transmission is not known. A potentiation of the adrenergic responses could be expected from uptake inhibitors (Hughes, 1972; Salt, 1972). For example, corticosteroids which inhibit extraneuronal uptake (Iversen & Salt, 1970; see Section 5.4) potentiate the response to sympathomimetic amines (Kalsner, 1969) and increase NA outflow during nerve stimulation in both vas deferens and portal vein (Hughes, 1972). It has been suggested that both corticosterone and cholesterol may potentiate the effects of circulating NA on tissues by this mechanism (Kalsner, 1969; Salt & Iversen, 1972). If neuronal uptake is inhibited, more NA is taken up by the extraneuronal mechanism, and vice versa, indicating that the two mechanisms are in dynamic balance (Hughes, 1972).

5.6.5 *Drugs which affect enzymic inactivation of NA*

Little work with the specific purpose of studying the effect of MAO and COMT inhibitors on adrenergic transmission has been performed. The inhibitors of MAO marsilid (isopropylisoniazide) and nialamide, either failed to alter the response to adrenergic stimulation (Kamijo, Koelle & Wagner, 1956; Day & Rand, 1963) or reduced it (Davey, Farmer & Reinert, 1963). Other inhibitors of MAO such as pheniprazine, tranylcypromine, and isonicotinic acid hydrazide were reported to potentiate adrenergic responses (Kamijo, Koelle & Wagner, 1956; Ryall, 1961; Bhargava, Kar & Parmar, 1965), but these drugs also have a cocaine-like activity inhibiting neuronal NA uptake (Kamijo, Koelle & Wagner, 1956; Iversen, 1967). Pargyline, which does not have cocaine-like activity (Iversen, 1967), increases NA overflow (Hughes, 1972) by affecting the presynaptic inhibitory feed-back mechanism (Blakeley, Powis & Summers, 1973); the small potentiation of the excitatory junction potentials recorded intracellularly in vas deferens (Bell, 1967) might be due to uptake inhibition rather than to inhibition of the enzyme activity. Inhibition of COMT by pyrogallol leads to a potentiation of the response of the cat nictitating membrane (Wylic, Archer & Arnold, 1960) and to a small potentiation of the response of the guinea-pig vas deferens (Bhargava, Kar & Parmar, 1965). Tropolone, a COMT inhibitor, in association with pargyline reduced the response of the rat portal vein and of rat vas deferens, but increased NA release by two to three fold (Hughes, 1972).

5.6.6 *Drugs which affect postsynaptic receptors*

The role of α- and β-adrenoceptors in the actions of CA is discussed in Section 5.3. The action of drugs which specifically block these receptors will now be discussed briefly, but fuller reviews on the subject are available (see Triggle, 1968; Gooman & Gilman, 1971; Raper & McCulloch, 1971).

α-Adrenoceptor blocking drugs. The ergot alkaloids were the first drugs which were found to antagonize α-adrenoceptor activity and to block sympathetic transmission to some smooth muscles (Dale, 1906), but they also have other non-specific effects. α-Adrenoceptor antagonists belong to two main chemical groups: the imidazolines (e.g. phentolamine) and the β-haloalkylamines (e.g. phenoxybenzamine).

Although drugs such as dibenamine, phenoxybenzamine and phentolamine specifically block the effect of NA on α-adrenoceptors, their efficacy in blocking adrenergic transmission depends on the ability of the drug to reach the receptor sites in the tissue. Differences in efficacy of drugs to antagonize nerve-mediated α-adrenergic effects have been found in different tissues. For example, contraction of blood vessels evoked by adrenergic stimulation is completely blocked by α-adrenoceptor blocking drugs (see Furchgott, 1952; Yates & Gillis, 1963; Paterson, 1965) and the inhibitory effect of adrenergic stimulation on the longitudinal muscle of the intestine after blockade of β-adrenoceptors is abolished by α-receptor blockers (see Day & Warren, 1968). In both of these organs the adrenergic neuromuscular distance is greater than 80 to 100 nm (see Burnstock, 1970). In contrast, contraction of the vas deferens in response to adrenergic nerve stimulation is resistant to these receptor blocking drugs being affected only by high concentrations where non-specific effects become apparent (Boyd, Chang & Rand, 1960; Birmingham & Wilson, 1963; Kuriyama, 1963; Ohling & Stromblad, 1963; Burnstock & Holman, 1964; Bentley & Smith, 1967; Holman, 1967; Hotta, 1969; Swedin, 1971; Ambache & Zar, 1971). The resistance of vas deferens contraction to α-adrenoceptor blockade has been related to the poor penetration of these drugs at the junctional receptors, since the minimum neuromuscular distance is 10 to 30 nm and no basement membrane is interposed between nerve varicosity and smooth muscle membrane (Burnstock & Holman, 1964; Bentley & Smith, 1967; Swedin, 1971; Furness & Iwayama, 1971, 1972; Furness, 1974). These drugs block the contraction produced by applied NA (Hotta, 1969; Bentley & Smith, 1967; Furness, 1974) at concentrations which do not affect nerve stimulation (Holman & Jowett, 1964). Contraction produced by tyramine, which releases NA from adrenergic terminals, is completely blocked by these drugs (Bentley & Smith, 1967). Finally the slow response, which follows the initial fast 'twitch' response in the nerve stimulated vas deferens and which is probably due to the effect of NA on nonjunctional receptors, is also completely blocked by α-adrenoceptor blocking drugs (Swedin, 1971).

Most α-adrenoceptors have pharmacological actions in addition to specific receptor blockade. For example, phenoxybenzamine in low concentrations increases the amount of NA release by blocking the negative feedback mechanism (see Section 5.2). Phenoxybenzamine also has an extraneuronal uptake blocking action and a weaker neuronal uptake blocking action (see Section 5.4). These actions may explain the potentiation of both inhibitory (see Burn & Kevin, 1965) and excitatory nerve responses by phenoxybenzamine (see Section 5.4). In some preparations, phenoxybenzamine has a strong blocking action on cholinergic transmission (Boyd, Burnstock, Campbell, Jowett, O'Shea & Wood, 1963).

In conclusion, the α-adrenoceptor blocking drugs are reliable pharmacological tools only for those smooth muscles with large neuromuscular distances.

β-*Adrenoceptor blocking drugs.* The discovery of a specific β-adrenoceptor blocking drug, dichloroisoproterenol (DCI) by Powell & Slater (1958) came much later than that of α-adrenoceptor blockers. Since then, other more powerful specific blockers, have been developed such as pronethalol, which was abandoned in clinical use because of its carcinogenic properties, propranol, sotalol, methoxamine, Kö 592, H13/62 and M13/57 (see Moran, 1967). Most of these drugs were reported to have a local anaesthetic action but usually in higher concentration than is necessary to block adrenergic responses (Gill & Vaughan Williams, 1964; Morales-Aguilera & Vaughan Willis, 1965; Blinks, 1967; Mylecharane & Raper, 1973).

These drugs also have cocaine-like properties, i.e. they inhibit neuronal uptake of NA being effective in decreasing order propranolol, pronethanol, DCI > Sotalol > Kö 592 which is devoid of this effect (Foo, Jowett & Stafford, 1968; Werner, Wagner & Schümann, 1971; Starke & Schumann, 1972; Mylecharane & Raper, 1973).

These drugs completely block β-adrenoceptors in some organs but not in others. For example propranolol and pronethanol inhibit both the positive chronotropic and inotropic effect of sympathetic stimulation in the heart (Blinks, 1967; Moran, 1967), but do not completely inhibit β-adrenoceptors in the guinea-pig colon (Costa & Furness, 1971). DCI and propranolol, but not butoxamine, inhibit the relaxation elicited by adrenergic stimulation in some intestinal preparations (Moran, 1967; Day & Warren, 1968). The differential effects in different tissues of β-adrenoceptor blocking drugs introduced recently (e.g. sotalol, proctalol) and the different relative potencies of agonists (e.g. salbutamol, soterenol), have been used as the basis for the suggestion that there is more than one type of β-adrenoceptor (see Furchgott, 1972). Propranolol and sotalol also have an adrenergic neuron blocking activity similar to guanethidine, the effect being reversed by amphetamine (Mylecharane & Raper, 1973). There seems to be little difference in the blockade of β-adrenoceptors activated either by applied NA or by nerve-released NA (Moran, 1967). This may be explicable again on the basis of neuromuscular distance since for the tissues having predominant β-inhibitory adrenoceptors it is greater than 100 nm (see Section 5.1).

5.6.7 *False adrenergic transmitters*

The mechanisms of synthesis, storage, release, uptake, metabolic degradation and postsynaptic action in adrenergic nerves are processes which are not entirely specific for NA. A large series of related compounds are capable of substituting for NA in some or all of these processes. Those compounds which can be stored in the adrenergic vesicles together with NA and which can be released by nerve stimulation fulfil the criteria for classification as 'false adrenergic transmitters' (Kopin, 1968a). The subject has been extensively reviewed by Muscholl (1972) and only the effect of false transmitters on adrenergic transmission will be considered here. The ability of a false adrenergic transmitter to alter adrenergic transmission depends on which process is affected. The final effects could be equivalent to one of the actions described above (Section 5.6) or more commonly to various combinations of them (see Kopin, 1968b). Some precursors of adrenergic false transmitters, including α-methyl-tyrosine, α-methyl-*m*-

tyrosine and tyramine, are taken up by adrenergic nerves and metabolized to α-methyl-NA, metaraminol and octapamine respectively and stored in the adrenergic vesicles. These analogues are all able to deplete NA stores, their end product displacing NA. Furthermore, depletion is also due to the inhibitory effect of these drugs on the biosynthetic enzyme T-OH (Day & Rand, 1964; Carlsson, 1966a; Muscholl & Sprenger, 1966; Maitre & Staehelin, 1967; Kopin, 1968b; Patil & Jacobowitz, 1968; Kilbinger, Lindman, Loffelholz, Muscholl & Patil, 1971; Thoenen & Tranzer, 1971; Muscholl, 1972).

In most tissues, little impairment of adrenergic transmission has been reported following injection of α-methyl-dopa (Goldberg, Da Costa & Ozaki, 1960; Varma & Benfey, 1963; Haefely, Hürlimann & Thoenen, 1966, 1967; Smith, 1966; Henning & Svensson, 1968; Mohammed, Gaffney, Yard & Gomez, 1968; Schmitt & Petillot, 1970; Kilbinger, Lindman, Loffelholz, Muscholl & Patil, 1971), but in other cases an impairment of adrenergic neurotransmission was reported (Kisin, 1967; Sugarman, Margolius, Gaffney & Mohammed, 1968; Mailk & Muscholl, 1969; Salmon & Ireson, 1970). These discrepancies are probably due to the similar potency of α-methyl-NA and NA in some tissues and the predominant effect of α-methyl-NA on β-adrenoceptors in other tissues (Muscholl & Sprenger, 1966; see Muscholl, 1972).

There is strong evidence that the hypotensive effect of α-methyl-dopa is central in origin and is not mediated by impaired adrenergic neurovascular transmission (see Henning, 1969; Muscholl, 1972); its effect remains following immunosympathectomy (Varma, 1967). On the contrary, α-methyl-*m*-tyrosine, which is taken up and metabolized in adrenergic nerves to α-methyl-*m*-tyramine and to metaraminol, impairs adrenergic transmission (see Muscholl, 1972). In fact, metaraminol displaces NA from storage sites and inhibits NA synthesis by preventing the β-hydroxylation of DA to NA; when released by nerve stimulation it shows a much lower potency on effector tissues than the physiological transmitter (Carlsson & Lindqvist, 1962; Crout, Alpers, Tatum & Shore, 1964; Haefely, Hürlimann & Thoenen, 1965, 1966; Jonsson & Ritzén, 1966; Schmitt & Petillot, 1970; Torchiana, Poster, Stone & Hanson, 1970).

In cases in which a compound related to NA has been found to be physiologically present in adrenergic nerves together with NA, as for instance for octopamine (Molinoff & Axelrod, 1969) the term 'false adrenergic transmitter' does not apply even if the physiological significance is not known. Octopamine has been found to displace NA from storage sites, to compete with DA for β-hydroxylation sites, is stored and released by nerve stimulation and impairs adrenergic transmission (Fischer, Horst & Kopin, 1965; Kopin, 1968a, b; Muscholl, 1972).

5.7 General discussion and models of adrenergic transmission

The schemes presented below represent a correlation and extension of previous models proposed for the anatomy of the sympathetic ground plexus (Hillarp, 1946, 1959; Malmfors, 1965), for the biochemistry of NA synthesis, storage, release and inactivation (Shore, 1962; Kopin, 1964; Iversen, 1967; Molinoff & Axelrod, 1971; von Euler, 1972), for neurovascular relationships (Somlyo & Somlyo, 1968; Holman, 1969, 1970; Ljung, 1970; Burnstock, Gannon & Iwayama, 1970; Folkow & Neil, 1971; Bevan & Su, 1973; Burnstock, 1975) and for the general definition of the autonomic neuromuscular junction (Bennett & Burnstock, 1968; Burnstock, 1970; Burnstock & Iwayama, 1971; Trendelenburg, 1972; Bennett, 1973a, d; Burnstock & Bell, 1974).

5.7.1 *Transmitter mechanisms at adrenergic neuromuscular junctions*

Figure 15 shows the essential features of the transmitter mechanisms in adrenergic neurons. Most of the NA is stored in the nerve terminals in vesicles thereby being biologically inactive. A large stable store (lNA) which is usually not saturated, is replenished by the binding of cytoplasmic NA (cNA) and by local synthesis of NA from tyrosine through dopa and DA. From this store, NA is transferred to a smaller store (sNA), which is also replenished by binding of cytoplasmic NA (cNA) and by local synthesis. It is usually assumed that both lNA and sNA are located in vesicles. The bulk of the cytoplasmic NA is inactivated by MAO and the deaminated metabolite leaks passively out of the nerve into the extracellular space and eventually the blood stream. The low concentration of cytoplasmic NA, due to this enzyme, facilitates the uptake of NA into the neuron through the axon membrane by increasing the concentration gradient.

At the arrival of the nerve impulse, NA is released from the small readily available pool (sNA) in discrete packets by exocytosis together with the proteins chromagranin and DβH and probably ATP.

sNA exerts a tonic inhibition on NA synthesis at the step of the hydroxylation of tyrosine. Thus, when the level of sNA is decreased by nerve stimulation, local synthesis of new NA is increased and more transmitter becomes available to cater for the increased nerve activity.

It seems likely that the proteins, once released by the process of exocytosis together with NA, are not taken up again into the neuron but are replaced by new proteins synthesized in the cell body and transported along the axons to the terminals by axonal flow.

The NA released by nerve stimulation diffuses into the neuromuscular

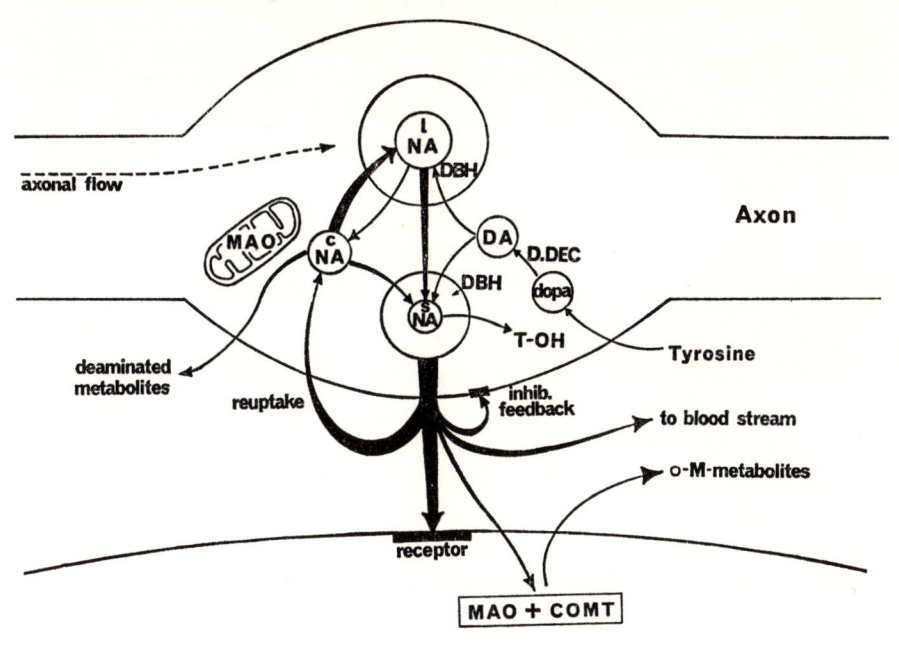

Figure 15. The basic mechanism involved in the synthesis, storage, uptake and release of NA from a varicosity of the adrenergic terminal axon innervating smooth muscle.

lNA: large storage pool of vesicular NA
sNA: small storage pool of vesicular NA
cNA: cytoplasmic NA
MAO: monoamine oxidase
COMT: catechol-*o*-methyl transferase
DA: dopamine
T-OH: tyrosine hydroxylase
D.DEC: dopa decarboxylase
DβH: dopamine-β-hydroxylase
o-M-metabolites: *o*-methylated-metabolites.

For further description see text.

gap binding temporarily to the receptors of the smooth muscle cell membrane and produce changes in conductivity which lead to a response. Part of the NA released is taken up again into the nerve where it is either incorporated into vesicular stores, or metabolized in the cytoplasm by MAO (and possibly also by COMT). Part of the released NA is taken up by smooth muscle cells or other extraneuronal tissues and metabolized by COMT and/or MAO; part diffuses into the blood stream, and part occupies α-adrenoceptors located on the nerve membrane which inhibit further release of NA.

5.7.2 *Maintenance of adrenergic transmission*

Maintenance of nerve transmission requires that a sufficient amount of NA is available for release at arrival of nerve impulses. Since NA is released together with DβH and chromagranin (see Section 5.2), maintenance of transmission requires a continuous replacement supply of NA, DβH and chromagranin.

The NA available for release originates from vesicular stores of NA which are replenished by axonal transport of NA to the terminals (see Section 3.1), by local synthesis of NA (see Section 4.1) and by reuptake of released NA (see Section 5.4).

Release of NA. Some NA is released spontaneously from adrenergic terminal varicosities as revealed by the spontaneous excitatory junction potentials (SEJPs) recorded in smooth muscle from vas deferens and other organs (see Section 5.2). The amount of NA released spontaneously has been estimated by Hughes (1972) to be 0.2 per cent of the total store of NA in the terminal per hour. During nerve activity, the amount of NA released per impulse increases with the frequency in the physiological range (see Section 5.2, and Hughes, 1972) and can be expressed as a fraction of the total store of NA (Table 3, Section 5.2). Taking the figures calculated by Hughes (1972) as an approximation of the fraction of NA released, at 2 Hz the fraction released per pulse is about 5×10^{-5} while at 16 Hz it is about 15×10^{-5}.

The mechanisms which lead to replacement of the NA lost by release will be considered.

Storage of NA. The total store of NA in a tissue varies with the density of innervation from less than $1\mu g\ g^{-1}$ in the intestine to about $20\mu g\ g^{-1}$ in the vas deferens.

Only a fraction of the total store of NA is releasable by nerve stimulation, since even continuous stimulation at high frequencies does not fully deplete NA stores. Furthermore, only a small fraction seems to be essential for transmission since NA stores can be depleted up to 99 per cent in some tissues without failure of transmission (see Section 5.6, and Antonaccio & Smith, 1974). When NA stores are reduced to 11 per cent of normal by inhibition of NA synthesis, transmission is maintained as long as this small pool of NA is maintained by local NA synthesis (Almgren 1973). Otherwise transmission is not sustained (Farnebo & Malmfors, 1971; Rubio & Langer, 1973) and in tissues in which depletion of NA by reserpine results in failure of transmission, recovery occurs at a time when only a small number of adrenergic vesicles is present (see Section 5.6).

Axonal flow. Axonal flow of NA supplies less than 1 per cent of the total store of NA in terminals per day (see Section 3.1). Therefore axonal flow does not appear to be an important source for replacement NA used in transmission, although it may be more important for the supply of replacement chromagranin and DβH. Banks and Mayor (1972) estimated that about 4000 vesicles per day are transported from the cell bodies to the terminal axon. If chromagranin, DβH and NA are released by exocytosis in the same proportions as they are bound in the vesicle (see Section 5.2), it follows that axonal flow would be unable to maintain a sufficient supply of chromagranin and DβH. The number of vesicles arriving at the terminals would only allow a release of 1 vesicle per second from the whole adrenergic branching terminals of one neuron. Even increasing this figure by a factor of 1000, which would be inconsistent with the number of vesicles found in non-terminal axons (see Banks & Mayor, 1972), the supply of vesicles by axonal flow would not allow maintenance of transmission. This difficulty could be explained by transport of chromagranin and DβH in a non-vesicular form and/or by local formation of vesicles from tubular reticulum (see Section 3.2).

Reuptake of NA. The reuptake of NA after release has been considered to be a major mechanism for maintenance of transmission (see Iversen, 1967, 1971). The percentage of NA taken up again after release has been estimated to be about 50 per cent (Table 6 and Section 5.5). However, despite the high efficiency of this uptake mechanism, its blockade does not result in the impairment of adrenergic transmission (Kalsner, 1972), although it does produce some depletion of the NA store (Bhatnagar & Moore, 1972). This suggests that reuptake of NA does not represent the major mechanism for maintenance of adrenergic transmission.

Synthesis of NA. The mechanism which appears to be the most important for maintenance of adrenergic transmission is synthesis of NA in the adrenergic terminal. NA in terminals is synthesized locally from tyrosine and the level of NA is kept constant by alterations in the rate of synthesis

ADRENERGIC NEURONS

Table 6

Preparation	Percentage of NA taken up by adrenergic nerves after release	Reference
Cat spleen	37	Thoenen, Hürlimann & Haefely, 1964
Cat hind leg blood vessels	54	Folkow, Häggendal & Lisander, 1967
Renal artery	39	Zimmerman & Gisslen, 1968
Cat nictitating membrane	37	Basset, Cairncross, Hacket & Story, 1969
Cat nictitating membrane	70	Langer, 1970
Guinea-pig uterine artery	63	Bell & Vogt, 1971
Dog spleen	66	De Potter, Chubb, Put & De Schaepdryver, 1971
Rat iris	30	Farnebo & Hamberger, 1971
Cat spleen	30	Kirpekar & Puig, 1971
Rabbit heart	50	Wennmalm, 1971
Rat vas deferens	72–79	Hughes, 1972
Rat portal vein	63–70	Hughes, 1972
Rabbit ear artery	30	Kalsner, 1972
Rabbit heart	50	Starke & Schümann, 1972

which follow fluctuations in the physiological frequencies of nerve activity, i.e. the fast regulatory mechanism of NA biosynthesis (see Section 4.1). The rate of synthesis of NA in the adrenergic terminals can be increased from the level of resting activity to a maximum of 5 per cent to 30 per cent of the total store synthesized per hour (Table 7). In 'silent' neurons there

Table 7

NA synthetized per h as percentage of total store		Reference
Range	5	Spector, Melmon & Sjoerdsma, 1962
	19	Montanari, Costa, Beaven & Brodie, 1963
Range	5 20	Bhagat and Friedman, 1969
Range	18 30	Folkow, Häggendal & Lisander, 1967
Average	15	Iversen, 1967
Range	6 30	Blakeley, Brown, Dearnaley & Harrison, 1968

is probably a low rate of synthesis just sufficient to compensate for the spontaneous release and basal metabolic degradation of NA. Low frequency (5 Hz), prolonged stimulation *in vitro* results in decline of transmission. Inhibition of NA synthesis by α-methyl-*p*-tyrosine leads to a faster decline of the transmission, confirming the importance of synthesis of NA in the maintenance of adrenergic transmission (see Section 5.6). Farnebo and Lidbrink (1972) reported that transmural stimulation of the rat iris at 40 Hz for 1 h produced a depletion of 50 per cent of NA which, after inhibition of NA synthesis, became 75 per cent. Almgren (1973) found that adrenergic stimulation of rat salivary gland at 5Hz for 5 h depleted 60 per cent of NA which after synthesis inhibition became 87 per cent. Intermittent stimulation allows recovery of transmission and therefore can be maintained for longer periods with little decrease of response and of NA stores (Weiner & Rabadjija, 1968a; van Orden, Robb, Bhatnagar & Burke, 1972).

Adrenergic transmission is maintained if the mechanisms of replacement of NA described above balance the NA lost by release. This suggests that it is essential for maintenance of transmission that the rate of release of NA does not exceed the rate of accumulation of NA from axonal flow plus the rate of NA taken up after release but not metabolized in the cytoplasm, plus the rate of local NA synthesis. Thus the store of NA in physiological conditions fluctuates around set values.

In order to visualize the relation between these mechanisms more clearly, the following equation is proposed:

$$R = S + U + F + \Delta st \qquad (1)$$

where R is the rate of release of NA at different frequencies, taking the values for fractional release, from Hughes (Hughes, 1972, see Table 3, Section 5.2); S represents the rate of local synthesis of NA as a fraction of the total store of NA; U is the rate of uptake of NA (since part of NA taken up by nerve after stimulation is metabolized by MAO and COMT, the value of U is an overestimation of the NA which re-enters the main NA store). About 50 per cent of released NA is taken back into the nerves (see Table 6). F represents the rate of NA axonal flow (about 0·04 per cent of total store/h). Δst represents the variation of the amount of NA stored. When Δst tends to zero, transmission can be maintained indefinitely, and this is the case of transmission from tonically active neurons; when Δst becomes negative, transmission declines as in the case of supraphysiological activation of adrenergic nerves; when Δst becomes positive, transmission is maintained and the total store increases as would occur during recovery from NA depletion.

Figure 16 is a schema of the rate of release at different frequencies, the rate of NA synthesized in different physiological conditions, the proportion taken up after release, the rate of NA transported to the terminals, and the rate of spontaneous release of NA; all rates are expressed as changes in percentage of the total store per hour.

The schema shows that the axonal flow plays a minor role in the maintenance

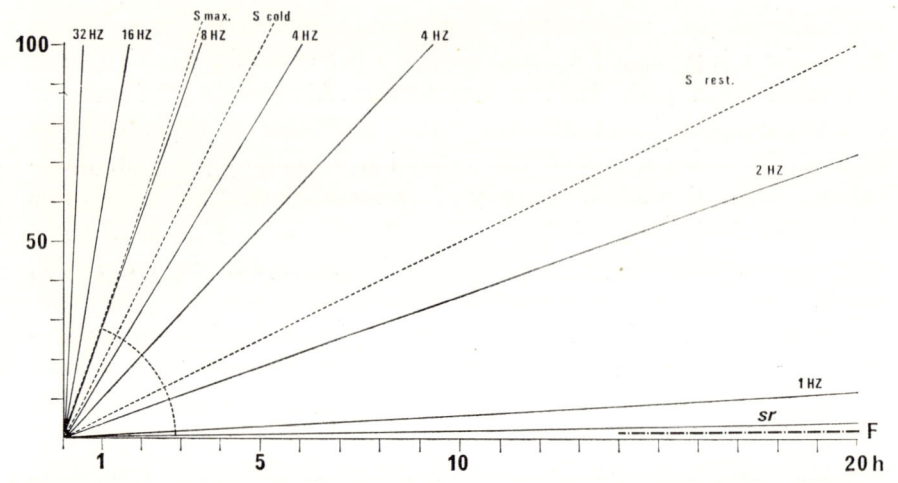

Figure 16. Graphical representation of the rates of release, synthesis, reuptake, and of axonal flow of NA in adrenergic terminal axons.
Ordinate: percentage of total storage of NA. Abscissa: time in hours (h). Continuous lines: rates of spontaneous release (*sr*) and of NA release at different impulse frequencies (1,2,4,8,16,32 Hz). Dashed lines: rates of NA synthesis during resting conditions (S resting) cold exposure (S cold) and maximal rate (S max) reported. The steeper of the two rates of release at 4 Hz represents the rate of release of NA assuming that the NA uptake mechanism which takes up 50 per cent of NA released is blocked (see Table 6). F: rate of axonal flow calculated from Geffen, Livett & Rush, 1970. The semcircular dotted line indicates the range of variability of rate of NA synthesis. Quantitative data on synthesis, uptake and release on which the schema is based were taken from authors listed in Tables, 3, 6 and 7. For complete description see text.

of NA store, and that the uptake mechanism alone can only prolong transmission but does not prevent its decline.

During physiological activity, when Δst tends to zero and recognizing that F is negligible, the net rate of release of NA (taking into account the rate of uptake of NA after release), should have the same slope as the rate of synthesis which occurs at that frequency. The schema shows a very good correlation between the rates of synthesis and release in the physiological range. Thus the rate of synthesis which occurs in the heart at resting conditions (5%/h) corresponds to a frequency of just over 2 Hz, while the rate of synthesis which occurs during cold exposure

(20%/h) corresponds to a frequency of about 6 Hz. Also the frequency which corresponds to the maximal rate of synthesis (30%/h) is 8 Hz which is in very good agreement with the range of physiological frequencies for adrenergic neurons (see Folkow & Neil, 1971 and Skok, 1973). The rate of synthesis therefore fluctuates in the range shown in the schema by the semicircular arrow, and follows fluctuations of discharge of the adrenergic neurons. Transmission is ensured as long as the two superimposing curves remain in this range. However, if the frequency increases over 8 Hz, the rate of synthesis cannot counteract the loss of NA and impairment of transmission during prolonged activity occurs, with depletion of NA stores. Dearnaley and Geffen (1966) obtained depletion of 7·7 per cent of the NA store following stimulation for 5 minutes at 30 Hz. From the schema, at 32 Hz depletion of the NA store would occur in about 30 minutes if no synthesis took place. In 5 min about 17 per cent of the store would be depleted; with a maximum synthesis of 30%/h, the degree of depletion could be about 14 per cent of the NA store, a figure very close to that observed experimentally. The diagram is also consistent with the findings of Folkow, Häggendal and Lisander (1967) that, while at 8 Hz the response of the cat nictitating membrane to adrenergic stimulation is sustained, after blockade of the reuptake of NA the response of the nictating membrane began to abate. In the diagram the curve of NA release at 8 Hz, in which uptake is working, is still inside the range of the capability of synthesis to counteract depletion, but after blockade of uptake the curve of 8 Hz would be steeper, and it would be out of range of NA synthesis.

Of course this schema gives only qualitative information and no quantitative implications can be drawn from it. The actual slope of the curves may be quite different. A factor that has not been taken into account, which would increase the slope of the frequency-release curves, is the amount of NA that is inactivated extraneuronally. The amount which overflows and which is measured, therefore underestimates the actual release (Hughes, 1972). However, there seems to be a good correlation between the amount released, measured as reduction of the store, and the amount recovered in the bath (6 per cent and 7·7 per cent respectively, Dearnaley & Geffen, 1966).

Factors which would decrease the slope of the NA release curves are: the probable reduction of the amount released per impulse by decreasing the total store of NA; the possible involvement of DβH in limiting transmission before NA stores are depleted.

By making gross approximations, it is possible to use equation 1 and the schema to establish the rate of synthesis which occurs in a neuron which is firing at a certain rate, or vice versa to find at which range of frequencies a neuron is capable of maintaining transmission indefinitely, by knowing the range of rate of NA synthesis of that neuron.

A final consideration about the maintenance of adrenergic transmission brings in the role of the neuromuscular distance. In organs with close neuromuscular junctions, smaller amounts of NA in the terminals are sufficient to maintain transmission, since very high peak concentrations occur at the receptors in these organs and therefore they are more suitable for prolonged transmission. Reduction of NA release in tissues with wide neuromuscular junctions more easily impairs transmission, because the peak concentration of NA at the receptors is just above threshold (see Section 5.6).

5.7.3 *Adrenergic responses and neuromuscular geometry*

The responses of a smooth muscle effector organ to activation of the adrenergic nerves involves a series of events i.e. release of NA from the adrenergic terminal, diffusion of NA across the junction, reaction of NA with the receptors and coupling to the mechanical response. Both the amplitude and time course of the mechanical response to adrenergic nerve stimulation will now be considered.

Amplitude of the adrenergic response. The relation between frequency of discharge and amplitude of response is given by a convex hyperbola as shown by Rosenblueth (Rosenblueth, 1950; see Johansson, Johansson, Ljung and Stage, 1972); thus by increasing the discharge frequency there is an initial steep increase of the response which reaches a plateau and then there is no further increase in amplitude.

The frequency-response curve varies with tissues. For example the frequency-response curve for the nictitating membrane or the rat iris is very steep so that the maximal response is achieved at low frequencies, while the frequency-response curve for muscle blood vessels is less steep and to achieve maximal effect requires higher frequencies (see Celander, 1954). The probable explanation is that the concentration of released NA at the receptors during transmission is inversely related to the neuromuscular distance (see Section 5.2). Therefore in tissues with short neuromuscular distances, such as the nictitating membrane, vas deferens, and iris, stimulation at low frequencies leads to a high concentration of NA at the receptors and therefore to a large response, while in tissues with larger neuromuscular distances, such as large blood vessels and intestine (see Section 5.1), stimulation at low frequencies is not able to build up at the receptors a concentration of NA sufficient to elicit the response. Thus, the vas deferens, and the nictitating membrane respond to a single pulse by contraction or by transient depolarization of the smooth muscle membrane or both (see Sections 5.5, 5.6 and Plate 12). In contrast, single pulses are ineffective in medium and large blood vessels and in the intestine and no hyperpolarization of the smooth muscle membrane or relaxation is observed; only frequencies higher than 5 Hz elicit a response and have a slight effect on membrane depolarization and produce relaxation (see Section 5.5, and Plate 13).

The response of smooth muscle organs *in vivo* to CA released by stimulation of the adrenal medulla could be regarded as a transmission with a very wide junction compared to the neuromuscular junctions. The frequency-response curves of various organs obtained by stimulating the splanchnic nerves to the adrenal medulla are flat compared with the frequency-response curves to stimulation of the adrenergic nerves supplying the same organs (Celander, 1954). When the neuromuscular distance increases, the frequency-response curve not only flattens and shifts to the right, but also the maximal response is decreased as shown in the figures of Celander (1954). In the pulmonary artery, the maximal response to nerve stimulation is much smaller than that to applied NA, probably because of the very long distance between the adrenergic terminals and medial smooth muscle (Bevan & Su, 1974).

Several authors reported a lack of correlation between the amount of NA released and the amplitude of the response; this has generally been attributed to failure of the effector muscle to respond rather than to failure of the adrenergic neuron to release NA (Langer, 1970; Farnebo & Malmfors, 1971; Hughes, 1972; Stjärne, 1973d). In fact, while the frequency-response curve is hyperbolic, the amount of NA released increases almost linearly with the frequency (see Hughes, 1972). At suprasphysiological frequencies, failure of transmission is also due to failure of the adrenergic neurons to release NA (Iversen, 1967). Another reason for the apparent lack of correlation of NA release and amplitude of response is that the amount of NA which overflows into the bath, depends on both the experimental design for collection and the proportion of released NA which is inactivated in the tissue before it reaches the bath. Furthermore the amount of NA collected does not necessarily reflect the amount of NA which was effective in eliciting the response. An example of this is the large twitch response of the vas deferens, which is probably due to release of NA from very few varicosities in close contact with the smooth muscle cells (Swedin, 1971; Furness & Iwayama, 1972; Furness, 1974), but does not correspond to a large amount of NA collected in the bath (Stjärne, 1973c). When all these considerations are taken into account there is still a good correlation between amplitude of response and frequency in the physiological range of 1 to 8 Hz (Haefley, Hürlimann & Thoenen, 1965; Blakeley & Brown, 1966; Kirpekar, Wakade, Dixon & Prat, 1969; Häggendal, Johansson, Jonason & Ljung, 1970).

The influence of neuronal uptake of CA in determining the concentration of NA at the receptor sites and therefore the amplitude of the response is also dependent on the neuromuscular geometry. It appears that neuronal uptake affects the amplitude of adrenergic responses only in tissues with close neuromuscular junctions (see Section 5.6). However, high concentration of NA at these junctions during stimulation tends to saturate the uptake mechanism and inhibit further release of NA via the prejunctional inhibitory mechanism (Section 5.2).

Time course of the adrenergic response. The *onset* of the adrenergic response is also a function of neuromuscular distance, e.g. Fig. 17 compares the rates of onset of adrenergic responses in large and small vessels, which have wide and close neuromuscular distances respectively.

The *offset* of the response reflects the time required for the termination of the effect of NA by the inactivating mechanism. In Sections 5.4 and 5.6, the mechanisms for inactivation of NA and their effect on adrenergic transmission were discussed separately. Here their relative roles will be considered. Only diffusion and both neuronal and extraneuronal uptake mechanisms will be discussed, since it has been shown earlier (Section 5.4) that inactivation of NA by metabolic degradation by MAO and COMT occurs inside both neuronal and extraneuronal tissues. Neuronal uptake of NA plays a major role in determining the time course of adrenergic responses. Blockade of such mechanisms delays the return of the smooth muscle to the original tension prior to nerve stimulation.

Diffusion and extraneuronal uptake mechanisms appear to influence the termination of the adrenergic response, but to a lesser extent than neuronal uptake. However, they may play a more important role in the slow responses produced by CA diffusing from adrenal medulla or from nerves with wide separation from the muscles. These mechanisms are interdependent and it is impossible at present to establish their relative contributions in different tissues (Hughes, 1972). Thus blockade of neuronal uptake by cocaine leads to an increase in extraneuronal uptake from 25 per cent to 75 per cent of NA released by nerve stimulation in the rabbit portal vein; conversely blockade of extraneuronal uptake by corticosterone leads to an increase of neuronal uptake from about 70 per cent to 90 per cent (Hughes, 1972). Another factor which influences the proportion of NA taken up by either of the two mechanisms is the local concentration of NA. Neuronal uptake works more efficiently in the lower range of concentration up to 10^{-6} g ml^{-1} becoming saturated at higher concentrations (Draskózcy & Trendelenburg, 1970), while extraneuronal uptake

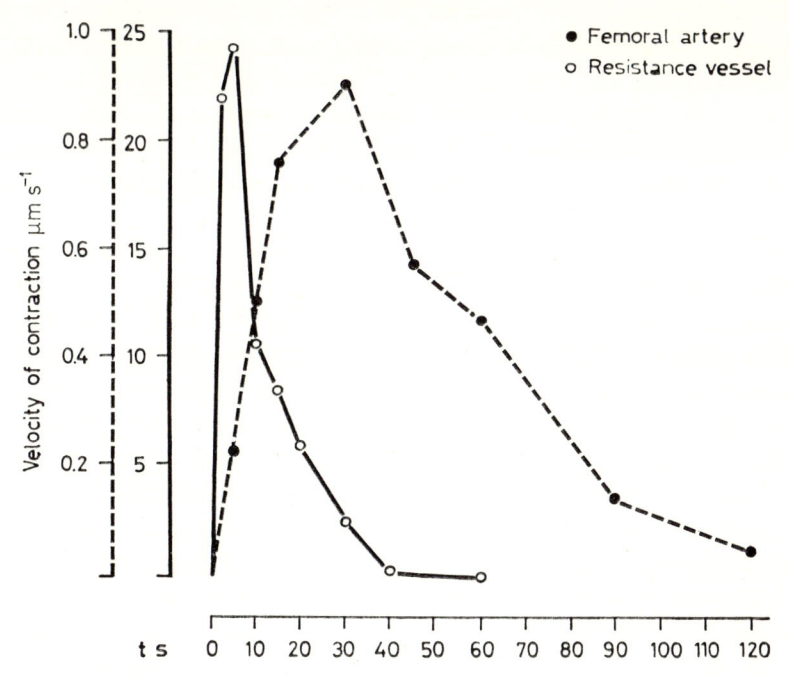

Figure 17. Comparison of the time course of the adrenergic response in two blood vessels with different neuromuscular relations. The response of a small resistance vessel to adrenergic stimulation of 15 Hz is prompt while the response of a large artery is sluggish. Smooth muscle cells are closer to the adrenergic terminal axons in the small than in the large arteries. Slightly modified from Gero and Gerova, 1971.

works from about 10^{-7} g ml^{-1} to higher concentrations (Lightman & Iversen, 1971). If follows that in smooth muscle relatively insensitive to NA, like those in the dog saphenous vein (Osswald, Guimarães & Coimbra, 1971), extraneuronal accumulation may occur at submaximally effective concentration of NA (Draskózcy & Trendelenburg, 1970).

Since the rate of diffusion depends on the rate of blood flow (see Section 5.4), changes in blood flow may alter the proportion of NA inactivated by neuronal and extraneuronal uptake. The relative importance of diffusion in the termination of the adrenergic response is also related to the density of innervation; in densely innervated tissues a great part of the NA released is recaptured by nerves before it escapes from the tissue, while in sparsely innervated tissues a larger proportion of NA escapes from the tissue by diffusion.

Blockade of all the inactivating mechanisms leads to an indefinitely persistent response. For example, the rate of relaxation of aorta muscle previously contracted by NA is extremely slow when diffusion of NA is reduced by using a mineral oil bath and neuronal uptake and extraneuronal metabolic degradation have been blocked (Kalsner & Nickerson, 1968; Osswald, Guimarães & Coimbra, 1971). Similarly, marked prolongation of contraction in aorta occurs under these conditions with NA released by from nerves (Gillis, 1971).

5.7.4 *Dual role of adrenergic nerves*

In most smooth muscles, the magnitude and time course of the response to exogenous NA differ markedly from the response to nerve-released NA. Densely innervated smooth muscles such as vas deferens and nictitating membrane are rather insensitive to applied NA *in vitro* (see Section 5.4), while they respond promptly to low frequency stimulation of their adrenergic nerve supply. Conversely, sparsely innervated smooth muscle such as the non-sphincteric parts of the intestine are very sensitive to applied NA but do not respond promptly to adrenergic nerve stimulation (Chapter 5.4 and above). Similarly, NA injected *in vivo* or circulating CA from the adrenals have little effect on nictitating membrane or vas deferens (Celander, 1954; Holman, 1967; Trendelenburg, 1972), while the non-sphincteric parts of the intestine are very sensitive to circulating CA (Furness & Burnstock, 1954). Following either degeneration of adrenergic nerves (Chapter 6) or blockade of neuronal uptake of NA by cocaine (Sections 5.4 and 5.6), densely innervated organs become sensitive to circulating CA (Cannon & Rosenblueth, 1949; Langer, 1966b).

The reason for these different responses to exogenously applied and nerve-released NA is probably that in densely innervated organs (nictitating membrane and vas deferens) there are close neuromuscular junctions (Chapter 5); thus high peak concentrations of NA released from nerves reach localized areas on the smooth muscle membrane, while applied NA is largely taken up by the dense network of adrenergic fibres before it reaches the receptors. In sparsely innervated tissues, neuromuscular distances are usually wider and, following stimulation of the adrenergic nerves, lower concentrations of NA reach the receptors; whereas applied or circulating CA reach receptor sites without being significantly inactivated by neuronal uptake.

These observations suggest that adrenergic fibres may have a dual role; not only do they provide nerve control of smooth muscle, its efficacy depending on the neuromuscular distance, but they also 'protect' a densely innervated organ from unwanted effects of circulating CA. It is noteworthy too, that densely innervated organs such as the vas deferens, nictitating membrane and iris, perform functions that are quite unrelated to general homeostatic functions which also involve the adrenal medulla. On the other hand, the sparsely innervated parts of the intestine are very sensitive to CA released from the adrenal medulla.

Blood vessels are concerned in homeostatic responses, which involve CA released from both adrenal medulla and adrenergic nerves. Their pattern of adrenergic innervation appears to be suitable for control by both circulatory CA (the non-innervated media behaves as a supersensitive smooth muscle following denervation or blockade of neural uptake by cocaine) and adrenergic nerves (see Folkow & Neil, 1971). Moreover, the absence of adrenergic nerves in the media of most blood vessels facilitates the diffusion of CA from the adrenals to their peripheral targets.

5.7.5 *Models of neuromuscular junctions*

Models of autonomic neuromuscular junctions are proposed (Figure 18) to illustrate the three main types found in different organs.

An essential and common feature of these models is that the effector unit is not the single smooth muscle cell but the muscle cell bundle, about 25 to 100 μm in diameter. Individual cells within the effector bundle are connected by low resistance electrical pathways, represented by 'nexuses' or 'gap junctions' of variable size which allow electrotonic spread of activity.

Smooth muscles with a dense adrenergic innervation and close neuromuscular junctions. This model (Figure 18a) is basically that proposed by Burnstock 1970 and Burnstock and Iwayama (1971) and applies to smooth muscles such as vas deferens, seminal vesicles, retractor penis, iris and nictitating membrane which are relatively densely innervated by adrenergic nerves. A variable proportion of the smooth muscle cells are 'directly-innervated' by one or more adrenergic terminals with close neuromuscular junctions (20 nm). The cells adjoining 'directly-innervated cells' are coupled electrotonically to them. Excitatory junction potentials (EJP's) elicited in

'Directly-innervated cell'

'Coupled cell' exhibits junction potentials carried by electrotonic coupling

'Indirectly-coupled cell' exhibits action potentials through low resistance pathways

Varicose adrenergic axon

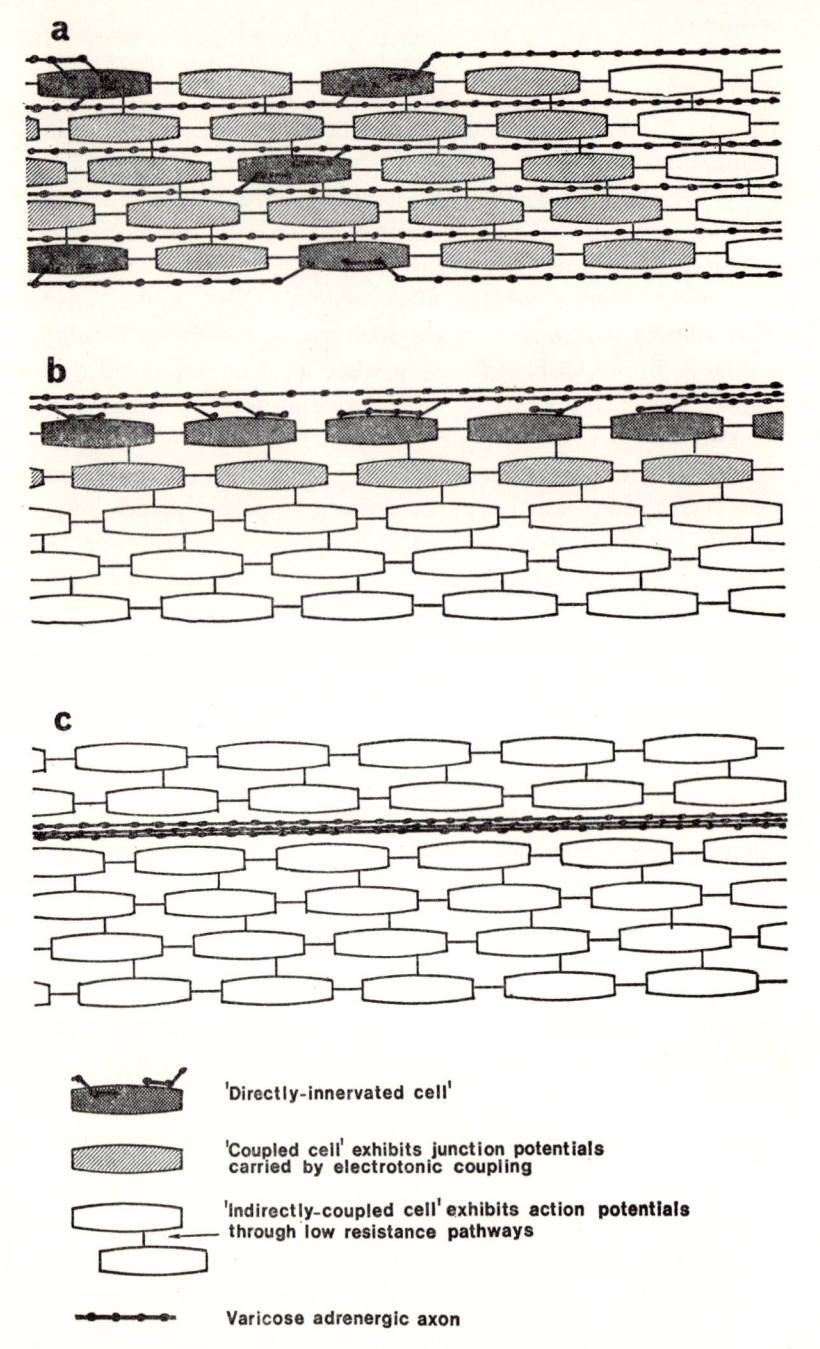

Figure 18. Schematic model of the three main types of adrenergic neuromuscular relationship.
(a) densely and homogenously innervated tissues such as vas deferens.
(b) asymmetrically innervated tissues such as blood vessels.
(c) sparsely innervated tissues such as non-sphincteric parts of the intestine.
For description see text.

'directly-innervated cells' can be recorded in these 'coupled cells'. When the muscle cells in an area of the effector bundle reach a threshold depolarization, an all-or-none action potential is initiated which propagates through the tissue. Thus 'indirectly-coupled cells' are invaded by the action potential and contract in response to adrenergic nerve stimulation.

The response to nerve stimulation is prompt, since the NA rapidly reaches high concentrations in the small neuromuscular space with little diffusion. This also results in a prompt termination of the response by reuptake of NA into the nerve, and in reduction of the amount of NA released through activation of the presynaptic α-adrenoceptors.

This neuromuscular arrangement is suitable for highly efficient adrenergic control of smooth muscle; these muscles readily respond to low frequency, or even single, nerve pulses. Circulating CA has little effect in these muscles because of inactivation by neuronal uptake of CA in the dense network of adrenergic nerves.

Smooth muscles with asymmetric adrenergic innervation. The innervation of most blood vessels is characterized by adrenergic nerves that are confined to the adventitial side of the media (Figure 18b). The minimal neuromuscular distance in small arteries and in veins is about 80 nm. Stimulation of the adrenergic nerves results in a prompt vasoconstriction. Excitatory junction potentials (EJP's) have been recorded from cells of the inner media remote from the nerve plexus (Bell, 1969; Speden, 1970), indicating that the muscle cells are electrically coupled (see Holman, 1969; Speden, 1970). However, the asymmetric disposition of the nerves leaves most of the medial smooth muscle devoid of adrenergic fibres and therefore of sites of NA inactivation, so that the muscle is more responsive to circulating CA.

These types of blood vessels therefore are under effective control by both adrenergic nerves and circulating CA.

Smooth muscles with low density of adrenergic nerves and large neuromuscular distances. In larger blood vessels such as elastic arteries, adrenergic innervation is sparser, neuromuscular distances are larger and the electrical coupling between smooth muscle cells and bundles seems likely to be less marked due to the presence of elastic laminae intermingled amongst the muscle bundles. The response to nerve stimulation is slow in onset and weaker compared to that of smaller arteries (see Figure 17); a higher frequency of stimulation is needed to elicit a response and no discrete junction potentials have been recorded from elastic arteries. These large blood vessels therefore appear to be mainly under humoral control by circulating CA while adrenergic nerve control is weak and sluggish. However, some of them (e.g. aorta) appear to be relatively insensitive to both circulating and nerve-released CA.

In the intestine, almost no adrenergic fibres are present in the longitudinal muscle coat and adrenergic fibres run in the nerve bundles of Auerbach's plexus separated by large distances from the longitudinal muscle cells. Adrenergic control of longitudinal intestinal muscle is not as effective as it is for densely innervated organs, since only high frequency pulses elicit responses. The sparse adrenergic innervation of the intestinal smooth muscle results in a high sensitivity to circulating CA.

This arrangement of adrenergic nerves (Figure 18c) provides a predominant humoral control by circulating CA and little nerve control.

5.8 Summary

The morphological relation between adrenergic nerves and smooth muscles, the mechanisms of noradrenaline release, reuptake, pre- and postjunctional effects, and inactivation, the electrophysiology and pharmacology and the general properties of adrenergic transmission are described in this chapter.

Adrenergic nerves become varicose and branch in effector organs to form the 'autonomic ground plexus'. The number of adrenergic terminal varicosities (a measure of the density of innervation) varies in different tissues; for example vas deferens, iris and nictitating membrane are densely innervated and the adrenergic terminal varicosities approach to within 10 to 30 nm of the smooth muscle membrane, while non-sphincteric parts of the intestine and ureter are sparsely innervated and have neuromuscular junctions wider than 100 nm. Muscular arteries have an asymmetrical adrenergic innervation, which is usually confined to the adventitial-medial border, and there is a range of neuromuscular distances from 80 nm to more than 1000 nm in some large arteries. In tissues with close neuromuscular junctions, some postjunctional specializations have been described. The motor unit in smooth muscle is represented by a bundle of smooth muscle cells electrically coupled through 'nexuses' or 'gap junctions' of variable size.

Noradrenaline is released from the storage vesicles of all varicosities of the adrenergic terminal axon together with the vesicular proteins chromagranin and dopamine-β-hydroxylase probably by a calcium-dependent process of exocytosis.

The fraction of the total noradrenaline store released per impulse increases with frequency in the physiological range, and decreases at supraphysiological frequencies. Local mechanisms also regulate the amount of noradrenaline released: noradrenaline and dopamine released from adrenergic nerves inhibit further release by acting on α-adrenoceptors located on adrenergic axon terminals; prostaglandins of the E type released from effector tissues inhibit noradrenaline release while prostaglandins of the F type enhance noradrenaline release. Angiotensin may modulate adrenergic transmission by enhancing noradrenaline release. Acetylcholine released from neighbouring cholinergic nerves may reduce noradrenaline release by acting on muscarinic receptors located on the adrenergic terminal axon.

Noradrenaline is released in discrete packages but it is not known whether the entire content of a vesicle represents the quantum or if there are several quanta of noradrenaline per vesicle. Following release of noradrenaline, high transient peaks of concentration occur near the muscle surface close to the site of release, in a range, depending on the neuromuscular distances, of 10^{-9}M for junctions of 4000 nm to 10^{-1}M for close junctions of 10 to 30 nm.

The effects of catecholamines on effector tissues are mediated by both α and β adrenoceptors. Several hypotheses concerning the nature of the receptors have been advanced.

Released noradrenaline is inactivated by various mechanisms namely: neuronal uptake, which is a specific and highly efficient amine pump that is inhibited by drugs such as cocaine and imipramine; extraneuronal uptake, which is a less specific amine pump present in smooth muscle and connective tissues and is inhibited by drugs such as some steroids, normetanephrine, metanephrine and phenoxybenzamine (species and tissue difference in this extraneuronal uptake have been reported). Noradrenaline, once it is taken up either neuronally or extraneuronally is catabolized by the enzymes monoamine oxidase (mainly inside the neurons) and catechol-*o*-methyl transferase (mainly in extraneuronal tissue). Some noradrenaline is inactivated by diffusing away from the effector tissue. Changes in blood flow influence the proportion of noradrenaline which diffuses away.

Adrenergic transmission has been studied by using intracellular recordings from smooth muscle cells. In tissues excited by sympathetic nerves, both spontaneous excitatory postsynaptic potentials and excitatory postsynaptic potentials induced by single nerve impulses have been described, while in tissues inhibited by sympathetic nerves only small and slow hyperpolarizations of the cell membrane occur following repetitive stimulation.

The effects of drugs which modify synthesis, storage, release, uptake and enzymatic degradation of noradrenaline on adrenergic transmission are described. Inhibition of synthesis, storage and release of noradrenaline and inhibition of post-junctional adrenoceptors impairs adrenergic transmission, while inhibition of uptake, enzymatic degradation and enhancement of release potentiate adrenergic transmission. 'False adrenergic transmitters....' reduce or inhibit adrenergic transmission.

A schema of the essential features of the transmitter mechanisms in adrenergic terminal axons is presented. Adrenergic transmission is maintained as long as noradrenaline loss by release is balanced by synthesis, reuptake and axonal flow of noradrenaline. Only a small part of the noradrenaline storage is essential for transmission. Synthesis of noradrenaline is the major mechanism for maintenance of adrenergic transmission, while axonal flow and reuptake play a minor role.

Neuronal uptake plays an important role in determining the time course of adrenergic responses by ensuring a prompt termination of the effect of noradrenaline. The neuromuscular geometry is a major factor in determining the amplitude and the time course of adrenergic responses, the safety factors for transmission, the susceptibility of transmission to drugs which affect synthesis, storage, uptake, prejunctional and postjunctional adrenoceptors, and metabolic degradation.

A dual role for adrenergic nerves is proposed: direct control of the effector tissues by release of noradrenaline; and indirect control by preventing unwanted effects of circulating catecholamines in densely-innervated tissues by the neuronal uptake mechanism.

Models of three main types of adrenergic neuromuscular arrangement are presented: densely, homogeneously innervated organs such as vas deferens, iris and nictitating membrane are under very effective nerve control, but little influenced by circulating catecholamines; arterial blood vessels with asymmetric innervation, are under effective control by both adrenergic nerves and circulating catecholamines; and sparsely innervated organs such as some large elastic arteries and the non-sphincteric parts of the longitudinal coat of the intestine are sensitive to circulating catecholamines, but are only weakly affected by adrenergic nerves.

6 GROWTH AND DEGENERATION OF ADRENERGIC NEURONS

6.1 Growth of adrenergic neurons and related cells in tissue culture

6.1.1 *Adrenergic neurons*

Neurons of mammalian sympathetic ganglia cultivated *in vitro* in suitable media survive, and their severed axons regrow (Matsumoto, 1920; see Murray, 1965 and Chamley, Mark, Campbell & Burnstock, 1972). Normal synthesis, storage, axonal flow, uptake and release of NA occur in these neurons; this has been demonstrated with fluorescence histochemistry, biochemistry and autoradiography (Goldstein, 1967; Sano, Odake & Yonezawa, 1967; Chamley, Mark, Campbell & Burnstock, 1972, Kopin & Silberstein, 1972). Cultured adrenergic neurons respond to nicotinic drugs (Larrabee, 1970).

When foetal neurons are cultured, they are able to develop to almost full maturity (Sano, Odake & Yonezawa, 1967). Growing axons cultured from foetal tissue first show a smooth weakly fluorescent appearance, but later individual fibres with the characteristic adult varicose appearance can be seen (Sano, Odake & Yonezawa, 1967). This development is comparable to that seen during ontogenesis (see Section 6.2). However, in neurons cultured from newborn tissue, some fibres are already varicose (Chamley, Mark, Campbell & Burnstock, 1972). The appearance of varicosities in cultured adrenergic neurons shows that these structures appear even if effector tissues are not present.

At least two types of adrenergic neuron have been identified in cultured mammalian sympathetic ganglia (Murray & Stout, 1947; Chamley, Mark, Campbell & Burnstock, 1972). A low percentage (5 to 10 per cent) of the total population of neurons are generally small with a few long smooth branching processes; they usually show an intense specific fluorescence for CA and exhibit strong migratory ability. The rest of the neuronal population is represented by larger neurons with less intense fluorescence and these do not usually migrate outside of the explant. The first type probably corresponds to the smaller neurons described in the sympathetic ganglia of both adult (see Section 3.1), and newborn animals (see Section 4.2).

Since the relative percentages of the two types do not change during further spontaneous or Nerve Growth Factor (NGF)-induced maturation (Chamley, Mark, Campbell & Burnstock, 1972), the smaller type is not likely to represent an immature form of the larger type.

6.1.2 *Interactions between adrenergic neurons and effector tissues*

Adrenergic axons in tissue culture grow towards explants of various organs (Levi-Montalcini, Meyer & Hamburger, 1954; Levi-Montalcini & Angeletti, 1961; Silberstein, Johnson, Jacobowitz & Kopin, 1971; Johnson, Silberstein, Hanbauer & Kopin, 1972). They grow preferentially and in larger numbers towards explants of tissues such as auricle and vas deferens which, in adults, are densely innervated by adrenergic nerves (Chamley, Goller & Burnstock, 1973; Chamley, Campbell & Burnstock, 1973). Nerve fibres appear to be attracted for up to about 1 mm, but only by a large clump of cells and not by single cells (Mark, Chamley & Burnstock, 1973; Chamley, Campbell & Burnstock, 1973). If the growing fibres encounter single smooth muscle cells they establish close, extensive and long-lasting associations, but not with fibroblasts. The association between adrenergic fibres and smooth muscle cells of iris, vas deferens and taenia coli shows morphological characteristics similar to neuromuscular junctions in the animal. Sympathetic neurons have recently been shown to form functional relations with effector cells *in vitro* (Purves, Hill, Chamley, Mark, Fry & Burnstock, 1974). Once contact is established, lasting contact of muscle cells with other adrenergic fibres is prevented. These findings indicate that the density of innervation depends on the effector tissue (Chamley, Goller & Burnstock, 1973; Chamley, Campbell & Burnstock, 1973).

It has been suggested that growth of adrenergic fibres *in vivo* is dependent on the concentration gradient of a chemical substance present in the effector organs, and that this substance is NGF, which has also been shown to stimulate the growth of adrenergic neurons *in vivo* (Levi-Montalcini, Meyer & Hamburger, 1954; Johnson, Silberstein, Hanbauer & Kopin, 1972; Chamley, Goller & Burnstock, 1973). The growth of adrenergic fibres is more marked in the presence of foetal tissues than of older tissues (Silberstein, Johnson, Jacobowitz & Kopin, 1971; Chamley, Goller & Burnstock, 1973), which is consistent with the presence of a higher amount of NGF in immature tissues than in adult tissues (Johnson, Silberstein, Hanbauer & Kopin, 1972).

The content of NGF in the effector tissue may therefore control the extent of the adrenergic innervation, while the properties of individual effector cells appear to determine the kind and number of neuromuscular junctions established (Johnson, Silberstein, Hanbauer & Kopin, 1972; Chamley, Goller & Burnstock, 1973; Chamley, Campbell & Burnstock, 1973).

6.1.3 *Chromaffin cells*

Little work has been carried out on *in vitro* culture of adrenal medullary chromaffin cells largely because of technical difficulties (Hoch, Ligiti & Camp, 1959; Coupland, 1965). However, extra-adrenal chromaffin cells, for example those from large preaortal paraganglia (Coupland & MacDougall, 1966) and sympathetic ganglia (Chamley, Mark & Burnstock, 1972), have been cultured successfully and maintain their ability to synthesize and store CA.

In cultures of sympathetic ganglia, chromaffin tissue is highly sensitive to glucocorticoid hormones, but not to NGF; the reverse holds for adrenergic neurons (Levi-Montalcini & Angeletti, 1968; Eränkö, Eränkö, Hill & Burnstock, 1972). Glucocorticoids induce the synthesis of A in NA-containing extra-adrenal chromaffin cells (Coupland & MacDougall, 1966) and produce a significant increase in the number of chromaffin cells in sympathetic ganglia with increase in CA content, and induce the appearance of A-storing vesicles (Eränkö, Eränkö, Hill & Burnstock, 1972). This suggests that glucocorticoids induce the methylating enzyme PNMT *in vitro* as well as *in vivo* (see Section 6.2). No detectable changes in the chromaffin cells in sympathetic ganglia were produced by NGF in tissue culture (Chamley, personal communication).

6.2 Ontogenesis of the sympatho-adrenal system

The early stages of the development of the sympatho-adrenal system in mammals have been described by several authors (Kuntz, 1910, 1920/21, 1953; Ranson & Billingsley, 1918; Waring, 1935/36; Wrete, 1940; Smitten, 1963; Coupland, 1965; Fernholm, 1971).

Sympathetic adrenergic neurons, adrenal and extra-adrenal chromaffin cells originate from a common ancestor, a primitive stem cell of the neuroectoderm. The first traces of the primordia of the sympatho-adrenal system are small aggregates of small cells with a basophilic nucleus, little cytoplasm and short processes; these cell aggregates extend along the lateral sides of the aorta in the anterior part of the thoracic region (Koelliker, 1879; Kohn, 1903; Keene & Hewer, 1927; Kuntz, 1953; Coupland, 1965;

Pick, 1970). Detailed accounts of the cytological differentiation of the primitive sympathetic cells at the ultrastructural level are available (Pick, Gerdin & Delemos, 1964; Coupland & Weackley, 1968; Hervonen, 1971; Eränkö, 1972b; Hervonen & Kanerva, 1972a; Papka, 1972).

The first appearance of CA in primitive mammalian sympathetic cells has been determined by the fluorescence histochemical method. Since it is known that embryological tissue is difficult to process for this method, extra care has to be taken in interpreting the results (de Champlain, Malmfors, Olson & Sachs, 1970). Specific fluorescence in primitive sympathetic cells, which indicates that they are capable of synthesizing and storing CA, appears at different times in different species, but it must be remembered that the rate and degree of maturation during foetal life varies with the species (Winckler, 1969; Schümann, 1969; de Champlain, Malmfors, Olson & Sachs, 1970). The first fluorescent cells appear in the primitive para-aortic sympathetic primordia at 11 days of gestation in the mouse, at 13 days in the rat, at 14 days in the rabbit and at 7 weeks in humans (de Champlain, Malmfors, Olson & Sachs, 1970; Read & Burnstock, 1970; Owman, Sjöberg & Swedin, 1971; Hervonen, 1971; Fernholm, 1972; Papka, 1972). At this stage, the cells are loosely packed with no desmosomal attachments. Organelles are weakly represented in the cytoplasm, although the number of free or membrane-bound ribosomes are increasing. No granular vesicles are present in the cytoplasm.

Shortly after this stage, large cells with long smooth axons show specific green fluorescence for CA, which can be clearly distinguished from smaller cells with a brighter yellowish fluorescence and a few short processes. These cells represent the immature adrenergic neuron (sympathetic neuroblast) and the immature chromaffin cell (pheochromoblast) respectively (Coupland, 1965; de Champlain, Malmfors, Olson & Sachs, 1970; Hervonen, Owman, Sjöberg & Swedin, 1971; Fernholm, 1972; Papka, 1972).

6.2.1 *Development of adrenergic neurons*

From the primordial cluster of primitive sympathetic cells or neuroblasts, immature adrenergic neurons migrate caudally and cranially forming a segmented column which represents the future paravertebral sympathetic chain. Immature adrenergic neurons also migrate ventrally in the preaortal region to form the prevertebral sympathetic ganglia (Paterson, 1890; Lutz, 1968; Fermholm, 1971).

The immature adrenergic neuron is significantly different from the mature neuron. The intensity of fluorescence in the cell bodies is low in early stages, increases rapidly during foetal life and decreases again towards the end of foetal life. In newborn animals, adrenergic neurons are still immature, showing different stages of development in the same ganglion. Most cell bodies show a low fluorescent intensity, but there are some with intense fluorescence which have processes (Eränkö, 1972b). Probably the latter correspond to the intensely fluorescent multipolar neurons described in the adult (Jacobowitz & Woodward, 1968; Jacobowitz, 1970; see also Section 3.1).

Shortly after the first appearance of fluorescence in the cell bodies of the immature adrenergic neurons, fluorescent axons start to grow towards the effector organs. The growing axons are smooth and show medium intensity fluorescence. When they first reach the effector organs, the terminal axon is still devoid of varicosities, but these develop rapidly in early postnatal life. The ductus arteriosus is one of the first effector organs to receive adrenergic fibres. During foetal life the ductus arteriosus becomes fully innervated; it becomes obliterated by constriction just after birth, possibly via an adrenergic mechanism (Boreus, Malmfors, McMurphy & Olson, 1969; Aronson, Gennser, Owman & Sjöberg, 1970; Folkow & Neil, 1971). The intestine receives adrenergic fibres during foetal life; fluorescent nerve trunks have been seen in the mesentery and in the gut wall at 18 to 20 days of gestation in the rat, at 14 days in the mouse, at 28 days in rabbit, at 8 to 10 weeks in humans. At birth, adrenergic innervation of the intestine shows a pattern similar to that seen in the adult (Gabella & Costa, 1969; Read & Burnstock, 1969a; de Champlain, Malmfors, Olson & Sachs, 1970; Fernholm, 1972; Gershon & Thompson, 1973). The heart also receives adrenergic fibres just before birth (Winckler, 1969; Friedman, Pool, Jacobowitz, Seagren & Braunwald, 1968; de Champlain, Malmfors, Olson & Sachs, 1970; Owman, Sjoberg & Swedin, 1971), but the iris, the vas deferens, the brown fat and the pineal gland all receive their first adrenergic nerves after birth (Håkanson, Lombard, des Gouttes & Owman, 1967; Machado, Wragg & Machado, 1968; Yamauchi & Burnstock, 1969; de Champlain, Malmfors, Olson & Sachs, 1970; Derry & Daniel, 1970; Furness, McLean & Burnstock, 1970; Owman, Sjöberg & Swedin, 1971).

The ability to take up and store NA is already present in the developing adrenergic neuron (Goldstein, 1967), and the first appearance of fluorescent fibres in developing organs almost coincides with the appearance of uptake of NA and binding to specific vesicular proteins (Glowinsky, Axelrod, Kopin & Wurtman, 1964; Iversen, de Champlain, Glowinski & Axelrod,

1967; Sachs, de Champlain, Malmfors & Olson, 1970; Mirkin, 1972). However, these properties appear in the developing adrenergic neuron before NA reaches a level detectable by fluorescence histochemistry (Read & Burnstock, 1969a; Machado, 1971; Gershon & Thompson, 1973). NA in the developing adrenergic neuron is stored, at least partially, in granular vesicles (Yamauchi & Burnstock, 1969; Hervonen, 1971; Machado, 1971; Eränkö, 1972b). No characteristic granular vesicles have been observed by electronmicroscopy in the primitive sympathetic cell, but they have been seen occasionally in the cell body and in the processes of early immature adrenergic neurons (Hervonen, 1971; Hervonen & Kanerva, 1972a; Papka, 1972). The number of vesicles in the cell body first increases then decreases, more vesicles being found in the axonal and dendritic processes (Eränkö, 1972b; Papka, 1972). In the developing adrenergic terminals, granular vesicles have been found as soon as the fibres reach the effector organs (Yamauchi & Burnstock, 1969; Machado 1971). The adrenergic granular vesicles in the cell bodies appear to originate from empty vesicles in the Golgi region which are progressively filled by dense material (Hervonen, 1971). In the terminals, the granular vesicles also seem to originate from empty vesicles budding off from the smooth endoplasmic reticulum and are subsequently filled by an increasing amount of dense material (Machado, 1971; see also Section 3.1). The size of the granular vesicles in the cell bodies of the developing adrenergic neuron is larger (60 to 110 nm) than those in the mature neuron (30 to 60 nm) (Hervonen, 1971; Eränkö, 1972b; Hervonen & Kanerva, 1972a; Papka, 1972). In the developing terminals, both small and large vesicles are present but the proportion of small vesicles increases with age (Yamauchi & Burnstock, 1969; Machado, 1971).

6.2.2 *Development of adrenergic transmission*

The functional state of the sympathetic nervous system during foetal life is little known. When the first adrenergic fibres reach the effector organs they are not yet capable of transmitting to the effector tissues (Furness, McLean & Burnstock, 1970). A period of maturation is required for both adrenergic terminals and effector tissues; the appearance of fibres containing NA in developing organs does not necessarily indicate the presence of functional transmission (Furness, McLean & Burnstock, 1970; Machado, 1971). Little is known about the onset of functional ganglionic transmission, i.e., between preganglionic cholinergic fibres and postganglionic adrenergic neurons. These junctions are present in newborn animals

(Halstead & Larrabee, 1972), and a 500-fold increase in the number of ganglionic synaptic junctions occurs during the first 8 days after birth, accompanied by an increase in the level of the presynaptic enzyme choline acetyltransferase (Black et al. 1971; Black et al. 1972a).

Sympathetic control of different effector organs appears at various times, probably related to their physiological roles (de Champlain, Malmfors, Olson & Sachs, 1970; Furness, McLean & Burnstock, 1970; Machado, 1971). Sympathetic control of the heart appears immediately before birth. A tonic pacemaker, probably located in the medulla oblongata feeds rhythmic impulses to the cardiac adrenergic nerves which provide excitatory control of heart beat (Corey, 1935; Bloor, 1964; Wekstein, 1965; Adolph, 1967; Schwieler, Douglas & Bouhuys, 1970). A further acceleration of heart beat of adrenergic origin appears two or three weeks after birth (Wekstein, 1965; Adolph, 1967). Less is known about the time of onset of adrenergic control of the vasculature. The baroreceptor reflexes are functional very soon after birth (Downing, 1960; Bloor, 1964). Adrenergic vasoconstrictor fibres to the vasculature of the hind limb of the dog become effective only after two weeks following birth (Boatman, Schaffer, Dixon & Brody, 1965). It has been suggested that the increase of blood pressure during postnatal development is due to increasing activity in the autonomic nervous system with the establishment of a sympathetic vasoconstrictor tone in the systemic circulation (Mott, 1961). The rabbit trachea relaxes in response to sympathetic stimulation one week after birth (Schwieler, Douglas & Bouhuys, 1970). The development of thermoregulation, which involves peripheral adrenergic nerves, i.e. brown fat thermogenesis (Himms-Hagen, 1972) and peripheral vasoconstriction (Folkow & Neil, 1971), is not well understood. Newborn animals placed in a cool environment are unable to maintain their body temperature (Adolph, 1957). The role of the sympathetic system is small during the first days after birth but as the animal approaches weaning age, the influence of the sympathetic system increases and remains significant in adult life (Antoshkina, 1939; Wekstein, 1965).

Adrenergic control of the gut appears at the end of foetal life (Gershon & Thompson, 1973). Burn (1968a) reported that stimulation of mesenteric nerves elicited relaxation of the gut in newborn kittens, but a cholinergic contraction in newborn rabbits. Relaxation of the gut in the rabbit was only observed after several days postnatal by Day and Rand (1961). However Gershon & Thompson (1973) recorded relaxations from the 30th day of gestation. Adrenergic nerves to the pineal gland seem to become functional in 8-day old rats (Machado, Wragg & Machado, 1968; Machado, 1971).

A significant delay between the arrival of adrenergic fibres and the onset of adrenergic neurotransmission occurs in the mouse vas deferens (Yamauchi & Burnstock, 1969; Furness, McLean & Burnstock, 1970). Excitatory junction potentials in response to nerve stimulation were not observed earlier than 18 days, although adrenergic varicose fibres were established in the muscle at twelve days. Junction potentials did not have the same time course as those recorded in adult tissue until thirty days, which corresponds with the onset of puberty in mice. Transmural stimulation of vas deferens from newborn rats produces a nerve-mediated contraction; the effect is readily blocked by α-adrenoceptor blocking drugs, since the neuromuscular junctions are less intimate and the characteristic resistance to blockade seen in the adult only develops after eleven days (Swedin, 1972; see Section 5.6).

6.2.3 *Factors which control the maturation of adrenergic neurons*

Little is known about the genetic basis for the determination of different types of autonomic neurons. It has been suggested that there is a primitive stem cell which gives rise to cholinergic, adrenergic or other types of cells; this stem cell has therefore a complete set of genes for enzymes of transmitters (Filogamo & Marchisio, 1971). However, the expression of a gene required for the synthesis of one neurotransmitter may restrict the expression of genes for other neurotransmitters (Amano, Richelson & Nirenberg, 1972). In fact, clones of cells capable of synthesizing either ACh or NA have been detected in mouse neuroblastoma, but no cell capable of synthesizing both neurotransmitters has been found. Gene expression may be reversible and different types of cells may differentiate along alternate pathways. Experiments on reinnervation of the nictitating membrane by preganglionic fibres have led to the suggestion that cholinergic neurons may become adrenergic (Ceccarelli *et al.* 1972), but this hypothesis needs further experimental confirmation.

Development and growth of the sympathetic nervous system is under the influence of a NGF which specifically stimulates the sympathetic neurons (Levi-Montalcini & Angeletti, 1961). NGF has been localized in serum, in sympathetic ganglia, in several organs and is present in sympathetic tissue during early embryogenesis (Winick & Greenberg, 1965; Levi-Montalcini & Angeletti, 1968). NGF is taken up by adrenergic terminals and transported by retrograde axonal flow to the cell body, where it exerts its major actions (Hendry *et al.*, 1974; Stöckel *et al.*, 1974). NGF induces an increase in size and rate of cell division of immature neurons, an increase in the rate of adrenergic axons, an increase in the content of NA in them by inducing the biosynthetic enzymes T-OH and DβH, and induces an early maturation of ganglionic transmission (Crain & Wiegand, 1961; Crain, Benitez & Vatter, 1964; Edwards, Fenton, Kakari, Large, Papadaki & Zaimis, 1966; Levi-Montalcini & Angeletti, 1968; Olson, 1967; Hendry & Iversen, 1971; Angeletti, Levi-Montalcini & Caramia, 1971; Thoenen, Angeletti, Levi-Montalcini & Kettler, 1971). Not all adrenergic neurons are affected equally by NGF injected into the newborn animal; for example, those supplying the vas deferens and the uterus are little affected (Edwards, Fenton, Kakari, Large, Papadaki & Zaimis, 1966; Olson, 1967; Levi-Montalcini, 1971; Hughes, Kirk, Kneen & Large, 1973). It has been suggested (Levi-Montalcini & Angeletti, 1968) that NGF could be placed in a class of biological substances between inducers and hormones, since it shares characteristics in common with both.

Hormones have been shown to influence sympathetic neurons, although it is not known whether they affect the neuron directly or indirectly. They may play an important role in the development and maturation of the adrenergic neuron as they do in other parts of the nervous system (Reiss, 1955). The adrenergic neuron during development, but probably not in mature stages, is sensitive to corticotrophic hormones (ACTH) and cortical hormones (Resti, 1962; Korochkin & Korochkina, 1970; Olson & Malmfors, 1970; Eränkö & Eränkö, 1972a; Costa, Eränkö & Eränkö, 1974b). In the newborn rat, treatment for 8 days with ACTH or glucocorticoids produced a significant increase in RNA synthesis and in the cellular volume and activity of neurons in sympathetic ganglia but not in enteric ganglia (Korochkin & Korochkina, 1970) and a significant increase in CA fluorescence of both cell bodies and terminals (Costa, Eränkö & Eränkö, 1974b). Insulin was reported to have a similar effect on sympathetic ganglia (Angeletti, Luzzi & Levi-Montalcini, 1966). Thyrotrophic hormone appears to produce an increase in cell volume and of Nissl substance (Atech, 1962). Angiotensin induces protein and CA synthesis in adrenergic terminals in the vas deferens (Roth & Hughes, 1972). Adrenergic neurons supplying the internal reproductive organs in females, but not in males, are specifically sensitive to the oestrogenic hormones (Sjöberg, 1967; Owman, Sjöberg, Sjöstrand & Swedin, 1970). Treatment of rats with 17-β-oestradiol produces a marked increase in NA content in adrenergic terminals in the uterus and vagina. A similar increase occurs during the first period of pregnancy, while towards the end of pregnancy there is a decrease in NA (Sjöberg, 1967; Rosengren & Sjöberg, 1968; Spratto & Miller, 1968a, b). Significant structural changes are produced in adrenergic terminals in the uterus and vagina by this hormone (Hervonen, Kanerva, Lietzen & Partanen, 1972). The diameter of the axons mainly in the intervaricose regions, is increased and a large number of tubular structures appear often with a dense material content. Adrenergic neurons supplying other organs are not affected by oestrogens (Brundin, 1965; Sjöberg, 1968).

The level of NA and the volume of smooth musculature of male internal reproductive organs is hormonal dependent. Castration leads to a reduction of both NA content and musculature volume while testosterone restores and increases both up to two to three times that of control tissue (Wakade & Kirpekar, 1973). Chronic treatment with testosterone produces a decrease of NA content, of fluorescence in nerve fibres and of adrenergic vesicles in guinea-pig seminal vesicles, which parallels the decrease in volume of smooth musculature (Greenberg, Long, Burke, Chapnik & Van Orden, 1973).

In the early stages of postnatal life, most adrenergic neurons receive functional preganglionic innervation (Crain, Benitez & Vatter, 1964). The more immature adrenergic neurons undergo intense mitotic activity (Levi-Montalcini & Booker, 1960) and there is an associated increase in ganglionic synapses (Black, Hendry & Iversen, 1971b; Black, Bloom, Hendry & Iversen, 1971). At the same time there is a fast rise of T-OH and DβH levels in adrenergic neurons which is prevented by preganglionic denervation or by ganglionic blockade (Black, Hendry & Iversen, 1971a, 1972b; Thoenen, Saner, Angeletti & Levi-Montalcini, 1972; Thoenen, Saner, Kettler & Angeletti, 1972; Black, 1973). This indicates that maturation of the adrenergic neuron depends, in part, on the preganglionic nerves and suggests that the mechanism may be similar to that involved in the slow regulation of NA synthesis by enzyme induction in the adult (Black, Hendry & Iversen, 1972b; Thoenen, Saner, Angeletti & Levi-Montalcini, 1972; Thoenen, Saner, Kettler & Angeletti, 1972) (see Section 4.1). The presence of presynaptic fibres is also required for normal increase in MAO in immature sympathetic neurons, while preganglionic activity is without effect on this enzyme in the adult; therefore, the regulation of these enzymes in the immature neurons may be different from that in the adult neuron (Black, Hendry & Iversen, 1972b). Furthermore, the maturation of preganglionic fibres appears to be dependent on the maturation of postganglionic adrenergic neurons. For example, increase of the preganglionic enzyme choline acetyltransferase, which normally occurs in coincidence with the phase of fast development of the ganglion in the newborn, is prevented by immunological or chemical adrenergic sympathectomy with consequent delay in the maturation (Thoenen, Saner, Kettler & Angeletti, 1972; Black, Hendry & Iversen, 1972a, b), and NGF treatment, which acts exclusively on the postganglionic neuron, produces a parallel increase in activity of choline acetyltransferase (Thoenen, Saner, Angeletti & Levi-Montalcini, 1972; Thoenen, Saner, Kettler & Angeletti, 1972).

6.2.4 *Development of chromaffin cells*

From the early stages that NA appears in the sympathetic primordia, the ancestors of both adrenergic neurons and of chromaffin cells are intermingled (de Champlain, Malmfors, Olson & Sachs, 1970; Fernholm, 1971, 1972; Hervonen, 1971; Owman, Sjöberg & Swedin, 1971; Papka, 1972). The primitive chromaffin cells (pheochromoblasts) show a steady increase of their fluorescence and migrate, following the adrenergic neurons, and reach definitive locations by the end of foetal life.

The main bulk of immature chromaffin tissue is located in the preaortic region; the more anterior part of it penetrates the cortical tissue and becomes surrounded by it, forming the *anlage* of the adrenal medulla (see Coupland, 1965; de Champlain, Malmfors, Olson & Sachs, 1970; Fernholm, 1971, 1972; Hervonen, 1971; Owman, Sjöberg & Swedin, 1971). The distal part of the preaortic immature chromaffin tissue represents the *anlage* of the organ of Zuckerkandl (Zuckerkandl, 1912) and together with all the other immature chromaffin cells scattered mainly in the abdominal and thoracic cavities will develop into the extra-adrenal chromaffin tissue (Coupland, 1965) (see Section 3.2). The developing chromaffin cells always appear around capillaries or small vessels and it has been suggested that this may be the reason for the faster maturation of the chromaffin cell in comparison to that of the adrenergic neuron (see Hervonen, 1971).

The CA contained in the early chromaffin tissue, both adrenal and extra-adrenal, is almost exclusively NA (see Brundin, 1966). In extra-adrenal chromaffin cells, NA remains the main CA during foetal and postnatal life, although small amounts of A and of PNMT have also been found (Brundin, 1966; Coupland, 1965; Gennser & Studnitz, 1969; Hervonen, 1971). In contrast, in the adrenal medulla, A appears early and increases rapidly becoming predominant over NA (Brundin, 1966; Comline & Silver, 1966; Lagerspetz & Hissa, 1968; Roffi, 1968a; von Studnitz, 1968). It is widely accepted that the transformation of NA-storing cells into A-storing cells in the adrenal medulla is induced by cortical glucocorticoid hormones by induction of PNMT (Kuntz, 1912; Coupland, 1953; Brundin, 1966; Eränkö, Lempinen & Räisänen, 1966; Wurtman & Axelrod, 1966a; Roffi, 1968a, b; Pohorecky & Wurtman, 1971). These hormones also have marked effects on extra-adrenal chromaffin cells. They prevent degeneration of the paraganglia which normally occurs in early postnatal life; they increase the size and content of CA of these paraganglia and they increase the number of granular vesicles in sympathetic ganglia of newborn rats (Lempinen, 1964; Eränkö, Heath & Eränkö, 1973; Costa, Eränkö & Eränkö, 1974a). The maturation and development of both medullary and extra-adrenal chromaffin tissue appears to be under significant control by cortical hormones (see Hervonen, 1971), while these tissues are barely sensitive to NGF (Crain & Wiegand, 1961; Olson, 1967). However, adrenal CA content is increased by NGF in kittens and rats (Edwards, Fenton, Large, Papadaki & Zaimis, 1966).

The origin of CA-storage vesicles in immature chromaffin tissue is comparable to that described for adrenergic neurons (Elfvin, 1967; Coupland & Weackley, 1968; Hervonen, 1971; and see Section 6.2).

The content of CA in the adrenal medulla increases faster than the content of ATP so that the CA/ATP ratio is significantly less than one in early stages and about one at birth (O'Brien, Da Prada; Pletscher, 1972) (see Section 3.1). This led the authors to postulate that the early vesicle contains mainly ATP and takes up CA later. There is a parallel increase in size of the CA-containtaining vesicles (Hervonen, 1971).

Few studies have been made on the onset of adrenal medullary functions and little information is available on the function of the extra-adrenal chromaffin tissue. A-storing cells of the adrenal medulla become functional, at different stages, early in lamb and late in calf, in conjunction with the development of preganglionic nerves, and they respond exclusively to nerve stimulation throughout their development (Comline & Silver, 1966). On the other hand, NA-storing cells are sensitive to asphyxia during foetal and early postnatal life and their innervation only becomes apparent after several weeks (Comline & Silver, 1966). Extra-adrenal chromaffin cells receive little innervation (see Section 3.2) and were shown to be sensitive to asphyxia in newborn rabbits (Brundin, 1966) and in foetal humans (see Hervonen, 1971). This suggests that the vascular tone of the foetus is under the control of circulating A and NA released from chromaffin tissue, and that only during postnatal life is this replaced by nervous control (Coupland, 1953; Comline & Silver, 1966; Hervonen, 1971).

6.3   **Growth and regeneration of adrenergic neurons in the adult**

Sympathetic postganglionic neurons continue to grow throughout life (Terni, 1922; Levi, 1925, 1946; de Castro, 1932; Weiss, 1961; Botar, 1966). The high regenerative ability of sympathetic neurons has been widely recognized (Langley, 1897-98; Langley & Anderson, 1904; Cajal, 1928; de Castro, 1932; Butson, 1950-51; Olson & Malmfors, 1970). Shortly after surgical or chemical damage to adrenergic axons, budding and sprouting occur. The axons show characteristic growth cones and thick beaded enlargements (Lawrentjew, 1925; de Castro, 1932; Olson, 1969) with a high content of NA (Blumcke & Niedorf, 1965; Dahlström; 1965; Olson, 1969). The appearance of growing non-terminal axons resembles that of developing axons (see Section 6.1) or axons growing in culture (see Section 6.1). After a variable time, depending on the type and location of the damage to adrenergic axons, the regenerating fibres reinnervate the organs to restore the original pattern of innervation and functional neuro-effector transmission is resumed (Tuckett, 1896; Langley, 1897-98; Machida, 1929; Brücke, 1931; Simeone, 1937; Butson, 1950-51; Guth, 1956).

Most current knowledge about the growth properties of adrenergic neurons in the adult come from studies on transplanted tissues and sympathetic ganglia (Cajal, 1928; de Castro, 1932; Ward, 1936; Malmfors & Olson, 1967; Olson, 1969; Olson & Malmfors, 1970; Björklund, Katzman, Stenevi & West, 1971; Björklund & Stenevi, 1971; Burnstock, Gannon, Malmfors & Rogers, 1971; Malmfors, Furness, Campbell & Burnstock, 1971). Each adrenergic neuron is capable of extending the area of effector tissue that is innervated (field of innervation) since a denervated tissue transplanted in the anterior eye chamber is reinnervated by adrenergic axons which already innervate the host iris. However, the capacity of the adrenergic neuron to grow appears to be limited because, although a sympathetic ganglion transplanted in the anterior eye chamber is capable of reinnervating the host iris, small pieces of ganglion only partially reinnervate it (Olson & Malmfors, 1970).

Adrenergic neurons can innervate tissues of different origin that are normally innervated by neurons from other ganglia, and the original pattern of innervation of the effector tissue is maintained suggesting that the pattern of innervation is specifically determined by the effector tissue Burnstock, Gannon, Malmfors & Rogers, 1971; Malmfors, Furness,

Campbell & Burnstock, 1971; Rogers, 1972). Conversely the same tissue can be reinnervated by adrenergic neurons from various ganglia (Olson & Malmfors, 1970) and even by central adrenergic neurons (Björklund & Steveni, 1971). Areas in which adrenergic nerves are not normally present are not reinnervated, but those tissues which are normally supplied by adrenergic nerves are reinnervated and nerve transmission restored. For example, the submandibular gland (but not the sublingual gland) and enteric cholinergic neurons (but not parasympathetic cholinergic neurons of other ganglia) are normally richly innervated by adrenergic nerves and are reinnervated following transplantation in the anterior eye chamber (Olson & Malmfors, 1970; Burnstock, Gannon, Malmfors & Rogers, 1971; Malmfors, Furness, Campbell & Burnstock, 1971).

The work of Olson & Malmfors (1970) has shown that adrenergic fibres from a transplanted ganglion do not grow into areas which are already innervated, suggesting that the presence of adrenergic fibres in the effector tissue may inhibit the growth of other fibres in the same area and may inhibit the formation of longlasting relationships with effector cells, a situation already described for nerve fibres innervating smooth muscle in culture (see Section 6.1). Preganglionic denervation delays the process of regeneration indicating that a normal preganglionic input is one of the requirements for the regeneration.

The growth of adrenergic neurons in the adult appears to be a process balanced between opposing factors: stimulation by preganglionic nerves and effector tissues; inhibition by adrenergic fibres innervating the same effector. Little is known about the mechanism of inhibition. The stimulating factor may be NGF, particularly since NGF has been shown to control the growth of adrenergic neurons during normal development and in culture (see Sections 6.1 and 6.2); it is neither cortisone nor testosterone (Olson & Malmfors, 1970). The finding that the adrenergic neurons which supply the vas deferens are less capable of reinnervating the iris than adrenergic neurons from paravertebral ganglia when transplanted in the anterior eye chamber could be explained by the low sensitivity of these neurons to NGF (see Section 6.1). It is interesting that adrenal chromaffin cells transplanted in the anterior eye chamber are capable of reinnervating the iris by giving rise to varicose processes which build up a terminal network of fibres indistinguishable from the normal adrenergic nerve ground plexus (Olson, 1970) and have small granular vesicles (Hökfelt, 1973), showing that under some conditions the chromaffin cells can develop neuronal properties (see Section 3.2).

6.4 Degeneration of adrenergic neurons

Little is known about degeneration of the sympathetic system throughout life or about the corresponding effects on the whole animal. An increasing number of neurons showing senescence alterations have been observed in aging sympathetic ganglia (de Castro, 1932; Amprino, 1938; Ehlers, 1951; Unger, 1951; Herman, 1952; Stohr, 1957; Botar, 1966). The number of terminals in effector tissues of old animals decreases in parallel with decrease of CA, and some terminals show signs of degeneration such as large swollen elongations (Frolkis, Bezrukov, Bogatskaya, Verhkratsky, Zamostian, Sheutchuk & Shtchegoleva, 1970). The various pathological alterations which have been described in sympathetic neurons may represent either a functional adaptation in old animals (Levi, 1946) or a toxic effect of diseases which affect the neurons (de Castro, 1932; Botar, 1966). More detailed experimental work is needed to correlate degenerative changes of adrenergic neurons to aging and to pathological situations. The findings that chemical and immunological agents are capable of inducing degenerative changes in adrenergic neurons supply invaluable tools for the study of the physiology of the adrenergic system (see Thoenen, 1972), and also opens up a promising field for the study of senescence and pathology in the autonomic nervous system.

6.4.1 *Surgical sympathectomy*

Both transection of postganglionic sympathetic nerves and the removal of the corresponding ganglia have been termed 'surgical sympathectomy' (Thoenen, 1972). Transection results in permanent degeneration of the adrenergic fibres peripheral to the section and in temporary degeneration of the cell body termed retrograde degeneration or 'chromatolysis' (see Botar, 1966; Dixon, 1970; Matthews & Raisman, 1972; Thoenen, 1972). Degeneration of peripheral adrenergic fibres with parallel disappearance of NA has been demonstrated by a large number of workers (see for references Cooper, 1966; Furness & Costa, 1971a; Garrett, 1971; Edvinsson, Owman, Rosengren & West, 1972; Thoenen, 1972). The time course of degeneration is variable probably depending on the axon length (Sachs & Jonsson, 1973). Taking the content of NA as a measure, the time for degeneration varies from 18 to 48 hours (Malmfors & Sachs, 1965b; Thoenen, 1972), while the general structure of sympathetic axons appears normal for several days (see de Castro, 1932 and Blumcke & Dengler, 1970).

The disappearance of transmitter, due to irreversible damage of the membrane amine-uptake process, occurs for every fibre at different times, although at a fast rate (Malmfors & Sachs, 1965; Sears & Gillis, 1967). The ultrastructure of degenerating adrenergic axons (Smith, Trendelenburg, Lanser & Tsai, 1966; Van Orden, Bensch, Langer & Trendelenburg, 1967; Kapeller & Mayor, 1969; Iwayama, 1970; Blumcke & Dengler, 1970; Garrett, 1971; Bray, Peyronnard & Aquayo, 1972; Knocke & Terworth, 1973) correlates well with the disappearance of NA and axon membrane properties.

Colchicine is a drug which disrupts microtubules and interferes with fast axonal flow (see Section 3.1). Applied locally to a nerve, colchicine produces a degeneration distal to the application and a retrograde degeneration in the cell body (Costa & Filogamo, 1970; Dahlström, 1970) without altering nerve conduction in the early stages (Keen & Livingston, 1970; Pilar & Landmesser, 1972). This suggests that the degeneration of the distal fibre is due to the interruption of the transport of substances necessary to the survival of the terminals.

6.4.2 *Drug-induced long-term structural changes in adrenergic neurons and chemical sympathectomy*

A number of drugs specifically affect some function of adrenergic neurons without long lasting alteration of the neurons (see Section 5.6), but a few substances produce specific degenerative changes in adrenergic neurons.

Guanacline. Chronic treatment for 12 weeks or more with the antihypertensive drug guanacline causes a massive deposition of lipoprotein granules resembling 'ageing pigment' in sympathetic neurons (Burnstock, Doyle, Gannon, Gerkins, Iwayama & Mashford, 1971); this appears to be associated with hypotension prolonged up to at least 6 months following cessation of treatment (Dawborn, Doyle, Ebringer, Howqua, Jerums, Johnson, Mashford & Parkin, 1969). Guanacline produces a slow dose-dependent sympathectomy, by a mechanism which is unknown.

6-Hydroxydopamine (6-OHDA). 6-OHDA causes a selective, reversible degeneration of axon terminals (but not cell bodies) of peripheral adrenergic neurons *in vitro* and *in vivo* (Tranzer & Thoenen, 1967b, 1968; Malmfors & Sachs, 1968; Knyihór, Ristovsky, Kálmán & Csillik, 1969; Furness, Campbell, Gillard, Malmfors, Cobb & Burnstock, 1970; Johnsson

& Sachs, 1970; Ausprunk, Berman & McNary, 1971; Cheah, Geffen, Jarrott & Ostberg, 1971; Goldman & Jacobowitz, 1971; Malmfors & Thoenen, 1971; Sachs, 1971, Hökfelt, Jonsson & Sachs, 1972; Thoenen & Tranzer, 1973).

The sensitivity to 6-OHDA degeneration varies from fibre to fibre and from ganglion to ganglion. The pelvic ganglia, for instance, are more resistant to 6-OHDA than the pre and paravertebral ganglia. The time course of degeneration is also variable and is probably related to the amount of 6-OHDA which is taken into the neuron to produce a degenerative effect. As a direct consequence of the degeneration of adrenergic terminals, neurotransmission is abolished (Furness, Campbell, Gillard, Malmfors, Cobb & Burnstock, 1970; Furness, 1971; Thoenen, 1972). The cell body shows the characteristic chromatolytic changes observed following surgical postganglionic denervation with decrease of the enzymes T-OH and MAO (Cheah, Geffen, Jarrott & Ostberg, 1971; Brimijoin & Molinoff, 1971). Recovery is from three weeks for the vas deferens to three months for the heart and depends probably on the length of the degenerated terminal fibre (see Malmfors & Thoenen, 1971; de Champlain, 1971). A faster recovery (less than a week) of adrenergic fibres to blood vessels after 6-OHDA degeneration has been reported (Finch, Haeusler, Kuhn & Thoenen, 1973).

6-OHDA injected into newborn animals results in a permanent massive destruction of pre and paravertebral sympathetic ganglia, indicating that immature adrenergic neurons are more sensitive to this drug (Sachs, de Champlain, Malmfors & Olson, 1970; Angeletti & Levi-Montalcini, 1970; Angeletti, 1971; Jaim-Etcheverry & Zieher, 1971b; Malmfors, 1971; Clark, Laverty & Phelan, 1972; Eränkö & Eränkö, 1972b; Finch, Haeusler & Thoenen, 1973; Papka, 1973). Some neurons survive after 6-OHDA. The adrenal chromaffin tissue shows a slight hyperplasia while the extra-adrenal chromaffin tissue is not affected (Chear, Geffen, Jarrott & Ostberg, 1971; Sachs, 1971; Angeletti, 1971; Tranzer & Richards, 1971; Eränkö & Eränkö, 1972b). Adrenergic neurons exposed to 6-OHDA in tissue culture also show degenerative changes (Hill, Mark, Eränkö, Eränkö & Burnstock, 1973).

The serotonin analogue, 5, 6-dihydroxytryptamine, which has been shown to induce chemical degeneration of serotonin terminals in the central nervous system (Baumgarten, Björklund, Lachenmayer, Nobin & Stenevi, 1971) is also capable of producing degeneration of the terminals of peripheral adrenergic neurons similar to that caused by 6-OHDA (Baumgarten, Göthert, Holstein & Schlossberge, 1972).

Guanethidine and Bretylium. Guanathidine, in addition to its adrenergic blocking action and NA-storage depleting properties, has been shown to cause severe specific damage to adrenergic neurons following chronic treatment (see Burnstock, Evans, Gannon, Heath & James, 1971; Eränkö & Eränkö, 1971c; Jensen-Holm & Juul, 1971; Angeletti & Levi-Montalcini, 1972; Evans, Gannon, Heath & Burnstock, 1972; Heath, Evans, Gannon, Burnstock & James, 1972; Juul & McIsaac, 1973). Low doses lead to long-term damage of the nerves supplying muscle of reproductive organs, while in high concentrations nearly total sympathectomy results (Burnstock, Evans, Gannon, Heath & James, 1971). Adrenergic neurons which supply the vas deferens and the sphincter urethrae are particularly sensitive to degeneration by guanethidine (Evans, Gannon, Heath & Burnstock, 1972). Since the toxic effect of guanethidine depends on its intraneuronal concentration and since guanethidine acts as a 'false transmitter', the damaging effect will be more marked in those neurons with lower transmitter turnover, which is the case for the neurons referred to above (see Heath, Evans, Gannon, Burnstock & James, 1972).

Guanethidine does not affect chromaffin cells in the adult (Burnstock, Evans, Gannon, Heath & James, 1971; Evans, Gannon, Heath & Burnstock, 1972), but in newborn animals, a large increase of chromaffin cells has been reported in sympathetic ganglia (Eränkö & Eränkö, 1971b). An irreversible sympathectomy is caused in the newborn by guanethidine (Eränkö & Eränkö, 1971c, d; Angeletti, Levi-Montalcini & Caramia, 1972).

Analysis of the mechanism of guanethidine action both in tissue culture (Hill, Mark, Eränkö, Eränkö & Burnstock, 1973) and *in vivo* (Heath, Evans & Burnstock, 1973; Heath, Hill & Burnstock, 1974) has shown that guanethidine causes 'retraction' rather than degeneration of terminal axons, and initial alteration of mitochondrial structure in the cell bodies. Bretylium, another adrenergic blocking agent has been reported to cause selective degenerative changes of mitochondria of the immature sympathetic neurons (Caramia, Angeletti, Levi-Montalcini & Caratelli, 1972).

6.4.3 *Immunosympathectomy*

The term immunosympathectomy has been taken to indicate destruction of the sympathetic cell population following treatment by a specific antiserum to NGF, although this destruction is not complete (Cohen, 1960; see Levi-Montalcini & Angeletti, 1966; Steiner & Schönbaum, 1972; Thoenen, 1972). The sensitivity of sympathetic neurons to the antiserum varies with the age of the animal (Levi-Montalcini & Angeletti, 1966; Klingman & Klingman, 1972a; Schucker, 1972). Before birth, the antiserum is effective but less than in the newborn in which it produces the maximum degenerative effects. The antiserum is less effective on animals several weeks old and is almost without effect in the adult. The 'sympathectomy' produced is not complete. The most sensitive ganglia are those of the paravertebral sympathetic chain followed next by the prevertebral ganglia. The adrenergic neurons supplying the internal male genital organs, the brown fat and the anococcygeus muscle are little affected by anti-NGF (Vogt, 1964; Zaimis & Berk, 1965; Levi-Montalcini & Angeletti, 1966; Derry & Daniel, 1970; Klingman & Klingman, 1972b; Levi-Montalcini, 1971; Gibson & Gillespie, 1973). The adrenal chromaffin cells are not affected by the antiserum at any stage of development, while its effect on extra-adrenal chromaffin cells is not known (Levi-Montalcini & Angeletti, 1966; Klingman & Klingman, 1972b).

6.4.4 *Secondary effects accompanying sympathectomy*

Degeneration contraction. Following sympathectomy, produced surgically or by 6-OHDA, degenerating adrenergic fibres lose their ability to take up and store NA. The NA which leaks from the damaged nerves acts on the effector tissues to produce a transient sympathomimetic effect or 'degeneration contraction', which lasts until the disappearance of the transmitter (Sears & Bárány, 1960; Coats & Emmelin, 1962; Malmfors & Sachs, 1965b; Langer, 1966a; Sears & Gillis, 1967; Ellemin, 1968; Lundberg, 1969; see Thoenen, 1972; and Trendelenburg & Emmelin, 1972). Degeneration contraction also occurs *in vitro* (Furness, 1971b; Geffen, 1972). Degeneration contraction caused by sympathectomy with 6-OHDA is preceded by the direct release of NA (Furness, 1971a; Lundberg, 1971; Trendelenburg & Wagner, 1971; Wagner & Trendelenburg, 1971).

Denervation supersensitivity. Following degeneration of adrenergic terminals, a rapid increase in the sensitivity of the denervated tissues to A and NA was observed, first by Budge (1855) and later by Langendorff (1900) in the denervated eye as 'paradoxical pupil dilatation'. This 'denervation supersensitivity' has been described and reviewed by several authors (see Trendelenburg, 1965; Thoenen, 1972), and is due mainly to the absence of NA uptake into neurons in the denervated tissue (Trendelenburg, 1963, 1972). Since this supersensitivity is mimicked by drugs which block the specific uptake mechanism of NA and related phenylethylamines (see Section 5.4), it has consequently also been termed 'cocaine-like'. A specific cocaine-like supersensitivity was demonstrated following degeneration *in vitro* of the adrenergic nerves to the vas deferens (Furness, 1971b). Following degeneration of the adrenergic nerves, a slow supersensitivity of the denervated tissues develops which is superimposed on the cocaine-like supersensitivity; this is non-specific and is mimicked by decentralization of the adrenergic neurons (Trendelenburg, 1963, 1972; Fleming, 1963; Thoenen, 1972; Fleming, McPhillis & Westfall, 1973). Since the first component of the denervation supersensitivity is related to a neuronal mechanism, it is also named 'prejunctional supersensitivity', while the second component which is related to a non-specific increase in sensitivity of the effector is termed 'postjunctional supersensitivity' (see Thoenen, 1972). Degeneration of the adrenergic terminals therefore produces both prejunctional and postjunctional supersensitivity and is mimicked by combined decentralization and treatment with cocaine (Thoenen, 1972). Denervation supersensitivity has also been described following chemical sympathectomy and immunosympathectomy (Brody, 1964, 1972; Zaimis, 1965; Lundberg, 1971; Trendelenburg & Wagner, 1971; see Steiner & Schönbaum, 1972; Trendelenburg, 1972). The supersensitivity following damage, but not degeneration, of nerves by low doses of guanethidine, leads to 'prejunctional supersensitivity' of the vas deferens which is maintained for at least 6 months following cessation of drug treatment (Evans, Iwayama & Burnstock, 1973).

Effects on whole animal. The dispensability of the sympathetic system in normal conditions was established by Cannon (Cannon, Newton, Bright, Menkin & Moore, 1929). It becomes indispensable in situations in which high adaptability is required, for example, under situations of stress (see Brody, 1972).

The effects produced by degeneration of adrenergic nerves are the product of the interactions of several factors which, on a theoretical basis, are likely to include: (a) the number, location and function of the specific adrenergic neurons that are damaged; (b) the physiological state of activity of the neurons and the pathological states which activate the neurons; (c) the nature of the compensatory reflex mechanisms available, and the side effects of these reflexes. For example, degeneration by 6-OHDA which acts predominantly on paravertebral and prevertebral ganglia, would produce effects totally different from those caused by the degeneration by guanethidine, which acts predominantly on the adrenergic nerves supplying the internal male genital organs or, in high doses, on the entire sympathetic system (see Burnstock, Evans, Gannon, Heath & James, 1971; Evans, Gannon, Heath & Burnstock, 1972). The effect of degeneration of neurons which are not physiologically active, for example the cardio-accelerator nerves (Folkow & Neil, 1971), would not be marked, as shown by the lack of effect of immunosympathectomy on resting heart rate (see Brody, 1964, 1972).

Finally, an example of compensatory effects masking effects due to sympathectomy, is the lack of change in blood pressure in resting conditions following immunosympathectomy (Dorr & Brody, 1966), despite the absence of adrenergic nerves to the blood vessels that are tonically active (Folkow & Neil, 1971). The compensatory mechanism is represented by reflex activation of the adrenal gland, with induction of CA synthetic enzymes and increase of CA synthesis and release (Axelrod, Mueller & Thoenen, 1970; Brody, 1972; see Axelrod, 1972) (see also Section 4.1). A possible side effect of sympathetic degeneration is the development of an undesirable supersensitivity to circulating CA in those organs which are normally densely innervated and under highly efficient nerve control, for example the vas deferens (see Sections 5 and 7).

6.5 Summary

Adrenergic neurons survive, grow and differentiate in tissue culture and are able to synthesize, transport from the cell body to the axons, take up, store and release noradrenaline. Cultured adrenergic neurons are also capable of forming functional neuromuscular junctions with smooth muscle, and develop a pattern and density of innervation which is similar to that seen *in vivo*.

Chromaffin tissue also survives in culture and is sensitive to corticosteroids which induce the appearance of adrenaline by inducing the enzyme, phenylethanolamine-N-methyl transferase.

Sympathetic adrenergic neurons, adrenal and extra-adrenal chromaffin cells have a common embryological origin in neural crest tissue. Primordial clusters of sympathetic cells migrate to form the sympathetic chains and axons grow towards the organs which are innervated at different stages. Immature adrenergic fibres show a smooth appearance and no varicosities are yet present. Fibres reach organs and become varicose, but functional transmission does not occur until a later stage.

The growth and maturation of adrenergic neurons is under the control of nerve growth factor (NGF), of preganglionic input and probably of hormones. Chromaffin cells appear to be controlled mainly by adrenocortical hormones. The sensitivity of adrenergic neurons to NGF and hormones is reduced in the adult, but preganglionic inputs still have an effect by inducing synthesis of proteins and new enzymes. In the adult, adrenergic fibres retain remarkable abilities for growth. They can increase the territory of innervation by branching when an organ has lost some of its adrenergic fibres; growth is inhibited by the presence of other adrenergic nerve fibres. The pattern of innervation appears to depend on the effector organ.

Transection of adrenergic axons leads to retrograde degeneration with chromatolysis of the cell body and degeneration of the part of the axon distal to transection. 6-Hydroxydopamine specifically destroys adrenergic terminals in the adult, but produces degeneration of the whole adrenergic neuron in newborn animals. Chronic treatment with high doses of guanethidine produces degeneration of adrenergic neurons; lower doses produce long term damage to those adrenergic neurons which supply the internal reproductive organs. Immunosympathectomy in the newborn produces degeneration of all adrenergic neurons with the exception of the neurons supplying the internal generative organs which are highly resistant. In the adult, immunosympathectomy is ineffective.

Degeneration of adrenergic terminals produces secondary effects including: short-lasting sympathomimetic effects due to the leakage of noradrenaline from degenerating axons i.e. 'degeneration contraction'; and a long lasting supersensitivity of denervated smooth muscle, due partly to the disappearance of the noradrenaline uptake mechanism, ('prejunctional supersensitivity') and partly to a change in sensitivity of the smooth muscle membrane ('postjunctional supersensitivity'). The general effect of sympathectomy on the physiology of the whole animal depends on the method used to achieve sympathectomy and on the number and location of the neurons which degenerate. Compensatory mechanisms in the sympathoadrenal system may mask some effects of sympathectomy.

7 GENERAL DISCUSSION AND CONCLUSIONS ABOUT ADRENERGIC NEURONS

From this extensive survey on the properties of adrenergic neurons and of adrenergic neurotransmission, some general features can be considered. While not intended as a general theory for adrenergic neurons, they supply possible guidelines for the interpretation of some of the data and findings in this enormous and continuously growing field. It is proposed here that the great variability of adrenergic neurons and their functions can be accounted for by a limited number of modulating factors.

7.1 Heterogeneity of peripheral adrenergic neurons

Peripheral adrenergic neurons differ in many aspects. A variety of morphological types of neurons has been described on the basis of size and number of processes. The length of their axons varies, and it has even been suggested that adrenergic neurons can be classified in the two categories, namely 'short' and 'long', on this basis (Sections 2.2, 3.1). The pattern of branching of adrenergic terminal axons varies with the tissues from a dense homogenously distributed network, to an asymmetrical distribution of fibres (Sections 3.1, 5.1, 5.7). The minimal distances between adrenergic terminal axons and the effector tissue vary from tens to several hundred nm (Section 5.1). The characteristics of response to the activation of adrenergic axons varies in different organs; in some tissues it is prompt and powerful, in others it is sluggish and weak (Section 5.7). The number of preganglionic fibres and the pattern of preganglionic innervation differ in sympathetic ganglia (Section 2.2) and so does the level of nerve activity of different adrenergic neurons, some being 'silent' unless activated reflexly while others are tonically active (Section 2.2). In accordance with these factors, there is good evidence that there are functionally distinct neural pathways through sympathetic ganglia (Bishop & Heinbecker, 1932; Skok, 1973). There are also groups of adrenergic neurons in sympathetic ganglia which differ in their sensitivity to ganglionic blocking or stimulating drugs (Mainland & Show, 1952; Hetzler, 1961). Adrenergic cell bodies have different contents of NA as shown by the different levels of fluorescence revealed by the fluorescence histochemical method for localizing CA (Section 3.1). The presence of different proportions of vesicles of different sizes in adrenergic neurons also suggests that there may be more than one type of adrenergic neuron (Section 3.1). The levels of activity of synthetic enzymes for NA varies in adrenergic neurons (Section 4.1). The sensitivity to drugs which affect storage and synthesis mechanisms as well as the toxic effects of drugs such as 6-OHDA, guanethidine and NGF-antiserum vary between different adrenergic neurons (Section 4.4). Hormones and NGF exert different effects in different ganglia, and these effects are also age-dependent (Sections 4.2 and 4.3).

All these features indicate that the peripheral adrenergic system is not homogenous. The question arises as to whether these differences can be accounted for by certain modulating factors.

7.2 Modulating factors in the heterogeneity of adrenergic neurons

Little is known about the genetic factors involved in the maturation and differentiation of adrenergic neurons, but it seems likely that the undifferentiated adrenergic neuron possesses potentialities for developing different properties depending on the conditions. The likely candidates as modulating factors of the undifferentiated cell include hormones, NGF (Sections 4.2, 4.3), and preganglionic nerve impulses (Sections 2.2, 4.1).

Although little is known about the effects of hormones during the development of neurons, the sensitivity of developing adrenergic neurons and of chromaffin tissue to corticosteroids and some sexual hormones (see Chapter 6) suggests that these hormones play a role in the development of the adrenergic system. NGF appears to be a powerful factor which influences the growth and morphology of the cell bodies, the extent of branching of adrenergic axon terminals, the levels of NA and the synthesis mechanisms (Section 6.2). The influence of NGF decreases during development, adrenergic neurons in adults being considerably less sensitive to NGF. The density and pattern of innervation of tissues appears to depend on the effector tissue. NGF and hormones in tissues are probably involved in determining innervation patterns, but other tissue properties are likely to be involved in establishing long lasting relations with adrenergic terminal axons (Chapter 6). An unknown factor has been postulated to explain the inhibition of adrenergic axon growth by other adrenergic nerves (Section 6.3).

DISCUSSION AND CONCLUSIONS ABOUT ADRENERGIC NEURONS

During postnatal life, the nerve impulses which impinge on adrenergic neurons play an important role in the determination of their physiological properties. The onset of nerve impulses affect the maturation of the adrenergic neuron (Section 6.2). The morphological changes induced by nerve impulses appear to be less prominent than those induced by NGF. Nevertheless, since nerve impulses are known to affect protein synthesis (Section 6.2). it is probable that slow changes occur following long term fluctuations of nerve impulse rate. Two types of functional mechanism of adaptation of the adrenergic neuron are activated by nerve impulses, which have been described as fast and slow control of NA metabolism (Section 4.1). Differences in NA content in each neuron can also be accounted for by differences in nerve impulse frequency (Section 4.1). The characteristics of every adrenergic neuron could be accounted for by specific hormonal and NGF effects during development, upon which are superimposed changes induced by preganglionic nerve activity throughout life.

The wide range of adrenergic neuron characteristics could thus be regarded as the modulation of a basic type of undifferentiated cell by the interaction of a limited number of factors, which make the adrenergic neurons highly adaptable to different functional requirements.

7.3 Functional requirements of different adrenergic neurons

Adrenergic neurons are part of neuronal pathways with different functions. Independent activation of sympathetic pathways has been shown in numerous instances. Cold exposure of animals activates adrenergic neurons to the heart but not those to salivary glands (Costa, Neff & Ngai, 1969). Adrenergic neurons innervating skin vessels, which are involved almost exclusively in thermoregulation, are activated by stimuli which differ from those which activate adrenergic neurons involved in blood pressure control (see Folkow, 1955; Walther, Iriki & Simon, 1970; Iriki, Walther, Pleschka & Simon, 1971; Kendrick, Öberg & Wennergren, 1972; Delius, Wallin & Hogbarth, 1973). Electrophysiological evidence for selective regional control of adrenergic nerve activity has been produced (Green & Heffron, 1966; Kendrick, Öberg & Wennergren, 1972; Iriki & Simon, 1973; Irisawa, Ninomiya & Wooley, 1973; Ninomiya, Irisawa & Nisimaru, 1973). Control of intestinal vasculature is exerted by adrenergic neurons which differ from those which affect intestinal motility (see Furness & Costa, 1974). Stimulation of aortic chemoreceptors produces differences in sympathetic nerve traffic to capacitance blood vessels in the mesentery and in the limb (Iizuka, Mark, Wendling, Schmid &

Eckstein, 1970). During early stages of haemorrhagic hypotension, the activity in cardiac splanchnic and renal sympathetic nerves may remain largely unaltered, while the increase of peripheral vascular resistance is probably due to circulating CA from adrenal glands (Chien, 1967; Bond, Lackey, Taxis & Green, 1970; Aars & Akre, 1973). The vas deferens musculature is innervated by adrenergic neurons located in the anterior pelvic (hypogastric) ganglia, while the adrenergic neurons innervating the vasculature of this organ are located in paravertebral sympathetic ganglia (Sjöstrand, 1965).

The view of a generalized activation of the sympatho-adrenal system (Cannon, 1929) does not cover the whole situation. Adrenergic neurons are the final link of different nerve pathways with independent functions. Consequently their morphological, biochemical and functional properties and their relations with the effector tissues would be expected to differ following modulation of the factors mentioned above.

7.4 Comparison of neurons with different functions

Adrenergic motorneurons to the cardiovascular system and adrenergic motorneurons to the vas deferens are selected for comparison. While adrenergic neurons to the cardiovascular system are involved in the general homeostasis of the body, adrenergic neurons to the vas deferens are involved in a specific and localized function, i.e. the emission of sperm. One major difference between these two groups of neurons is that cardiovascular adrenergic neurons are more sensitive to NGF than vas deferens neurons (Section 6.2), while sexual hormones affect vas deferens neurons but not the others (Section 6.2, 6.3). This suggests that the development of adrenergic neurons with different functions may be controlled by different factors. From these basic differences, other variations arise as a result of different nerve activity impinging on the two groups of neurons. Adrenergic neurons to muscular arteries are continuously firing since they are controlled by tonically active vasomotor centres (Folkow & Neil, 1971), while adrenergic neurons to vas deferens are likely to be silent except during emission of sperm.

The continuous preganglionic activity initiated early in development on adrenergic neurons supplying muscular arteries might be expected to produce a faster maturation of the adrenergic neurons and plastic changes such as growth of dendrites and increases in NA biosynthetic enzymes (Sections 4.1, 6.2, 6.3). Conversely, in adrenergic neurons to the vas deferens, maturation is slower and few dendritic processes are present

(Sections, 2.2, 3.1). Turnover rate of NA is higher in cardiovascular neurons (Spector, Tarver & Berkowitz, 1972) than in vas deferens neurons (Swedin, 1970). Consequently cardiovascular motorneurons are more sensitive to NA depletion by drugs than vas deferens neurons (Swedin, 1970). Toxic drugs, such as guanethidine are more effective on vas deferens neurons than on cardiovascular neurons probably because lower turnover of NA leads to a higher intraneuronal concentration of guanethidine (Section 6.4).

Since the physiological functions of these two types of adrenergic neuron are different, reflex stimuli which affect one may not affect the other. For example, fall in blood pressure due to α-adrenoceptor blockade would result in a reflex activation of cardiovascular adrenergic neurons but not of vas deferens adrenergic neurons. Trans-synaptic induction of enzymes would be expected in ganglia containing cardiovascular adrenergic neurons such as the superior cervical ganglion, but not in the hypogastric ganglion which supplies the vas deferens. Weiner, Cloutier, Bjur & Pfeiffer (1972) found this to be the case.

The kind of control these two groups of neurons exert on their effector tissues is different. Muscular arteries are under both nerve and humoral control, while vas deferens is under exclusive nerve control (Section 5.7). The distribution of adrenergic terminals in the two groups of neurons is different. The vas deferens has a dense homogeneously distributed innervation, which ensures an exclusive nerve control by 'protecting' the vas deferens from circulating CA. Blood vessels have a dense network of adrenergic nerve terminals at the adventitial-medial border, which ensures effective nerve transmission, but leaves the media exposed to the effect of circulating CA (Section 5.7).

There seems to be a good correlation between the physiological role of adrenergic neurons and their structural, biochemical and functional properties as well as their relations with effector tissues.

7.5 Speculations on the functional organization of the sympathoadrenal system

Adrenergic and adreno-medullary systems share a common feature in that their effects are mediated by the release of a catecholamine which acts as the chemical signal for the response of the effector organ.

The most important factor that determines the time course and the magnitude of the response is the geometry of the relationship between the source of CA release (whether chromaffin cells or adrenergic neurons) and the effector tissue (Sections 5.1, 5.7). The range of distances between CA-containing cells and effector cells varies enormously: at one extreme, the chromaffin cell has targets as far as the most peripheral tissues; at the other extreme, adrenergic neurons supplying the vas deferens have terminal axons deeply embedded in smooth muscle cells with membrane separation of as little as 10 to 30 nm (Section 5.1). Between these extremes there are intermediate situations, for example small groups of chromaffin cells with local target organs, such as the carotid body and adrenergic neurons with wide neuromuscular distances such as those in large elastic arteries and non-sphincteric parts of the intestine.

The wider the neuroeffector junction, the less localized are the CA effects, the longer the latency of the response and the slower its offset. The frequencies required to elicit maximal responses of the effector tissue are higher, the maximal response is smaller and the number of effector cells effected is larger. Thus the effect is more widespread and is involved in more general, non-specialized functions. Conversely, the narrower the neuroeffector junction, the more localized is the response. This response is prompter, i.e. the latency is shorter, the onset is faster and the frequency at which the maximal response to nerve stimulation is elicited is lower.

Integration of sympatho-adrenal effects occurs at different levels from the periphery to higher centres of the CNS. For example, scattered, non-innervated chromaffin cells respond to a variety of secretory stimuli (Section 3.2) and the cell itself integrates all the inputs. As an example of the other extreme of the scale, little integration occurs at the level of the adrenergic neurons which supply the vas deferens (Section 2.2) and emission of sperm is part of a complex pattern of activation involving high centres.

The dispensability of the various sympatho-adrenal functions (i.e. how essential they are for the physiology of the whole organism) also varies. The more general the function, the more dispensable it appears to be, for example, those components of the sympatho-adrenal system involved in body homeostasis (Cannon, 1929). The more specific and localized the function, the more it is irreplaceable such as the adrenergic motor innervation of the male internal genital organs, which in this respect resembles a somatic innervation.

The wide range of relations between CA-releasing cells and effector tissues in adult mammals suggests that there is a temporal sequence in their appearance during development. Isolated groups of CA-containing cells appear first, some of which develop into chromaffin cells which are functional at early stages, while some develop into neurons which become functional at a later stage (Section 6.2). As a consequence of growth of cell processes, the neuroeffector distances decrease; the sooner they stop growing, the wider will be the neuroeffector separation. Close neuromuscular junctions are the last to appear (Section 6.2). A similar trend may be found during phylogenesis of vertebrates judging from the large number of scattered chromaffin-like cells in lower vertebrates (Coupland, 1965) and the sparse adrenergic innervation of organs in primitive vertebrates such as the cyclostomes (Burnstock, 1969; Costa, unpublished results).

A major consequence of the existance of a range of cells (from chromaffin cells to adrenergic neurons) which release CA that act on the same receptors, is that undesirable interaction between systems with different functions could occur. For example, reflex activation of adrenal-medulla during fear, would lead to undesirable effects of the resulting circulating CA on organs such as the vas deferens and iris. The CA uptake mechanism of adrenergic neurons plays an important role in this respect (see Section 5.7), i.e. CA-uptake may represent a mechanism which 'protects' an effector organ from unwanted effects of CA originating from other sources. It has been shown that NA released from adrenergic nerves have little, or no, remote effect on other organs (Rosenblueth, 1950; Celander, 1954; Langer & Vogt, 1971), unless these organs have been adrenergically denervated or treated with cocaine (Cannon & Bacq, 1931; Liu & Rosenblueth, 1935; Partington, 1936; le Compte, 1941; Cannon & Rosenblueth, 1949; Celander, 1954). CA released from the adrenal glands affect some tissues such as blood vessels and the non-sphincteric parts of the intestine. However, circulating CA has little effect on iris muscle, nictitating membrane, pilomotor muscles or blood vessels of the skin (see White, Okelberry & Whitelaw, 1936; Fatherree, Adson & Allen, 1940; Celander, 1954; Folkow & Neil, 1971, Furness & Burnstock, 1974), although A released from the adrenals and taken up by adrenergic nerves can reach up to 8% of the total store of CA in the nictitating membrane (Langer & Vogt, 1971). Following denervation, densely innervated tissues that were previously insensitive to circulating CA, become sensitive (Hartman, McCordock & Loder, 1923; Freeman, 1935; White, Okelberry & Whitelaw, 1936; Bender, 1938; White & Smithwick, 1944; Cannon & Rosenblueth, 1949; Langer, 1966b, see Section 5.7).

The view put forward by Cannon (1929), according to which sympathetic neurons and adrenal CA secretion have the same actions needs to be modified. Thus, in mammals multiple sympatho-adrenal functions are superimposed, but independent control of certain effector tissues is ensured by their particular pattern and density of adrenergic innervation.

7.6 Summary

The peripheral adrenergic system represents an heterogeneous system. The morphological, biochemical and functional differences between adrenergic neurons are tentatively accounted for by the interaction of some basic factors including hormones, nerve growth factor; preganglionic nerve impulses and unknown inhibitory factors in adrenergic neurons which limit overgrowth, and in effector cells which determine the kind of long lasting contacts. Each adrenergic neuron is part of a neuronal pathway which can be activated independently and perform different functions, for example, the adrenergic neurons to skin blood vessels and those to muscle blood vessels, or the adrenergic neurons to intestinal blood vessels and those to intestinal musculature.

A comparison is made between the adrenergic neurons supplying the cardiovascular system and those controlling the male internal genital organs describing the known properties as a function of the factors mentioned above.

The good correlation between some of the properties of adrenergic neurons and their relations with effector tissues and their physiological role allows some speculations on the sympatho-adrenal system. Taking the distance between the point of release of CA and the effector tissue, a range of cases is described. Chromaffin CA-releasing cells represent one extreme of the scale; adrenergic neurons supplying the vas deferens represent the other end. The roles of these cells range from general, non-localized, with sluggish control of effector tissues, to specific, well localized and prompt.

In mammals, during development all stages of this range appear and in the adult they coexist. The independence of different functional groups of CA releasing cells (adrenergic neurons and adrenal and extra-adrenal chromaffin cells) is ensured by the protective role of adrenergic fibres in some organs from unwanted effects of circulating CA. The density and pattern of adrenergic innervation are therefore determining factors for the type of adrenergic function.

REFERENCES

AARS, H. (1971) Effects of noradrenaline on activity in single aortic baroreceptor fibres. *Acta physiol. scand.* **83**, 335–343.

AARS, H. and AKRE, S. (1973) Nervous and humoral control of vascular resistance during acute hemorragic hypotension in rabbits. *Acta physiol. scand.* **87**, 404–410.

ABERCROMBIE, G. F. and DAVIES, B. N. (1963) The action of guanethidine with particular reference to the sympathetic nervous system. *Brit. J. Pharmacol.* **20**, 171–177.

ADOLPH, E. F. (1957) Ontogeny of physiological regulations in the rat. *Quart. Rev. Biol.* **32**, 89–137.

ADOLPH, E. F. (1967) Ranges of heart rates and their regulations at various ages (rat). *Amer. J. Physiol.* **212**, 595–602.

AHLQUIST, R. P. (1948) A study of the adrenotropic receptors. *Amer. J. Physiol.* **153**, 586–600.

AHLQUIST, R. P. (1966) The adrenergic receptor. *J. pharm. Sci.* **55**, 359–367.

ALCOCER, C., ARÉCHIGA, H., AGUILAR, H. U. and GUEVARA, R. (1972) Mecanismos eferentes de regulación de la actividad en la vía olfatoria del gato. *Bol. Esp. Med. Mexico.* **27**, 119–120.

ALDRICH, T. B. (1901) A preliminary report on the active principle of the suprarenal gland. *Amer. J. Physiol.* **5**, 457–461.

ALLISON, D. J. and POWIS, D. A. (1971) Adrenal catecholamines secretion during stimulation of the nasal mucous membrane in the rabbit. *J. Physiol. (Lond.).* **217**, 327–339.

ALMGREN, O. (1971) Influences of synthesis and membrane pump. Inhibition on the nerve impulse induced disappearance of NA from rat salivary glands. *Acta physiol. scand.* **83**, 515–528.

ALMGREN, O. (1973) Functional significance of synthesis of noradrenaline in adrenergic nerves of rat salivary glands. *J. Pharm. Pharmacol.* **26**, 23–29.

ALMGREN, O., ANDÉN, N-E., JONASON, J., NORBERG, K-A. and OLSON, L. (1966) Cellular localization of monoamine oxidase in rat salivary glands. *Acta physiol. scand.* **67**, 21–26.

ALOUSI, A. and WEINER, N. (1966) The regulation of norepinephrine synthesis in sympathetic nerves: effect of nerve stimulation, cocaine, and catecholamine-releasing agents. *Proc. nat. Acad. Sci. U.S.A.* **56**, 1491–1496.

AMANO, T., RICHELSON, E. and NIRENBERG, M. (1972) Neurotransmitter synthesis by neuroblastoma clones. *Proc. nat. Acad. Sci. U.S.A.* **69**, 258–263.

AMBACHE, N. and ZAR, M. A. (1971) Evidence against adrenergic motor transmission in the guinea-pig vas deferens. *J. Physiol. (Lond.).* **216**, 359–389.

AMPRINO, R. (1938) Modifications de la structure des neurones sympathiques pendant l'accroissement et la sénescence. Recherches sur le ganglion cervical supérieur. *C.R. Ass. Anat.* **33**, 3–18.

ANDÉN, N-E, and HENNING, M. (1966) Adrenergic nerve function, noradrenaline level and noradrenaline uptake in cat nictitating membrane after reserpine treatment. *Acta physiol. scand.* **67**, 498–504.

ANDÉN, N-E. and HENNING, M. (1968) Effect of reserpine on the urinary excretion and the tissue levels of noradrenaline in the rat. *Acta. physiol. scand.* **72**, 134–138.

ANDÉN, N-E., HENNING, M. and OBIANWU, H. (1968) Effect of epsilon aminocaproic acid on adrenergic nerve function and tissue monoamine level. *Acta Pharmacol.* **26**, 113–129.

ANDRES, K. H. (1971) Structure of cutaneous receptors. *Proc. Int. Union Physiol. Sci.* **8**, 136–137.

ANGELETTI, P. U. (1971) Chemical sympathectomy in newborn animals. *Neuropharmacology.* **10**, 55–59.

ANGELETTI, P. U. and LEVI-MONTALCINI, R. (1970) Sympathetic nerve cell destruction in newborn mammals by 6-hydroxydopamine. *Proc. nat. Acad. Sci., U.S.A.* **65**, 114–121.

ANGELETTI, P. U. and LEVI-MONTALCINI, R. (1972) Selective lesions of sympathetic neurons by some adrenergic blocking agents. *Proc. nat. Acad. Sci., U.S.A* **69**, 86–88.

ANGELETTI, P. U., LEVI-MONTALCINI, R. and CARAMIA, F. (1971) Ultrastructural changes in sympathetic neurons of newborn and adult mice treated with nerve growth factor. *J. Ultrastruct. Res.* **36**, 24–36.

ANGELETTI, P. U., LEVI-MONTALCINI, R. and CARAMIA, F. (1972) Structural and ultrastructural changes in developing sympathetic ganglia induced by guanethidine. *Brain Res.* **43**, 515–525.

ANGELETTI, P. U., LUZZI, A. and LEVI-MONTALCINI, R. (1966) Effetti dell'insulina sulla sintesi di RNA e di lipidi nei gangli sensitivi embrionali. *Ann. Ist. Super. Sanitá.* **2**, 420–422.

ANTONACCIO, M. J. (1970) Effects of chronic pretreatment with small doses of reserpine upon adrenergic nerve function. *Fed. Proc.* **29**, 679.

ANTONACCIO, M. J. and SMITH, G. B. (1974) Effects of chronic pretreatment with small doses of reserpine upon adrenergic nerve function. *J. Pharmacol exp. Ther.* **188**, 654–667.

ANTOSHKINA, E. D. (1939) Ueber die Ausbildung der Warmeregulierung im Laufe der Ontogenese. I and II. *Fiziol. Zh. SSSR.* **26**, 1–15, 16–29.

ARONSON, S., GENNSER, G., OWMAN, ch. and SJÖBERG, N-O. (1970) Innervation and contractile response of the human ductus arteriosus. *Eur. J. Pharmacol.* **11**, 178-186.

ARMSTRONG, J. M. and BOURA, A. L. A. (1970) Relative susceptibility of peripheral sympathetic nerves to adrenergic neurone blockade by bethanedine. *Brit. J. Pharmacol.* **40**, 551P-552P.

ARMSTRONG, J. M. and BOURA, A. L. A. (1973) Effects of clonidine and guanethidine on peripheral sympathetic nerve function in the pithed rat. *Brit. J. Pharmacol.* **47**, 850-852.

ATECH, Y. L. (1962) Réactions structurales des neurones des ganglions de la chaîne sympathique du rat thyroidectomisé à l'hormone thyréotrope hypophysaire. *C.R. Soc. Biol., Paris.* **156**, 1322-1324.

AUSPRUNK, D. H., BERMAN, H. and MCNARY, WM. F. JR. (1971) Perivascular nerve degeneration produced in the hamster cheek pouch by 6-hydroxydopamine. *Eur. J. Pharmacol.* **15**, 292-299.

AUSTIN, L., LIVETT, B. G. and CHUBB, I. W. (1967) Increased synthesis and release of noradrenaline and dopamine during nerve stimulation. *Life Sci.* **6**, 97-104.

AVAKIAN, O. V. and GILLESPIE, J. S. (1968) Uptake of noradrenaline by adrenergic nerves, smooth muscle and connective tissue in isolated perfused arteries and its correlation with the vasoconstrictor response. *Brit. J. Pharmacol.* **32**, 168-184.

AXELROD, J. (1962) Purification and properties of phenylethanolamine-N-methyl transferase. *J. biol. Chem.* **237**, 400-401.

AXELROD, J. (1966) Methylation reactions in the formation and metabolism of catecholamines and other biogenic amines. *Pharmacol. Rev.* **18**, 95-113.

AXELROD, J. (1971) Noradrenaline: Fate and control of its biosynthesis. *Science.* **173**, 598-606.

AXELROD, J. (1972) Dopamine-β-hydroxylase: Regulation of its synthesis and release from nerve terminals. *Pharmacol. Rev.* **24**, 233-243.

AXELROD, J., ALBERS, R. W. and CLEMENTE, C. D. (1959) Distribution of catechol-O-methyl transferase in the nervous system and other tissues. *J. Neurochem.* **5**, 68-72.

AXELROD, J., MUELLER, R. A. and THOENEN, H. (1970) Neuronal and hormonal control of tyrosine hydroxylase and phenylethanolamine N-methyltransferase activity. In 'New Aspects of Storage and Release Mechanisms of Catecholamines'. H. J. Schümann and G. Kroneberg, eds. Springer Verlag. N.Y., Heidelberg, Berlin, pp. 212-219.

AXELSSON, J. (1971) Catecholamine functions. *Amer. Rev. Physiol.* **33**, 1-30.

BACQ, Z. M. (1933) Recherches sur la physiologie du système nerveux autonome. III les propriétés biologiques et physicochimiques de la sympathine comparées à celles de l'adrénaline. *Arch. int. Physiol.* **36**, 167-246.

BACQ, Z. M. and FREDERICQ, H. (1934) Sensibilization à l'excitation sympathique par la cocaine: méthode indirecte pour déterminer la nature de la substance formeé par l'excitation des fibres sympathiques adrénergiques. *C.R. Soc. Biol. Paris.* **117**, 76–79.

BANKS, P. and HELLE, K. B. (1971) Chromogranins in sympathetic nerves. *Phil. Trans. Roy. Soc. Lond. Ser. B.* **261**, 305–310.

BANKS, P. and MAYOR, D. (1972) Intra-axonal transport in noradrenergic neurons in the sympathetic nervous system. *Biochem. Soc. Symp.* **36**, 133–149.

BANKS, P., MANGNALL, D. and MAYOR, D. (1969) The re-distribution of cytochrome oxidase, noradrenaline and adenosine triphosphate in adrenergic nerves constricted at two points. *J. Physiol. (Lond.).* **200**, 745–762.

BANKS, P., MAYOR, D., MITCHELL, M. and TOMLINSON, D. (1971) Studies on the translocation of noradrenaline-containing vesicles in post-ganglionic sympathetic neurons *in vitro*. Inhibition of movement by colchicine and vinblastine and evidence for the involvement of axonal tubules. *J. Physiol. (Lond.).* **216**, 625–639.

BARGER, G. and DALE, H. H. (1910) Chemical structure and sympathomimetic action of amines. *J. Physiol. (Lond.).* **41**, 19–59.

BARKER, D. (1948) The innervation of the muscle-spindle. *Quart. J. micr. Sci.* **89**, 143–186.

BARONDES, S. H. (1969) Axoplasmic transport. *In* 'Handbook of Neurochemistry'. Vol. II. A. Lajtha, ed. Plenum Press, New York, Lond., pp. 435–446.

BASSET, J. R., CAIRNCROSS, K. D., HACKET, N. B. and STORY, M. (1969) Studies on the peripheral pharmacology of fenazoxine, a potent antidepressive drug. *Brit. J. Pharmacol.* **37**, 69–78.

BAUMGARTEN, H. G., BJÖRKLUND, A., LACHENMAYER, L., NOBIN, A. and STENEVI, U. (1971) Long-lasting selective depletion of brain serotonin by 5,6-dihydroxytryptamine. *Acta physiol. scand.* Suppl. **373**, 1–15.

BAUMGARTEN, H. G., GÖTHERT, M., HOLSTEIN, A. F. and SCHLOSSBERGE, H. C. (1972) Chemical sympathectomy induced by 5,6-dihydroxytryptamine. *Z. Zellforsch. Mikrosk. Anat.* **128**, 115–134.

BEACHAM, W. S. and PERL, E. R. (1964) Background and reflex discharge of sympathetic preganglionic neurons in the spinal cat. *J. Physiol. (Lond.).* **172**, 400–416.

BEANI, L., BIANCHI, C. and CREMA, A. (1969) The effect of catecholamines and sympathetic stimulation on the release of acetylcholine from the guinea-pig colon. *Brit. J. Pharmacol.* **36**, 1–17.

BELL, C. (1967) Effects of cocaine and of monoamine oxidase and catechol-*o*-methyl transferase inhibitors on transmission to the guinea-pig vas deferens. *Brit. J. Pharmacol.* **31**, 276–286.

BELL, C. (1969a) Fine structural localization of acetylcholinesterase at a cholinergic vasodilator nerve-arterial smooth muscle synapse. *Circulation Res.* **24**, 61–70.

BELL, C. (1969b) Transmission from vasoconstrictor and vasodilator nerves to single smooth muscle cells of the guinea-pig uterine artery. *J. Physiol. (Lond.).* **205**, 695–708.

BELL, C. (1972) Mechanism of enhancement by angiotensin II of sympathetic adrenergic transmission in the guinea pig. *Circulation Res.* **31**, 348–355.

BELL, C. and BURNSTOCK, G. (1971) Cholinergic vasomotor neuroeffector junctions. *In* 'Symposium on the Physiology and Pharmacology of Vascular Neuroeffectors Systems'. Interlaken, 1969. J. A. Bevan, R. F. Furchgott, R. A. Maxwell and A. P. Somlyo, eds. Publ. S. Karger, Basel, pp. 37–46.

BELL, C. and VOGT, M. (1971) Release of endogenous noradrenaline from an isolated muscular artery. *J. Physiol. (Lond.).* **215**, 509–520.

BELLEAU, B. (1960) Relationship between agonists, antagonists and receptor sites. *In* 'Adrenergic mechanisms'. Ciba Foundations symposium. J. & A. Churchill Ltd., London. pp. 223–245.

BELLEAU, B. (1967) Stereochemistry of adrenergic receptors: newer concepts on the molecular mechanism of action of catecholamines and antiadrenergic drugs at the receptor level. *Ann. N.Y. Acad. Sci.* **139**, 580–605.

BELMONTE, C., SIMON, J., GALLEGO, R. and BARON, M. (1972) Sympathetic fibers in the aortic nerve of the cat. *Brain Res.* **43**, 25–35.

BENELLI, G., DELLA BELLA, D. and GANDINI, A. (1964) Angiotensin and peripheral sympathetic nerve activity. *Brit. J. Pharmacol. (Chemother).* **22**, 211–219.

BENNETT, M. R. (1973a) 'Autonomic neuromuscular transmission.' Cambridge University Press, Cambridge.

BENNETT, M. R. (1973b) An electrophysiological analysis of the release of noradrenaline at sympathetic nerve terminals. *J. Physiol. (Lond.).* **229**, 515–532.

BENNETT, M. R. (1973c) An electrophysiological analysis of the uptake of noradrenaline at sympathetic nerve terminals. *J. Physiol. (Lond.).* **229**, 533–546.

BENNETT, M. R. (1973d) Structure and electrical properties of the autonomic neuromuscular junction. *Phil. Trans. Roy. Soc. Lond. Ser. B.* **265**, 25–34.

BENNETT, M. R. and BURNSTOCK, G. (1968) Electrophysiology of the innervation of intestinal smooth muscle. *In* 'Handbook of Physiology, Sect. 6, Alimentary Canal'. Vol. IV. Motility., pp. 1709–1732. American Physiological Society, Washington, D.C.

BENNETT, M. R., BURNSTOCK, G. and HOLMAN, M. E. (1966a) Transmission from perivascular ihhibitory nerves to the smooth muscle of the guinea-pig taenia coli. *J. Physiol. (Lond.).* **182**, 527–540.

BENNETT, M. R., BURNSTOCK, G. and HOLMAN, M. E. (1966b) Transmission from intramural inhibitory nerves to the smooth muscle of the guinea-pig taenia coli. *J. Physiol. (Lond.).* **182**, 541–558.

BENNETT, M. R. and MERRILLEES, N. C. R. (1966) An analysis of the transmission of excitation from autonomic nerves to smooth muscle. *J. Physiol. (Lond.).* **185**, 520–535.

BENNETT, M. R. and ROGERS, D. C. (1967) A study of the innervation of the taenia coli. *J. Cell. Biol.* **33**, 573–596.

BENNETT, T., BURNSTOCK, G., COBB, J. L. S. and MALMFORS, T. (1970) An ultrastructural and histochemical study of the short term effects of 6-hydroxydopamine on adrenergic nerves in the domestic fowl. *Brit. J. Pharmacol.* **38**, 802–809.

BENTLEY, G. A. (1965) Potentiation of responses to noradrenaline and reversal of sympathetic nerve blockade in the guinea-pig vas deferens. *Brit. J. Pharmacol.* **25**, 243–256.

BENTLEY, G. A. and SMITH, G. (1967) Effects of alpha-adrenergic receptor blocking drugs on the response of vas deferens and arterial muscle to sympathetic drugs and stimulation. *Circulation Res.* **21** Suppl. III, 101–110.

BERNARD, C. (1851) Influence du grand sympathique sur la sensibilité et sur le calorification. *C.R. Soc. Biol. Paris.* **3**, 163–169.

BERNEIS, K. H., PLETSCHER, A. and DA PRADA, M. (1969) Metal-dependent aggregation of biogenic amines: a hypothesis for their storage and release. *Nature (Lond.).* **224**, 281–283.

BERNEIS, K. H., PLETSCHER, A. and DA PRADA, M. (1970) Phase separation in solutions of noradrenaline and adenosine triphosphate: influence of bivalent cations and drugs. *Brit. J. Pharmacol.* **39**, 382–389.

BERTLER, Å., CARLSSON, A. and ROSENGREN, E. (1956) Release by reserpine of catechol-amines from rabbit's hearts. *Naturwissenschaften.* **22**, 521.

BEVAN, J. A., BEVAN, R. D., PURDY, R. E., ROBINSON, C. P., SU, C. and WATERSON, J. G. (1972) Comparison of adrenergic mechanisms in an elastic and a muscular artery of the rabbit. *Circulation Res.* **30**, 541–542.

BEVAN, J. A., CHESHER, G. B. and SU, C. (1969) Release of adrenergic transmitter from terminal nerve plexus in artery. *Agents & Actions.* **1**, 20–26.

BEVAN, J. A. and PURDY, R. E. (1973) Variations in adrenergic innervation and contractile responses of the rabbit saphenous artery. *Circulation Res.* **32**, 746–751.

BEVAN, J. A. and SU, C. (1971) Distribution theory of resistance of neurogenic vasoconstriction to alpha-receptor blockade in the rabbit. *Circulation Res.* **27**, 179–187.

BEVAN, J. A. and SU, C. (1973) Sympathetic mechanisms in blood vessels nerve and muscle relationship. *Ann. Rev. Pharmacol.* **14**, 269–285.

BHAGAT, B. (1967) The influence of sympathetic nervous activity on cardiac catecholamine levels. *J. Pharmacol. exp. Ther.* **157**, 74–80.

BHAGAT, B. and FRIEDMAN, E. (1969) Factors involved in maintenance of cardiac catecholamine content: relative importance of synthesis and reuptake. *Brit. J. Pharmacol.* **37**, 24–33.

BHAGAT, B. and ZEIDMAN, H. (1970) Increased retention of norepinephrine-^{3}H in vas deferens during nerve stimulation. *Amer. J. Physiol.* **219**, 692–696.

BHARGAVA, K. P., KAR, K. and PARMAR, S. S. (1965) Independent cholinergic and adrenergic mechanism in the guinea-pig isolated nerve vas deferens preparation. *Brit. J. Pharmacol.* **24**, 641–690.

BHATNAGAR, R. K. and MOORE, K. E. (1971a) Effects of electrical stimulation, α-methyltyrosine and desmethylimipramine on the norepinephrine contents of neuronal cell bodies and terminals. *J. Pharmacol. exp. Ther.* **178**, 450–463.

BHATNAGAR, R. K. and MOORE, K. E. (1971b) Maintenance of noradrenaline in neuronal cell bodies and terminals: effect of frequency of stimulation. *J. Pharm. Pharmacol.* **23**, 625–627.

BHATNAGAR, R. K. and MOORE, K. E. (1972) Regulation of norepinephrine contents in neuronal cell bodies and terminals during and after cessation of preganglionic stimulation. *J. Pharmacol. exp. Ther.* **180**, 265–276.

BIRMINGHAM, A. T. and WILSON, A. B. (1963) Preganglionic and postganglionic stimulation of the guinea-pig isolated vas deferens preparation. *Brit. J. Pharmacol.* **21**, 569–580.

BIRMINGHAM, A. T. and WILSON, A. B. (1965) An analysis of the blocking action of Dimethylphenylpiperazinium iodide on the inhibition of isolated small intestine produced by stimulation of the sympathetic nerves. *Brit. J. Pharmacol.* **24**, 375–386.

BISBY, M. A., CRIPPS, H. and DEARNALEY, F. (1971) Effects of nerve stimulation on the subcellular distribution of noradrenaline in the cat spleen. *J. Physiol. (Lond.).* **214**, 13–14P.

BISBY, M. A. and FILLENZ, M. (1971) The storage of endogenous noradrenaline in sympathetic nerve terminals. *J. Physiol. (Lond.).* **215**, 163–179.

BISBY, M. A., FILLENZ, M. and SMITH, A. D. (1973) Evidence for the presence of dopamine-β-hydroxylase in both populations of noradrenaline storage vesicles in sympathetic nerve terminals of the rat vas deferens. *J. Neurochem.* **20**, 245–248.

BISCOE, T. J. (1971) Carotid body: structure and function. *Physiol. Rev.* **51**, 437–496.

BISHOP, G. H. and HEINBECKER, P. (1932) A functional analysis of the cervical sympathetic nerve supply to the eye. *Amer. J. Physiol.* **100**, 519–532.

BJÖRKLUND, A. (1968) Monoamine-containing fibres in the pituitary neuro-intermediate lobe of the pig and rat. *Z. Zellforsch. Mikrosk. Anat.* **89**, 573–589.

BJÖRKLUND, A., CEGRELL, L., FALCK, B., RITZÉN, M. and ROSENGREN, E. (1970) Dopamine-containing cells in sympathetic ganglia. *Acta physiols. Scand.* **78**, 334–338.

BJÖRKLUND, A., KATZMAN, R., STENEVI, V. and WEST, K. A. (1971) Development and growth of axonal sprouts from noradrenaline and 5-hydroxytryptamine neurones in the rat spinal cord. *Brain Res.* **31**, 21–33.

BJÖRKLUND, A., OWMAN, ch. and WEST, K. A. (1972) Peripheral sympathetic innervation and serotonin cells in the habenular region of the rat brain. *Z. Zellforsch. Mikrosk. Anat.* **127**, 570–579.

BJÖRKLUND, A. and STENEVI, V. (1971) Growth of central catecholamine neurones into smooth muscle grafts in the rat mesencephalon. *Brain Res.* **31**, 1–20.

BLACK, I. B. (1973) Development of adrenergic neurons *in vivo*: inhibition of ganglionic blockade. *J. Neurochem.* **20**, 1265–1276.

BLACK, I. B., BLOOM, F. E., HENDRY, I. A. and IVERSEN, L. L. (1971) Growth and development of a sympathetic ganglion: maturation of transmitter enzymes and synapse formation in the mouse superior cervical ganglion. *J. Physiol. (Lond.).* **215**, 24–25P.

BLACK, I. B., HENDRY, I. A. and IVERSEN, L. L. (1971a) Differences in the regulation of tyrosine hydroxylase and dopa decarboxylase in sympathetic ganglia and adrenals. *Nature: New Biol.* **231**, 27–29.

BLACK, I. B., HENDRY, I. A. and IVERSEN, L. L. (1971b) Trans-synaptic regulation of growth and development of adrenergic neurones in a mouse sympathetic ganglion. *Brain Res.* **34**, 229–240.

BLACK, I. B., HENDRY, I. A. and IVERSEN, L. L. (1972a) The role of post-synaptic neurones in the biochemical maturation of presynaptic cholinergic nerve terminals in a mouse sympathetic ganglion. *J. Physiol. (Lond.).* **221**, 149–160.

BLACK, I. B., HENDRY, I. A. and IVERSEN, L. L. (1972b) Effects of surgical decentralization and nerve growth factor on the maturation of adrenergic neurons in a mouse sympathetic ganglion. *J. Neurochem.* **19**, 1367–1377.

BLACKMAN, J. G., CROWCROFT, P. J., DEVINE, C. E., HOLMAN, M. E. and YONEMURA, K. (1969) Transmission from pre-ganglionic fibres in the hypogastric nerve to peripheral ganglia of male guinea-pig. *J. Physiol. (Lond.).* **201**, 723–743.

BLACKMAN, J. G. and PURVES, R. D. (1969) Intracellular recordings from ganglia of the thoracic sympathetic chain of the guinea-pig. *J. Physiol. (Lond.).* **203**, 173–198.

BLACKELEY, A. G. H. and BROWN, G. L. (1966) Release and turnover of the adrenergic transmitter. *In* 'Mechanism of Release of Biogenic Amines'. U.S. Von Euler, S. Rosell and B. Uvnäs, eds. Pergamon Press, Oxford. pp. 185–187.

BLAKELEY, A. G. H., BROWN, G. L., DEARNALEY, D. P. and HARRISON, V. (1968) The effect of nerve stimulation on the synthesis of ^{3}H-tyrosine in the isolated blood-perfused cat spleen. *J. Physiol. (Lond.).* **200**, 59–60P.

BLAKELEY, A. G. H., BROWN, G. L. and GEFFEN, L. B. (1969) Uptake and re-use of sympathetic transmitter in the cat spleen. *Proc. Roy. Soc. Lond. Ser. B.* **174**, 51–68.

BLAKELEY, A. G. H., POWIS, G. and SUMMERS, R. J. (1973) The effects of pargyline on overflow of transmitter and uptake of noradrenaline in the cat spleen. *Brit. J. Pharmacol.* **47**, 719–728.

BLASCHKO, H. (1939) The specific action of l-dopa decarboxylase. *J. Physiol. (Lond.).* **96**, 50–51P.

BLASCHKO, H. (1954) Metabolism of epinephrine and norepinephrine. *Pharmacol. Rev.* **6**, 23–28.

BLASCHKO, H. (1972a) Catecholamines 1922–1971. *In* 'Catecholamines'. H. Blaschko and E. Muscholl, eds. Springer-Verlag, Berlin, Heidelberg, New York, pp. 1–15.

BLASCHKO, H. (1972b) Introduction and Historical Background. *In* 'Monoamine oxidases New Vistas. Advances in Biochemical Psychopharmacology 5.' E. Costa and M. Sadler, eds. Raven Press, New York, pp. 1–10.

BLINKS, J. R. (1967) Evaluation of the cardiac effects of several beta adrenergic blocking agents. *Ann. N.Y. Acad. Sci.* **139**, 673–685.

BLOOM, F. E. (1972) Electron microscopy of catecholamine-containing structures. *In* 'Catecholamines'. H. Blaschko and E. Muscholl, eds. Springer-Verlag, Berlin, Heidelberg, New York, pp. 46–78.

BLOOM, B. M. and GOLDMAN, I. M. (1966) The nature of catecholamine-adenine mononucleotide interactions in adrenergic mechanisms. *Adv. Drug Res.* **3**, 121–169.

BLOOR, C. M. (1964) Aortic baroreceptor threshold and sensitivity in rabbits at different ages. *J. Physiol. (Lond.).* **174**, 163–171.

BLÜMCKE, S. and DENGLER, H. J. (1970) Noradrenalin content and ultrastructure of adrenergic nerves of rabbit iris after sympathectomy and hypoxia. *Virchows Arch. Abt. B. Zell. Path.* **6**, 281–293.

BLÜMCKE, S. and NIEDORF, H. R. (1965) Fluoreszenmikroskopische und Electronenmikroskopische untersuchungen an Regenerierenden adrenergischen Nervenfasern. *Z. Zellforsch. Mikrosk. Anat.* **68**, 724–732.

BOADLE-BIBER, M. C., HUGHES, J. and ROTH, R. H. (1969) Angiotensin accelerates catecholamine biosynthesis in sympathetically innervated tissues. *Nature (Lond.).* **222**, 987–988.

BOATMAN, D. L., SHAFFER, R. A., DIXON, R. L. and BRODY, M. J. (1965) Function of vascular smooth muscle and its sympathetic innervation in the newborn dog. *J. clin. Invest.* **44**, 241–246.

BÖCK, P. (1970) Die Feinstruktur des paraganglionären gewebes im plexus suprarenalis des Meerschweinchem. *Z. Zellforsch. Mikrosk. Anat.* **105**, 389–404.

BÖCK, P. (1973) Das Glomus caroticum der maus. *Embr. Cell biol.* **48**, 1–84.

BOGDANSKI, D. F. and BRODIE, B. B. (1966) Role of sodium and potassium ions in storage of norepinephrine by sympathetic nerve endings. *Life Sci.* **5**, 1563–1569.

BOHR, D. F. (1973) Vascular smooth muscle updated. *Circulation Res.* **32**, 665–672.

BOLTON, T. B. (1972) The depolarizing action of acetylcholine or carbachol in intestinal smooth muscle. *J. Physiol. (Lond.).* **220**, 647–671.

BOND, R. F., LACKEY, G. F., TAXIS, J. A. and GREEN, H. D. (1970) Factors governing cutaneous vasoconstriction during hemorrhage. *Amer. J. Physiol.* **219**, 1210–1215.

BONDAREFF, W. (1965) Submicroscopic morphology of granular vesicles in sympathetic nerves of rat pineal body. *Z. Zellforsch. Mikrosk. Anat.* **67**, 211–218.

BORÉUS, L. O., MALMFORS, T., MCMURPHY, D. M. and OLSON, L. (1969) Demonstration of adrenergic receptor function and innervation in the ductus arteriosus of the human fetus. *Acta physiol. scand.* **77**, 316–321.

BOTAR, J. (1966) 'The Autonomic Nervous System.' Publ. Akadémiai kiadó, Budapest.

BOULLIN, D. J. (1967) The action of extracellular cations on the release of the sympathetic transmitter from peripheral nerves. *J. Physiol. (Lond.).* **189**, 85–99.

BOURA, A. L. A. and GREEN, A. F. (1965) Adrenergic neurone blocking agents. *Ann. Rev. Pharmacol.* **5**, 183–212.

BOWER, E. A. (1966) The activity of post-ganglionic sympathetic nerves to the uterus of the rabbit. *J. Physiol. (Lond.).* **183**, 748–767.

BOWMAN, W. C., RAND, M. J. and WEST, G. B. (1971) 'Textbook of Pharmacology.' 4th ed. Blackwell Scientific Publications, Oxford and Edinburgh.

BOYD, H., BURNSTOCK, G., CAMPBELL, G., JOWETT, A., O'SHEA, J. and WOOD, M. (1963) The cholinergic blocking action of adrenergic blocking agents in pharmacological analysis of autonomic innervation. *Brit. J. Pharmacol.* **20**, 418–435.

BOYD, H., CHANG, V. and RAND, M. J. (1960) The anticholinesterase activity of some antiadrenaline agents. *Brit. J. Pharmacol.* **15**, 525–531.

BOYD, I. A. and MARTIN, A. R. (1956) The end-plate potential in mammalian muscle. *J. Physiol. (Lond.).* **132**, 74–91.

BOYD, J. A. (1960) Origin, development and distribution of chromaffin cells. *In* 'Adrenergic Mechanisms'. Ciba Foundation Symposium. G. E. W. Wolstenholme and M. O'Connor, eds. J. & A. Churchill Ltd., London, pp. 63–82.

BOZLER, E. (1948) Conduction, automacity and tonus of visceral muscles. *Experientia (Basel).* **4**, 213–218.

BRALET, J., BELAY, A., LALLEMANT, A. M. and BRALET, A. M. (1973) Evolution de la radioactivité spécifique de la noradrenaline -^{3}H dans différents organes du rat *in vivo* aprés inhibition de la tyrosine hydroxylase. *Biochem. Pharmacol.* **21**, 1107–1115.

BRALET, J. and ROCHETTE, L. (1973) Influence du traitement par la clonidine sur la synthese du catécholamines dans le coeur, les glandes sous-maxillaires et la surrenale du rat. *Biochem. Pharmacol.* **22**, 3173–3180.

BRAY, G. M., PEYRONNARD, J. M. and AGUAYO, A. J. (1972) Reactions of unmyelinated nerve fibres to injury. An ultrastructural study. *Brain Res.* **42**, 297–309.

BRETTSCHNEIDER, H. (1962) Elektronenmikroskopische Untersuchungen über die Innervation der glatten Muskulatur des Darmes. *Z. Zellforsch. Mikrosk. Anat.* **68**, 333–360.

BRIMIJOIN, S. and MOLINOFF, P. B. (1971) Effects of 6-hydroxydopamine on the activity of tyrosine hydroxylase and dopamine-β-hydroxylase in sympathetic ganglia of the rat. *J. Pharmacol. exp. Ther.* **178**, 417–424.

BRIMIJOIN, S. and TRENDELENBURG, U. (1971) Reserpine-induced release of norepinephrine from isolated spontaneously beating guinea-pig atria. *J. Pharmacol. exp. Ther.* **176**, 149–159.

BRODY, M. J. (1964) Cardiovascular responses following immunological sympathectomy. *Circulation Res.* **15**, 161–167.

BRODY, M. J. (1972) Physiological status and pharmacologic responses of the immuno-sympathectomized animal. *In* 'Immunosympathectomy'. G. Steiner and E. Schönbaum, eds. Elsevier Pub. Co., Amsterdam, London, New York, pp. 147–158.

BRONK, D. W. (1939) Synaptic mechanisms in sympathetic ganglia. *J. Neurophysiol.* **2**, 380–401.

BROWN, G. L. (1965) The release and fate of the transmitter liberated by adrenergic nerves. *Proc. Roy. Soc. Lond. Ser. B.* **162**, 1–19.

BROWN, G. L. and GILLESPIE, J. S. (1957) The output of sympathetic transmitter from the spleen of the cat. *J. Physiol. (Lond.).* **138**, 81–102.

BRÜCKE, H. V. (1931) Recovery of normal tonus in the course of regeneration of the cervical sympathetic nerve. *J. Comp. Neurol.* **53**, 225–262.

BRUNDIN, J. (1965) Distribution and function of adrenergic nerves in the rabbit fallopian tube. *Acta physiol. scand.* **66**, Suppl. 259, 1–57.

BRUNDIN, J. (1966) Studies on the preaortal paraganglia of newborn rabbits. *Acta physiol. scand.* **70**, Suppl. 290, 1–54.

BUDGE, J. L. (1855) 'Über die Bewegung der Iris; für Physiologen und Ärzte.' Braunschwieg, Vieweg.

BÜLBRING, E. (1957) The actions of humoral transmitters on smooth muscle. *Brit. med. Bull.* **13**, 172–175.

BÜLBRING, E. and KURIYAMA, H. (1963) The effect of adrenaline on the smooth muscle of guinea-pig taenia coli in relation to the degree of stretch. *J. Physiol. (Lond.).* **169**, 198–212.

BÜLBRING, E. and TOMITA, T. (1969a) Increase of membrane conductance by adrenaline in the smooth muscle of guinea-pig taenia coli. *Proc. Roy. Soc. Lond. Ser. B.* **172**, 89–102.

BÜLBRING, E. and TOMITA, T. (1969b) Suppression of spontaneous spike generation by catecholamines in the smooth muscle of the guinea-pig taenia coli. *Proc. Roy. Soc. Lond. Ser. B.* **172**, 103–119.

BÜLBRING, E. and TOMITA, T. (1969c) Effect of calcium, barium and manganese on the action of adrenaline in the smooth muscle of the guinea-pig taenia coli. *Proc. Roy. Soc. Lond. Ser. B.* **172**, 121–136.

BURN, J. H. (1968a) The development of the adrenergic fibre. *Brit. J. Pharmacol.* **32**, 575–582.

BURN, J. H. (1968b) The mechanism of the release of noradrenaline. *In* 'Adrenergic Neuro-transmission'. Ciba Foundation Study Group No. 33. G. E. W. Wolstenholme and M. O'Connor, eds. J. & A. Churchill Ltd., London, pp. 16–25.

BURN, J. H. and KEVIN, K. F. N. G. (1965) The action of pempidine and antiadrenaline substances at the sympathetic postganglionic termination. *Brit. J. Pharmacol.* **24**, 675–688.

BURN, J. H. and RAND, M. J. (1959) Sympathetic postganglionic mechanism. *Nature (Lond.).* **184**, 163–165.

BURN, J. H. and RAND, M. J. (1965) Acetylcholine in adrenergic transmission. *Ann. Rev. Pharmacol.* **5**, 163–182.

BURN, J. H. and ROBINSON, J. (1954) Effect of denervation on amine oxidase in structures innervated by the sympathetic. *Brit. J. Pharmacol.* **7**, 304–318.

BURN, J. H. and TAINTER, M. L. (1931) An analysis of the effect of cocaine on the action of adrenaline and tyramine. *J. Physiol. (Lond.).* **71**, 169–193.

BURNSTOCK, G. (1958) Some observations concerning the electrophysiological basis of the action of acetylcholine and adrenaline on smooth muscle. *J. Physiol. (Lond.).* **143**, 19–20P.

BURNSTOCK, G. (1960) Membrane potential changes associated with stimulation of smooth muscle by adrenalin. *Nature (Lond.).* **186**, 727–728.

BURNSTOCK, G. (1969) Evolution of the autonomic innervation of visceral and cardiovascular systems in vertebrates. *Pharmacol. Rev.* **21**, 247–324.

BURNSTOCK, G. (1970) Structure of smooth muscle and its innervation. *In* 'Smooth Muscle'. E. Bülbring, A. Brading, A. Jones and T. Tomita, eds. Edward Arnold Publ. Ltd., London, pp. 1–69.

BURNSTOCK, G. (1972) Purinergic nerves. *Pharmacol. Rev.* **24**, 509–581.

BURNSTOCK, G. (1975) Innervation of vascular smooth muscle: histochemistry and electronmicroscopy *Clin. exp. Pharmacol. Physiol. Suppl. 1 (In press.)*

BURNSTOCK, G. and BELL, C. (1974) Peripheral autonomic transmission. *In* 'The Vertebrate Peripheral Nervous System'. J. I. Hubbard, ed. Plenum Press, New York. pp. 277–327.

BURNSTOCK, G. and COSTA, M. (1973) Inhibitory innervation of the gut. *Gastroenterology,* **64**, 141–144.

BURNSTOCK, G., DOYLE, A. E., GANNON, B. J., GERKINS, J. F., IWAYAMA, T.

and MASHFORD, M. L. (1971) Prolonged hypotension and ultrastructural changes in sympathetic neurones following guanacline treatment. *Eur. J. Pharmacol.* **13**, 175–187.

BURNSTOCK, G., EVANS, B., GANNON, B. J., HEATH, J. and JAMES, V. (1971) A new method for destroying adrenergic nerves in adult animals using guanethidine. *Brit. J. Pharmacol.* **43**, 295–301.

BURNSTOCK, G., GANNON, B. J. and IWAYAMA, T. (1970) Sympathetic innervation of vascular smooth muscle in normal and hypertensive animals. *In* Symposium on 'Hypertensive Mechanisms'. *Circulation Res.* **27**, Suppl. II, 5–24.

BURNSTOCK, G., GANNON, B. J., MALMFORS, T. and ROGERS, D. C. (1971) Changes in the physiology and fine structure of the taenia of the guinea-pig caecum following transplantation into the anterior eye chamber. *J. Physiol. (Lond.).* **219**, 139–154.

BURNSTOCK, G. and HOLMAN, M. E. (1960) Autonomic nerve-smooth muscle transmission. *Nature (Lond.).* **187**, 951–952.

BURNSTOCK, G. and HOLMAN, M. E. (1961) The transmission of excitation from autonomic nerve to smooth muscle. *J. Physiol. (Lond.).* **155**, 115–133.

BURNSTOCK, G. and HOLMAN, M. E. (1962a) Spontaneous potentials at sympathetic nerve endings in smooth muscle. *J. Physiol. (Lond.).* **160**, 446–460.

BURNSTOCK, G. and HOLMAN, M. E. (1962b) Effect of denervation and of reserpine treatment on transmission at sympathetic nerve endings. *J. Physiol. (Lond.).* **160**, 461–469.

BURNSTOCK, G. and HOLMAN, M. E. (1963) Smooth muscle: autonomic nerve transmission. *Ann. Rev. Physiol.* **25**, 61–90.

BURNSTOCK, G. and HOLMAN, M. E. (1964) An electrophysiological investigation of the actions of some autonomic blocking drugs on transmission in the guinea-pig vas deferens. *Brit. J. Pharmacol.* **23**, 600–612.

BURNSTOCK, G. and HOLMAN, M. E. (1966) Junction potentials at adrenergic synapses. *Pharmacol. Rev.* **18**, 481–493.

BURNSTOCK, G., HOLMAN, M. E. and KURIYAMA, H. (1964) Facilitation of transmission from autonomic nerve to smooth muscle of guinea-pig vas deferens. *J. Physiol. (Lond.).* **172**, 31–49.

BURNSTOCK, G., HOLMAN, M. E. and PROSSER, C. L. (1963) Electrophysiology of smooth muscle. *Physiol. Rev.* **43**, 482–527.

BURNSTOCK, G. and IWAYAMA, T. (1971) Fine structural identification of autonomic nerves and their relation to smooth muscle. *Progr. Brain Res.* **34**, 389–404.

BURNSTOCK, G. MCCULLOCH, M., STORY, D. and WRIGHT, M. (1972) Factors affecting the extraneuronal inactivation of noradrenaline in cardiac and smooth muscle. *Brit. J. Pharmacol.* **46**, 243–253.

BURNSTOCK, G., MCLEAN, J. R. and WRIGHT, M. (1971) Noradrenaline uptake by non-innervated smooth muscle. *Brit. J. Pharmacol.* **43**, 180–189.

BURNSTOCK, G. and ROBINSON, P. M. (1967) Localization of catecholamines and acetylcholinesterase in autonomic nerves. *Circulation Res.* Vol. 21, Suppl. III, 43–55.

BUTSON, A. R. C. (1950/51) Regeneration of the cervical sympathetic. *Brit. J. Surg.* **38**, 223–239.

CAIRNE, A. B., KOSTERLITZ, H. W. and TAYLOR, D. W. (1961) Effect of morphine on some sympathetically innervated effectors. *J. Pharmacol. exp. Ther.* **17**, 539–551.

CAJAL, S. R. (1905) Las células del gran sympático del hombre adulto. *Trab. Lab. Invest. biol. Univ. Madr.* **4**, cuadernos 1 & 2.

CAJAL, S. R. (1928) 'Degeneration and Regeneration of the Nervous System.' Vol. I. Hafner Publ. Co., London & New York.

CALLINGHAM, B. A. and BURGHEN, A. S. V. (1966) The uptake of isoprenaline and noradrenaline by the perfused rat heart. *Mol. Pharmacol.* **2**, 37–42.

CAMPBELL, G. (1970) Autonomic nervous supply to effector tissues. In 'Smooth Muscle'. E. Bülbring, A. Brading, A. Jones and T. Tomita, eds. Edward Arnold Publ. Co. London, pp. 451–495.

CAMPBELL, G. R., UEHARA, Y., MARK, G. and BURNSTOCK, G. (1971) Fine structure of smooth muscle cells grown in tissue culture. *J. Cell Biol.* **49**, 21–34.

CANNON, W. B. (1914) The interrelation of emotions as suggested by recent physiological researches. *Amer. J. Psychol.* **25**, 256–282.

CANNON, W. B. (1929) Bodily changes in pain, hunger, fear and rage. Ch. J. Branford Company, Boston.

CANNON, W. B. and BACQ, Z. M. (1931) Studies on the conditions of activity in endocrine organs. XXVI. A hormone produced by sympathetic action on smooth muscle. *Amer. J. Physiol.* **96**, 392–412.

CANNON, W. B., NEWTON, H. F., BRIGHT, E. M., MENKIN, V. and MOORE, R. M. (1929) Some aspects of the physiology of animals surviving complete exclusion of sympathetic nerve impulses. *Amer. J. Physiol.* **89**, 84–107.

CANNON, W. B. and ROSENBLUETH, A. (1933) Studies on conditions of activity in endocrine organs. XXIX. Sympathin E and sympathin I. *Amer. J. Physiol.* **104**, 557–574.

CANNON, W. B. and ROSENBLUETH, A. (1937) 'Autonomic Neuroeffector Systems.' Macmillan, New York.

CANNON, W. B. and ROSENBLUETH, A. (1949) 'The Supersensitivity of Denervated Structures. A Law of Denervation.' Macmillan Company, New York.

CANNON, W. B. and URIDIL, J. E. (1921) Studies on the conditions of activity in endocrine glands. VIII. Some effects on the denervated heart of stimulating the nerves of the liver. *Amer. J. Physiol. (Lond.).* **58**, 353–364.

CARAMIA, F., ANGELETTI, P. U., LEVI-MONTALCINI, R. and CARATELLI, L. (1972) Mitochondrial lesions of developing sympathetic neurons induced by bretylium tosylate. *Brain Res.* **40**, 237–246.

CARLSSON, A. (1966a) Pharmacological depletion of catecholamine stores. *Pharmacol. Rev.* **18**, 541–549.
CARLSSON, A. (1966b) Drugs which block the storage of 5-hydroxytryptamine and related amines. *In* 'Handbuch der experimentellen Pharmakologie.' Vol. 19, pp. 529–592. V. Erspamer, ed. Springer-Verlag, Berlin, Heidelberg, Göttingen.
CARLSSON, A., FOLKOW, B. and HÄGGENDAL, J. (1964) Some factors influencing the release of noradrenaline into the blood following sympathetic stimulation. *Life Sci.* **3**, 1335–1341.
CARLSSON, A., FUXE, K., HAMBERGER, B. and LINDQVIST, M. (1966) Biochemical and histochemical studies on the effects of imipramine-like drugs and (+) amphetamine on central and peripheral catecholamine neurons. *Acta physiol. scand.* **67**, 481–497.
CARLSSON, A. and LINDQVIST, M. (1962) *In vivo* decarboxylation of α-methyl dopa and α-methyl metatyrosine. *Acta Physiol. scand.* **54**, 87–94.
CARLSSON, A., ROSENGREN, E., BERTLER, A. and NILSSON, J. (1957) Effect of reserpine on the metabolism of catecholamines. *In* 'Psychotropic Drugs'. S. Garattini and V. Ghetti, eds. Elsevier Publ. Co., Amsterdam. pp. 363–372.
CASS, R. and SPRIGGS, T. L. B. (1961) Tissue amine levels and sympathetic blockade after guanethidine and bretylium. *Brit. J. Pharmacol.* **17**, 442–450.
CECCARELLI, B., CLEMENTI, F. and MANTEGAZZA, P. (1972) Adrenergic re-innervation of smooth muscle of nictitating membrane by preganglionic sympathetic fibres. *J. Physiol. (Lond.).* **220**, 211–228.
CEGRELL, L. (1970) Monoaminergic mechanisms in the pancreatic α-cells. *In* 'The Structure and Metabolism of the Pancreatic Islets'. S. Falkmer, B. Melman and T-B. Täljedal, eds. Pergamon Press, New York.
CELANDER, O. (1954) The range of control exercised by the sympathico-adrenal system. *Acta. physiol. scand.* **32**, Suppl. 116, 1–132.
CHALAZONITIS, A. and GONELLA, J. (1971) Activité électrique spontanée des neurones du ganglion stellaire du chat. Étude par micro-électrode extracellulaire. *J. Physiol. (Paris).* **63**, 599–609.
CHAMLEY, J. H., CAMPBELL, G. R. and BURNSTOCK, G. (1973) An analysis of the interactions between sympathetic nerve fibres and smooth muscle cells in tissue culture. *Dev. Biol.* **33**, 344–361.
CHAMLEY, J., GOLLER, I. and BURNSTOCK, G. (1973) Selective growth of sympathetic nerve fibers to explants of normally densely innervated autonomic effector organs in tissue culture. *Dev. Biol.* **31**, 362–379.
CHAMLEY, J. H., MARK, G. E., CAMPBELL, G. R. and BURNSTOCK, G. (1972) Sympathetic ganglia in culture. I. Neurons. *Z. Zellforsch. Mikrosk. Anat.* **135**, 287–314.
CHAMLEY, J. H., MARK, G. E. and BURNSTOCK, G. (1972) Sympathetic ganglia in culture. II. Accessory cells. *Z. Zellforsch. Mikrosk. Anat.* **135**, 315–327.

CHANG, C. C., COSTA, E. and BRODIE, B. B. (1965) Interaction of guanethidine with adrenergic neurons. *J. Pharmacol. exp. Ther.* **147**, 303-312.

CHEAH, T. B., GEFFEN, L. B., JARROTT, B. and OSTBERG, A. (1971) Action of 6-hydroxydopamine on lamb sympathetic ganglia, vas deferens and adrenal medulla: a combined histochemical, ultrastructural and biochemical comparison with the effects of reserpine. *Brit. J. Pharmacol.* **42**, 543-557.

CHIDSEY, C. A. and HARRISON, D. C. (1963) Studies on the distribution of exogenous norepinephrine in the sympathetic neurotransmitter store. *J. Pharmacol. exp. Ther.* **140**, 217-223.

CHIEN, S. (1967) Role of the sympathetic nervous system in hemorrhage. *Physiol. Rev.* **47**, 214-288.

CHRIST, D. D. and NISHI, S. (1971) Site of adrenaline blockade in the superior cervical ganglion of the rabbit. *J. Physiol. (Lond.).* **213**, 107-117.

CHUBB, I. W., DE POTTER, W. P. and DE SCHAEPDRYVER, A. F. (1972) Tyramine does not release noradrenaline from splenic nerves by exocytosis. *Naunyn Schmied Arch. exp Path. Pharmak.* **274**, 281-286.

CLARK, D. W. J., LAVERTY, R. and PHELAN, E. L. (1972) Long lasting peripheral and central effects of 6-OHDA in rats. *Brit. J. Pharmacol.* **44**, 233-243.

CLARKE, D. E., JONES, C. J. and LINLEY, P. A. (1969) Histochemical fluorescence studies on noradrenaline accumulation by Uptake$_2$ in the isolated rat heart. *Brit. J. Pharmacol.* **37**, 1-9.

CLOUTIER, G. and WIENER, N. (1973) Further studies on the increased synthesis of norepinephrine during nerve stimulation of guinea pig vas deferens preparation: effect of tyrosine and 6-7 dimethyl tetrahydropteridin. *J. Pharmacol. exp. Ther.* **186**, 75-85.

COATS, D. A. and EMMELIN, N. (1962) The short term effects of sympathetic ganglionectomy on the cats salivary secretion. *J. Physiol. (Lond.).* **162**, 282-288.

COHEN, S. (1960) Purification of a nerve-growth promoting protein from the mouse salivary gland and its neuro-cytotoxic antiserum. *Proc. Nat. Acad. Sci. U.S.A.* **46**, 302-311.

COHEN, M. I. and GOOTMAN, P. M. (1970) Periodicities in efferent discharge of splanchnic nerve of the cat. *Amer. J. Physiol.* **218**, 1092-1101.

COLLINS, G. G. S. and WEST, G. B. (1968a) The release of ^{3}H-dopamine from the isolated rabbit ileum. *Brit. J. Pharmacol.* **34**, 514-522.

COLLINS, G. G. S. and WEST, G. B. (1968b) Some pharmacological actions of diethyldithiocarbamate on rabbit and rat ileum. *Brit. J. Pharmacol.* **32**, 402-409.

COMLINE, R. S. and SILVER, M. (1966) Development of activity in the adrenal medulla of the foetus and new-born animal. *Brit. med. Bull.* **22**, 16-20.

CONSOLO, S., GARATTINI, S., LADINSKI, H. and THOENEN, H. (1972) Effect of chemical sympathectomy on the content of acetylcholine, choline and choline acetyltransferase activity in the rat spleen and iris. *J. Physiol. (Lond.).* **220**, 639–646.

COOPER, T. (1966) Surgical sympathectomy and adrenergic function. *Pharmacol. Rev.* **18**, 611–618.

COOTE, J. H., DOWNMAN, C. B. B. and WEBER, W. V. (1969) Reflex discharges into thoracic white rami elicited by somatic and visceral afferent excitation. *J. Physiol. (Lond.).* **202**, 147–159.

COQUIL, J. F., GORIDIS, C., MACK, G. and NEFF, N. H. (1973) Monoamine oxidase in rat arteries: evidence for different forms and selective localization. *Brit. J. Pharmacol.* **48**, 590–599.

COREY, E. L. (1935) Observations on cardiac activity in the fetal rat. *J. exp. Zool.* **72**, 127–145.

CORRODI, H., FUXE, K., HAMBERGER, B. and LJUNGDAHL, A. (1970) Studies on central and peripheral noradrenaline neurons using a new dopamine-β-hydroxylase inhibitor. *Eur. J. Pharmacol.* **12**, 145–155.

CORRODI, H. and MALMFORS, T. (1966) The effect of nerve activity on the depletion of the adrenergic transmitter by inhibitions of noradrenaline synthesis. *Acta physiol. scand.* **67**, 352–357.

COSTA, E., NEFF, N. H. and NAGI, S. H. (1969) Regulation of metaraminol efflux from rat heart and salivary gland. *Brit. J. Pharmacol.* **36**, 153–160.

COSTA, M. and ERÄNKÖ, O. (1973) Histochemical correlates of cold-induced trans-synaptic induction in the rat superior cervical ganglion. *Histochem. J.* **6**, 35–53.

COSTA, M., ERÄNKÖ, L. and ERÄNKÖ, O. (1974a) Effect of hydrocortisone on the extra-adrenal intensely fluorescent chromaffin and non-chromaffin cells in stretch preparations of the newborn rat. *Z. Zellforsch. Mikrosk. Anat.* (In press.)

COSTA, M., ERÄNKÖ, O. and ERÄNKÖ, L. (1974b) Hydrocortisone–induced increase in the histochemically demonstrable catecholamine content of sympathetic neurons of the newborn rat. *Brain Res.* **67**, 457–466.

COSTA, M. and FILOGAMO, G. (1970) Effetti della colchicina nelle fibre adrenergiche dei plessi nervosi intestinali. *Boll. soc. ital. Biol. sper.* **46**, 865–867.

COSTA, M. and FURNESS, J. B. (1971) The innervation of the proximal colon of the guinea-pig. *Proc. Aust. Phys. Pharm. Soc.* **2**, 29–30.

COSTA, M. and FURNESS, J. B. (1973) Observations on the anatomy and amine histochemistry of the nerves and ganglia which supply the pelvic viscera and on the associated chromaffin tissue in the guinea-pig. *Z. Anat. EntwGesch.* **140**, 85–108.

COSTA, M. and GABELLA, G. (1971) Adrenergic innervation of the alimentary canal. *Z. Zellforsch. Mikrosk. Anat.* **122**, 357–377.

COTÉ, M. G., PALAIC, D. and PANISSET, J. C. (1970) Change in the number of vesicles and the size of sympathetic nerve terminals following nerve stimulation. *Rev. canad. Biol.* **29**, 111–114.

COTTEN, M. DE V. (1972) Regulation of catecholamine metabolism in the sympathetic nervous tissue. *Pharmacol. Rev.* **24**, 161–434.

COUPLAND, R. E. (1953) On the morphology and adrenaline-noradrenaline content of chromaffin tissue. *J. Endocrin.* **9**, 194–203.

COUPLAND, R. E. (1965) 'The Natural History of the Chromaffin Cell.' Longmans Green & Co. Ltd., London.

COUPLAND, R. E. (1972) The Chromaffin System. *In* 'Catecholamines'. H. Blaschko and E. Muscholl, eds. Springer-Verlag, Berlin, Heidelberg, New York, pp. 16–39.

COUPLAND, R. E. and MACDOUGALL, J. D. (1966) Adrenaline formation in noradrenaline-storing chromaffin cells *in vitro* induced by corticosterone. *J. Endocrin.* **36**, 317–324.

COUPLAND, R. E. and WEACKLEY, B. S. (1968) Developing chromaffin tissue in the rabbit: an electron microscopic study. *J. Anat.* **102**, 425–455.

COYLE, J. T. and WOOTEN, G. F. (1972) Rapid axonal transport of tyrosine hydroxylase and dopamine-β-hydroxylase. *Brain Res.* **44**, 701–704.

CRAIN, S. M. (1968) Development of functional neuromuscular connections between separate explants of fetal mammalian tissues after maturation in culture. *Anat. Rec.* **160**, 466.

CRAIN, S. M., BENITEZ, H. and VATTER, A. E. (1964) Some cytological effects of salivary nerve-growth factor on tissue cultures of peripheral ganglia. *Ann. N.Y. Acad. Sci.* **118**, 206–231.

CRAIN, S. M. and WIEGAND, R. G. (1961) Catecholamine levels of mouse sympathetic ganglia following hypertrophy produced by salivary nerve-growth factor. *Proc. Soc. exp. Biol. Med.* **107**, 663–665.

CRIPPS, H. and DEARNALEY, D. P. (1970) The effect of cocaine on the noradrenaline overflow from the isolated blood–perfused cat spleen. *J. Physiol. (London).* **209**, 37–38P.

CRIPPS, H. and DEARNALEY, D. P. (1971) Evidence suggesting uptake of noradrenaline at adrenergic receptors in the isolated blood-perfused cat spleen. *J. Physiol. (Lond.).* **216**, 55–56P.

CRIPPS, H., DEARNALEY, D. P. and HOWE, P. R. C. (1972) Enhancement of noradrenaline depletion in the cat spleen by phenoxybenzamine and phentolamine. *Brit. J. Pharmacol.* **46**, 358–361.

CROUT, R. J. (1964) The uptake and release of H^3-norepinephrine by the guinea-pig heart *in vivo*. *Naunyn Schmied Arch. exp. Path. Pharmak.* **248**, 85–98.

CROUT, J. R., ALPERS, H. S., TATUM, E. L. and SHORE, P. A. (1964) Release of metaraminol (Aramine) from the heart by sympathetic nerve stimulation. *Science.* **145**, 828–829.

CROWCROFT, P. J., HOLMAN, M. E. and SZURSZEWSKI, J. H. (1971) Excitatory

input from the distal colon to the inferior mesenteric ganglion in the guinea-pig. *J. Physiol. (Lond.).* **219**, 443-461.
CROWCROFT, P. J. and SZURSZEWSKI, J. H. (1971) A study of the inferior mesenteric and pelvic ganglia of guinea-pigs with intracellular electrodes. *J. Physiol. (Lond.).* **219**, 421-441.
DAHLSTRÖM, A. (1965) Observations on the accumulation of noradrenaline in the proximal and distal parts of peripheral adrenergic nerves after compression. *J. Anat.* **99**, 677-689.
DAHLSTRÖM, A. (1966) 'The Intraneuronal Distribution of Noradrenaline and the Transport and Life-span of amine storage Granules in the Sympathetic Adrenergic Neuron.' Pub. Ivor Haeggströms Tryckeri Ab, Stockholm.
DAHLSTRÖM, A. (1968) Effect of colchicine on transport of amine storage granules in sympathetic nerves of rat. *Eur. J. Pharmacol.* **5**, 111-113.
DAHLSTRÖM, A. (1970) Effect of mitosis inhibitors on the transport of amine storage granules in monoaminergic neurons in the rat. *Acta physiol. scand. Suppl.* **357**, 6.
DAHLSTRÖM, A. (1971) Axoplasmic transport (with particular respect to adrenergic neurons). *Phil. Trans. Roy. Soc. Lond. Ser. B.* **261**, 325-358.
DAHLSTRÖM, A. and FUXE, K. (1964) A method for the demonstration of adrenergic nerve fibres in peripheral nerves. *Z. Zellforsch. Mikrosk. Anat.* **62**, 602-607.
DAHLSTRÖM, A. and HÄGGENDAL, J. (1966a) Some quantitative studies of the noradrenaline content in the cell bodies and terminals of a sympathetic adrenergic neuron system. *Acta physiol. scand.* **67**, 271-277.
DAHLSTRÖM, A. and HÄGGENDAL, J. (1966b) Studies on the transport and life span of amine storage granules in a peripheral adrenergic neuron system. *Acta physiol. scand.* **67**, 278-288.
DAHLSTRÖM, A. and HÄGGENDAL, J. (1969) Recovery of noradrenaline in adrenergic axons of rat sciatic nerves after reserpine treatment. *J. Pharmac. Pharmacol.* **21**, 633-638.
DAHLSTRÖM, A., HÄGGENDAL, J. and HÖKFELT, T. (1966) The noradrenaline content of the varicosities of sympathetic adrenergic nerve terminals in the rat. *Acta physiol. scand.* **67**, 289-294.
DAHLSTRÖM, A. and JONASON, J. (1968) Dopa-decarboxylase activity in sciatic nerves of the rat after constriction. *Eur. J. Parmacol.* **4**, 377-383.
DAHLSTRÖM, A., JONASON, J. and NORBERG, K-A. (1969) Monoamine oxidase activity in rat sciatic nerves after constriction. *Eur. J. Pharmacol.* **6**, 248-254.
DAIRMAN, W., GORDON, R., SPECTOR, S., SJOERDSMA, A. and UDENFRIEND, S. (1968) Increased synthesis of catecholamines in the intact rat following administration of α-adrenergic blocking agents. *Mol. Pharmacol.* **4**, 457-464.
DAIRMAN, W. and UDENFRIEND, S. (1971) Decrease in adrenal tyrosine

hydroxylase and increase in norepinephrine synthesis in rats given L-dopa. *Science.* 171, 1022-1024.

DALE, H. H. (1906) On some physiological actions of ergot. *J. Physiol. (Lond.).* 34, 163-206.

DALE, H. H. (1933) Nomenclature of fibres in the autonomic system and their effects. *J. Physiol. (Lond.).* 80, 10-11P.

DALE, H. H. (1937) Acetylcholine as a chemical transmitter of the effects of nerve impulses. I. History of ideas and evidence. Peripheral autonomic actions. Functional nomenclature of nerve fibres. *J. Mt Sinai Hosp.* 4, 401-415.

DA PRADA, M., BERNEIS, K. H. and PLETSCHER, A. (1971) Storage of catecholamines in adrenal medullary granules: formation of aggregates with nucleotides. *Life Sci.* 10, 639-646.

DAVEY, M. J., FARMER, J. B. and REINERT, H. (1963) The effect of nialamide on adrenergic functions. *Brit. J. Pharmacol.* 20, 121-134.

DAVIES, B. N. (1963) Effect of prolonged activity on the release of the chemical transmitter at post-ganglionic sympathetic nerve endings. *J. Physiol. (Lond.).* 167, 52P.

DAVIES, B. N. and WITHRINGTON, P. G. (1968) The release of noradrenaline by the sympathetic post-ganglionic nerves to the spleen of the cat in response to low frequency stimulation. *Arch. int. Pharmacodyn, Ther.* 171, 185-196.

DAWBORN, J. K., DOYLE, A. E., EBRINGER, A., HOWQUA, J., JERUMS, G., JOHNSTON, C. I., MASHFORD, M. L. and PARKIN, J. D. (1969) Persistent postural hypotension due to guanacline. *Pharmacol. Clin.* 2, 1.

DAY, M. D. and OWEN, D. A. A. (1968) The interaction between angiotensin and sympathetic vasoconstriction in the isolated artery of the rabbit ear. *Brit. J. Pharmacol.* 34, 499-507.

DAY, M. D. and RAND, M. J. (1961) Effect of guanethidine in revealing cholinergic sympathetic fibres. *Brit. J. Pharmacol.* 17, 245-260.

DAY, M. D. and RAND, M. J. (1963) Evidence for a competitive antagonism of guanethidine by dexamphetamine. *Brit. J. Pharmacol.* 20, 17-28.

DAY, M. D. and RAND, M. J. (1964) Some observations on the pharmacology of α-methyldopa. *Brit. J. Pharmacol.* 22, 72-86.

DAY, M. D. and WARREN, P. R. (1968) A pharmacological analysis of the responses to transmural stimulation in isolated intestinal preparations. *Brit. J. Pharmacol.* 32, 227-240.

DEARNALEY, D. P. and GEFFEN, L. B. (1966) Effect of nerve stimulation on the noradrenaline content of the spleen. *Proc. Roy. Soc. Lond. Ser. B.* 166, 303-315.

DE CASTRO, F. (1932) Sympathetic ganglia, normal and pathological. *In* 'Cytology and Cellular Pathology of the Nervous System'. Vol. I, Section VII.' W. Penfield, ed. Paul B. Hoeber Inc., New York, pp. 317-379.

DE CHAMPLAIN, J. (1971) Degeneration and regrowth of adrenergic nerve fibres in the rat peripheral tissues after 6-hydroxydopamine. *Canad. J. Physiol. Pharmacol.* **48**, 345-355.

DE CHAMPLAIN, J., MALMFORS, T., OLSON, L. and SACHS, CH. (1970) Ontogenesis of peripheral adrenergic neurons in the rat: pre- and postnatal observations. *Acta physiol. scand.* **80**, 276-288.

DE GROAT, W. C. and SAUM, W. R. (1971a) Adrenergic inhibition in mammalian parasympathetic ganglia. *Nature (Lond.).* **231**, 188-189.

DE GROAT, W. C. and SAUM, W. R. (1971b) Sympathetic-inhibition of the urinary bladder and of pelvic parasympathetic ganglionic transmission in the cat. *Fed. Proc.* **30**, 656.

DE GROAT, W. C. and SAUM, W. R. (1972) Sympathetic inhibition of the urinary bladder and of pelvic ganglionic transmission in the cat. *J. Physiol. (Lond.).* **220**, 297-314.

DE JONGH, D. K. and VAN PROOSDIJ-HARTZEMA, E. G. (1955) Investigations into experimental hypertension. IV. The acute effects of Rauwolfia alkaloids on the blood pressure of rats with experimental hypertension. *Acta physiol. pharm. néerl.* **4**, 175-179.

DE JONGH, D. K. and VAN PROOSDIJ-HARTZEMA, E. G. (1958) Augmentation of blood pressure in rabbits by reserpine. *Acta physiol. pharm. néerl.* **7**, 364-365.

DE LA LANDE, I. S., FREWIN, D. and WATERSON, J. G. (1967) The influence of sympathetic innervation on vascular sensitivity to noradrenaline. *Brit. J. Pharmacol.* **31**, 82-93.

DE LA LANDE, 1. S., JELLETT, L. B., LAZNER, M. A., PARKER, D. A. S and WATERSON, J. G. (1974) Histochemical analysis of the diffusion of noradrenaline across the artery wall. *Aust. J. Exp. Biol. Med. Sci.* **52**, 193-200.

DE LA LANDE, I. S. and JOHNSON, S. M. (1972) Evidence for inactivation of noradrenaline by extraneuronal monoamine oxidase in the rabbit ear artery. *Aust. J. exp. Biol. med. Sci.* **50**, 119-121.

DE LA LANDE, I. S., PATON, W. D. and WAND, B. (1968) Output of sympathetic transmitter in the isolated rabbit ear. *Aust. J. Exp. Biol. Med. Sci.* **46**, 727-738.

DEL CASTILLO, J. and KATZ, B. (1954) Quantal components of the end-plate potential. *J. Physiol. (Lond.).* **124**, 560-573.

DELIUS, W., WALLIN, G. and HAGBARTH, K-E. (1973) Role of sympathetic nerve impulses in regulation of peripheral circulation. *J. Clin. Lab. Inv.* **31**, Suppl. 128, 47-50.

DE POTTER, W. P. (1971) Noradrenaline storage particles in splenic nerve. *Phil. Trans. Roy. Soc. Lond. Ser. B.* **261**, 313-317.

DE POTTER, W. P. (1973) Release of amines from sympathetic nerves. *In* 'Frontiers in Catecholamine Research'. E. Usdin and S. Snyder, eds. Pergamon Press, Oxford.

DE POTTER, W. P. and CHUBB, I. W. (1971) The turnover rate of noradrenergic vesicles. *Biochem. J.* **125**, 375–376.

DE POTTER, W. P., CHUBB, I. W. and DE SCHAEPDRYVER, A. F. (1972) Pharmacological aspects of peripheral noradrenergic transmission. *Arch. int. Pharmacodyn. Thér.* **196**, 258–287.

DE POTTER, W. P., CHUBB, I. W., PUT, A. and DE SCHAEPDRYVER, A. F. (1971) Facilitation of the release of noradrenaline and dopamine-β-hydroxylase at low stimulation frequencies by α-blocking agents. *Arch. int. Pharmacodyn. Thér.* **193**, 191–197.

DE POTTER, W. P., DE SCHAEPDRYVER, A. F., MOERMAN, E. J. and SMITH, A. D. (1969) Evidence for the release of vesicle proteins together with noradrenaline upon stimulation of the splenic nerve. *J. Physiol. (Lond.).* **204**, 102–104P.

DERMIETZEL, R. (1971) Electronenmikroskopische untersuchung über die innervation der pars pylorica des Mäusemagens. *Z. Mikrosk. anat. Forsch. 84,* 225–256.

DE ROBERTIS, E. (1964) 'Histophysiology of Synapses and Neurosecretion.' Pergamon Press, Oxford.

DE ROBERTIS, E. and PELLEGRINO DE IRALDI, A. (1961). A plurivesicular component in adrenergic nerve endings. *Anat. Rec.* **139**, 299.

DE ROBERTIS, E., PELLEGRINO DE IRALDI, A., RODRIGUEZ, G. and GOMEZ, C. J. (1961) On the isolation of nerve endings and synaptic vesicles. *J. biophys. biochem. Cytol.* **9**, 229–235.

DERRY, D. M. and DANIEL, H. (1970) Sympathetic nerve development in the brown adipose tissue of the rat. *Canad. J. physiol. pharm.* **48**, 160–168.

DEVINE, C. E. (1967) Electron microscope autoradiography of rat arteriolar axons after noradrenaline infusion. *Proc. Univ. Otago med. Sch.* **45**, 7–8.

DEWEY, M. M. and BARR, L. (1962) Intercellular connection between smooth muscle cells: the nexus. *Science.* **137**, 670–672.

DIXON, J. S. (1970) Some fine structural changes in sympathetic neurons following axon section. *Acta anat.* **76**, 473–487.

DOMINO, E. F. and RECH, R. H. (1957) Observations on the initial hypertensive response to reserpine. *J. Pharmacol. exp. Ther.* **121**, 171–182.

DORR, L. D. and BRODY, M. J. (1966) Preliminary observations on the role of the sympathetic nervous system in development and maintenance of experimental renal hypertension. *Proc. Soc. exp. Biol. med.* **123**, 155–158.

DOUGLAS, W. W. (1968) Stimulus-secretion coupling: the concept and clues from chromaffin and other cells. *Brit. J. Pharmacol.* **34**, 451–474.

DOWNING, S. E. (1960) Baroreceptor reflexes in new-born rabbits. *J. Physiol. (Lond.).* **150**, 201–213.

DRASKÓCZY, P. R. and LYMAN, C. P. (1967) Turnover of catecholamines in active and hibernating ground squirrels. *J. Pharmacol. exp. Ther.* **155**, 101–111.

DRASKÓCZY, P. R. and TRENDELENBURG, U. (1970) Intraneuronal and extraneuronal accumulation of sympathomimetic amines in the isolated nictitating membrane of the cat. *J. Pharmac. exp. Ther.* **174**, 290–306.

DUBOCOVICH, M. and LANGER, S. Z. (1973) Effects of flow-stop on the metabolism of NA released by nerve stimulation in the perfused spleen. *Naunyn Schmied. Arch. exp Path. Pharmak.* **278**, 179–194.

DUNANT, Y. (1967) Organisation topographique et fonctionnelle du ganglion cervical supérieur chez le rat. *J. Physiol. (Paris).* **59**, 17–38.

DUNCAN, D. and YATES, R. (1967) Ultrastructure of the carotid body of the cat as revealed by various fixatives and the use of reserpine. *Anat. Rec.* **157**, 667–682.

EBBESSON, S. O. E. (1968) Quantitative studies of superior cervical sympathetic ganglia in a variety of primates including man. I. The ratio of preganglionic fibers to ganglionic neurons. *J. Morph.* **124**, 117–132.

ECCLES, J. C. (1935) Action potential of the superior cervical ganglia. *J. Physiol. (Lond.).* **85**, 179–206.

ECCLES, J. C. (1944) The nature of synaptic transmission in a sympathetic ganglion. *J. Physiol. (Lond.).* **103**, 27–54.

ECCLES, J. C. (1964) 'The Physiology of Synapses.' Springer-Verlag, Berlin.

ECCLES, J. C. and MAGLADERY, J. W. (1937) The excitation and response of smooth muscle. *J. Physiol. (Lond.).* **90**, 31–67.

ECCLES, R. M. (1955) Intracellular potentials recorded from a mammalian sympathetic ganglion. *J. Physiol. (Lond.).* **130**, 572–584.

ECCLES, R. M. and LIBET, B. (1961) Origin and blockade of the synaptic responses of curarized sympathetic ganglia. *J. Physiol. (Lond.).* **157**, 484–503.

EDVINSSON, L., NIELSON, K. C., OWMAN, CH. and SPORRONG, B. (1972) Cholinergic mechanisms in pial vessels. Histochemistry, electron microscopy and pharmacology. *Z. Zellforsch. Mikrosk. Anat.* **134**, 311–325.

EDVINSSON, L., OWMAN, CH., ROSENGREN, E. and WEST, K. A. (1972) Concentration of noradrenaline in pial vessels, choroid plexus, and iris during two weeks after sympathetic ganglionectomy or decentralization. *Acta physiol. scand.* **85**, 201–206.

EDWALL, L. and SCOTT, D. JR. (1971) Influence of changes in micro-circulation on the excitability of the sensory unit in the tooth of the cat. *Acta physiol. scand.* **82**, 555–556.

EDWARDS, D. C., FENTON, E. L., KAKARI, S., LARGE, B. J., PAPADAKI, L. and ZAIMIS, E. (1966) Effects of nerve growth factor in newborn mice, rats and kittens. *J. Physiol. (Lond.).* **186**, 10–12P.

EHINGER, B. and FALCK, B. (1970) Uptake of some catecholamines and their precursors into neurons of the cat ciliary ganglion. *Acta physiol. scand.* **78**, 132–141.

EHINGER, B., FALCK, B., PERSSON, H., ROSENGREN, A. M. and SPORRONG, B. (1971) Acetylcholine in adrenergic terminals in the cat iris. *J. Physiol. (Lond.).* **209**, 557–565.

EHINGER, B., FALCK, B., PERSSON, H. and SPORRONG, B. (1968) Adrenergic and cholinesterase-containing neurons of the heart. *Histochemie.* **16,** 197–205.

EHINGER, B., FALCK, B. and SPORRONG, B. (1970) Possible axo-axonal synapses between adrenergic and cholinergic nerve terminals. *Z. Zellforsch. Mikrosk. Anat.* **107,** 508–521.

EHLERS, P. (1951) Über Altersveränderungen an Grenzstrang-Ganglien vom Meerschweinchen. *Anat. Anz.* **98,** 24–34.

EISENFELD, A. J., AXELROD, J. and KRAKOFF, L. (1967) Inhibition of the extraneuronal accumulation and metabolism of norepinephrine by adrenergic blocking agents. *J. Pharmacol. exp. Ther.* **156,** 107–113.

ELDRED, E., SCHNITZLEIN, N. and BUCHWALD, J. (1960) Response of muscle spindles to stimulation of the sympathetic trunk. *Exp. Neurol.* **2,** 13–25.

ELFVIN, L-G. (1958) The ultrastructure of unmyelinated fibers in the splenic nerve of the cat. *J. Ultrastruct. Res.* **1,** 428–454.

ELFVIN, L-G. (1961) Electron-microscopic investigation of filament structures in unmyelinated fibers of cat splenic nerve. *J. Ultrastruct. Res.* **5,** 51–64.

ELFVIN, L-G. (1963) The ultrastructure of the superior cervical sympathetic ganglion of the cat. I. The structure of the ganglion cell processes as studied by serial sections. *J. Ultrastruct. Res.* **8,** 403–440.

ELFVIN, L-G. (1967) The development of the secretory granules in the rat adrenal medulla. *J. Ultrastruct. Res.* **17,** 45–62.

ELFVIN, L-G. (1968) A new granule-containing nerve cell in the inferior mesenteric ganglion of the rabbit. *J. Ultrastruct. Res.* **22,** 37–44.

ELFVIN, L-G. (1971a) Ultrastructural studies on the synaptology of the inferior mesenteric ganglion of the cat. I. Observations on the cell surface of the postganglionic perikarya. *J. Ultrastruct. Res.* **37,** 411–425.

ELFVIN, L-G. (1971b) Ultrastructural studies on the synaptology of the inferior mesenteric ganglion of the cat. II. Specialized serial neuronal contacts between preganglionic end fibers. *J. Ultrastruct. Res.* **37,** 426–431.

ELFVIN, L-G. (1971c) Ultrastructural studies on the synaptology of the inferior mesenteric ganglion of the cat. III. The structure and distribution of the axodendritic and dendrodendritic contacts. *J. Ultrastruct. Res.* **37,** 432–448.

ELLIOTT, T. R. (1904) On the action of adrenalin. *J. Physiol. (Lond.).* **31,** XX–XXI.

ELLIOTT, T. R. (1905) The action of adrenalin. *J. Physiol. (Lond.).* **32,** 401–467.

EMMELIN, N. (1968) Degeneration activity after sympathetic denervation of the submaxillary gland and the eye. *Experientia (Basel).* **24,** 44–45.

EMMELIN, N. and TRENDELENBURG, U. (1972) Degeneration activity after parasympathetic or sympathetic denervation. *Ergebn. Physiol.* **66,** 147–211.

ENERO, M. A. and LANGER, S. Z. (1973) Influence of reserpine-induced depletion of NA on the negative feed-back mechanism for transmitter release during nerve stimulation. *Brit. J. Pharmacol.* **49,** 214–225.

ENERO, M. A., LANGER, S. Z., ROTHLIN, R. P. and STEFANO, F. J. E. (1972) The role of the alpha adrenoreceptor in regulating noradrenaline overflow by nerve stimulation. *Brit. J. Pharmacol.* **44**, 672–688.

ENGELMAN, K. and PORTNOY, B. (1970) A sensitive double-isotope derivative assay for norepinephrine and epinephrine. *Circulation Res.* **26**, 53–57.

ERÄNKÖ, O. (1960) Cell types in the adrenal medulla. *In* 'Adrenergic Mechanisms'. CIBA Foundation Symposium. G. E. W. Wolstenholme and M. O'Connor, eds. J. & A. Churchill Ltd., London, pp. 103–110.

ERÄNKÖ, O. (1972a) Light and electron microscopic histochemical evidence of granular and non-granular storage of catecholamines in the sympathetic ganglion of the rat. *Histochem. J.* **4**, 213–224.

ERÄNKÖ, L. (1972b) Histochemical and electron microscopical observations on catecholamines in developing sympathetic ganglion. *Acta Instituti Anatomica Univ. Helsinkiensis. Suppl.* **5**.

ERÄNKÖ, O. and ERÄNKÖ, L. (1971a) Small, intensely fluorescent granule-containing cells in the sympathetic ganglion of the rat. *Progr. Brain Res.* **34**, 39–51.

ERÄNKÖ, O. and ERÄNKÖ, L. (1971b) Effect of guanethidine on nerve cells and small intensely fluorescent cells in sympathetic ganglia of newborn and adult rats. *Acta pharmacol. toxicol.* **30**, 403–416.

ERÄNKÖ, O. and ERÄNKÖ, L. (1971c) Histochemical evidence of chemical sympathectomy by guanethidine in newborn rats. *Histochem. J.* **3**, 451–456.

ERÄNKÖ, O. and ERÄNKÖ, L. (1971d) Loss of histochemically demonstrable catecholamines and acetylcholinesterase from sympathetic nerve fibres of the pineal body of the rat after chemical sympathectomy with 6-hydroxydopamine. *Histochem. J.* **3**, 357–363.

ERÄNKÖ, L. and ERÄNKÖ, O. (1972a) Effect of hydrocortisone on histochemically demonstrable catecholamines in the sympathetic ganglia and extra-adrenal chromaffin tissue of the rat. *Acta physiol. scand.* **84**, 125–133.

ERÄNKÖ, L. and ERÄNKÖ, O. (1972b) Effect of 6-hydroxydopamine on the ganglion cells and the small intensely fluorescent cells in the superior cervical ganglion of the rat. *Acta physiol. scand.* **84**, 115–124.

ERÄNKÖ, O., ERÄNKÖ, L., HILL, C. E. and BURNSTOCK, G. (1972) Hydrocortisone-induced increase in the number of small intensely fluorescent cells and their histochemically demonstrable catecholamine content in cultures of sympathetic ganglia of the rat. *Histochem. J.* **4**, 49–58.

ERÄNKÖ, O. and HÄRKÖNEN, M. (1963) Histochemical demonstration of fluorogenic amines in the cytoplasm of sympathetic ganglion cells of the rat. *Acta physiol. scand.* **58**, 285–286.

ERÄNKÖ, O. and HÄRKÖNEN, M. (1965) Monoamine-containing small cells in the superior cervical ganglion of the rat and an organ composed of them. *Acta physiol. scand.* **63**, 511–512.

ERÄNKÖ, O., HEATH, J. and ERÄNKÖ, L. (1973) Effect of hydrocortisone on the ultrastructure of the small, granule-containing cells in the superior cervical ganglion of the newborn rat. *Experientia (Basel).* **29**, 457–459.

ERÄNKÖ, L., HILL, C., ERÄNKÖ, O. and BURNSTOCK, G. (1972) Lack of toxic effect of guanethidine on nerve cells and small intensely fluorescent cells in cultures of sympathetic ganglia of newborn rats. *Brain Res.* **43**, 501–513.

ERÄNKÖ, O., LEMPINEN, M. and RÄISÄNAN, L. (1966) Adrenaline and noradrenaline in the organ of Zuckerkandl and adrenals of newborn rats treated with hydrocortisone. *Acta physiol. scand.* **66**, 253–254.

ERÄNKÖ, O., RECHARDT, L., ERÄNKÖ, L. and CUNNINGHAM, A. (1970) Light and electron microscopic histochemical observations on cholinesterase-containing sympathetic nerve fibres in the pineal body of the rat. *Histochem. J.* **2**, 479–489.

ERULKAR, S. D. and WOODWARD, J. K. (1968) Intracellular recording from mammalian superior cervical ganglion *in situ. J. Physiol. (Lond.).* **199**, 189–203.

ESTERHUIZEN, A. C., GRAHAM, J. D. P., LEVER, J. D. and SPRIGGS, T. L. B. (1968) Catecholamine and acetylcholinesterase distribution in relation to noradrenaline release. An enzyme histochemical and autoradiographic study on the innervation of the cat nictitating membrane. *Brit. J. Pharmacol.* **32**, 46–56.

EVANS, B., GANNON, B. J., HEATH, J. W. and BURNSTOCK, G. (1972) Long-lasting damage to the internal male genital organs and their adrenergic innervation in rats following chronic treatment with the antihypertensive drug guanethidine. *Fertil Steril.* **23**, 657–667.

EVANS, B., IWAYAMA, T. and BURNSTOCK, G. (1973) Long-lasting supersensitivity of the rat vas deferens to norepinephrine following chronic guanethidine administration. *J. Pharmacol. exp. Ther.* **185**, 60–69.

EVANS, C. A. N. and SAUNDERS, N. R. (1967) The distribution of acetylcholine in normal and in regenerating nerves. *J. Physiol. (Lond.).* **192**, 79–92.

EVANS, D. H. L. and EVANS, E. M. (1964) The membrane relationships of smooth muscles: an electronmicroscope study. *J. Anat.* **98**, 37–44.

FALCK, B. (1962) Observations on the possibilities of the cellular localization of monoamines by a fluorescence method. *Acta physiol. scand.* Suppl. **197**, 1–25.

FALCK, B., MCHEDLISHUILI, G. I. and OWMAN, CH. (1965) Histochemical demonstration of adrenergic nerves in cortex-pia of rabbit. *Acta pharmacol toxicol.* **23**, 133–142.

FALCK, B. and OWMAN, CH. (1968) 5-hydroxytryptamine and related amines in endocrine cell systems. *Advanc. Pharmacol.* **6**, 211–231.

FALCK, B., OWMAN, CH. and ROSENGREN, E. (1966) Changes in the pineal stores of 5-hydroxytryptamine after inhibition of its synthesis or break-down. *Acta physiol. scand.* **67**, 300–305.

FALCK, B., OWMAN, CH. and SJÖSTRAND, N. O. (1965) Peripherally located adrenergic nerves innervating the vas deferens and the seminal vesicle of the guinea-pig. *Experientia (Basel).* **21**, 98–100.

FARNEBO, L-O. and HAMBERGER, B. (1970) Release of norepinephrine from isolated rat iris by field stimulation. *J. Pharmacol. exp. Ther.* **172**, 332–341.

FARNEBO, L-O. and HAMBERGER, B. (1971) Drug-induced changes in the release of ^{3}H-noradrenaline from field stimulated rat iris. *Brit. J. Pharmacol.* **43**, 97–106.

FARNEBO, L-O. and LIDBRINK, P. (1972) Synthesis of noradrenaline in isolated rat iris during field stimulation. *Acta physiol. scand. Suppl.* **371**, 29–36.

FARNEBO, L-O. and MALMFORS, T. (1971) ^{3}H-noradrenaline release and mechanical response in the field stimulated mouse vas deferens. *Acta physiol. scand. Suppl.* **371**, 1–18.

FATHERREE, T. J., ADSON, A. W. and ALLEN, E. V. (1940) The vasoconstrictor action of epinephrine on the digital arterioles of man before and after sympathectomy. *Surgery.* **7**, 75–94.

FERNHOLM, M. (1971) On the development of the sympathetic chain and the adrenal medulla in the mouse. *Z. Anat. EntwGesch.* **133**, 305–317.

FERNHOLM, M. (1972) On the appearance of monoamines in the sympathetic systems and the chromaffin tissue in the mouse embryo. *Z. Anat. EntwGesch.* **135**, 350–361.

FERREIRA, S. H., MONCADA, S. and VANE, J. R. (1973) Some effects of inhibiting endogenous prostaglandin formation on the responses of the cat spleen. *Brit. J. Pharmacol.* **47**, 48–58.

FERREIRA, S. H. and VANE, J. R. (1967) Prostaglandins: their disappearance from and release into the circulation. *Nature (Lond.).* **216**, 868–873.

FERRY, C. B. (1966) Cholinergic link hypothesis in adrenergic neuroeffector transmission. *Physiol. Rev.* **46**, 420–456.

FILLENZ, M. (1971) Fine structure of noradrenaline storage vesicles in nerve terminals of the rat vas deferens. *Phil. Trans. Roy. Soc. Lond. Ser. B.* **261**, 319–323.

FILOGAMO, G. and MARCHISIO, P. C. (1971) Acetylcholine system and neural development. *Neurosciences Research.* **4**, 29–64.

FINCH, L., HAEUSLER, G., KUHN, H. and THOENEN, H. (1973) Rapid recovery of vascular adrenergic nerves in the rat after chemical sympathectomy with 6-hydroxydopamine. *Brit. J. Pharmacol.* **48**, 59–72.

FINCH, L., HAEUSLER, G. and THOENEN, H. (1973) A comparison of the effect of chemical sympathectomy by 6–HDA in newborn and adult rats. *Brit. J. Pharmacol.* **47**, 249–260.

FISCHER, J. E., HORST, W. D. and KOPIN, I. J. (1965) β-Hydroxylated sympathomimetic amines as false neurotransmitters. *Brit. J. Pharmacol.* **24**, 477–484.

FISCHER, J. E. and SNYDER, S. (1965) Disposition of norepinephrine-H^3 in sympathetic ganglia. *J. Pharmacol. exp. Ther.* **150**, 190–195.

FLEMING, W. W. (1963) A comparative study of supersensitivity to norepinephrine and acetylcholine produced by denervation, decentralization and reserpine. *J. Pharmacol. exp. Ther.* **141**, 173-179.

FLEMING, W. W. and TRENDELENBURG, U. (1961) Development of supersensitivity to norepinephrine after pretreatment with reserpine. *J. Pharmacol. exp. Ther.* **133**, 41-51.

FLEMING, W. W., MCPHILLIPS, J. J. and WESTFALL, D. P. (1973) Postjunctional supersensitivity and subsensitivity of excitable tissues to drugs. *Ergebn. Physiol.* **68**, 56-119.

FOLKOW, B. (1952) Impulse frequency in sympathetic vasomotor fibres correlated to the release and elimination of the transmitter. *Acta physiol. scand.* **25**, 49-76.

FOLKOW, B. (1955) Nervous control of the blood vessels. *Physiol. Rev.* **35**, 629-663.

FOLKOW, B. and HÄGGENDAL, J. (1970) Some aspects of the quantal release of the adrenergic transmitter. *In* 'Bayer-Symposium II', Springer-Verlag, Berlin, Heidelberg, New York. 91-97.

FOLKOW, B., HÄGGENDAL, J. and LISANDER, B. (1967) Extent of release and elimination of noradrenaline at peripheral adrenergic nerve terminals. *Acta physiol. scand. Suppl.* **307**, 1-38.

FOLKOW, B. and NEIL, E. (1971) 'Circulation.' Oxford University Press, New York, London, Toronto.

FONNUM, F. (1967) The 'compartmentation' of choline acetyl transferase within the synaptosome. *Biochem. J.* **103**, 262-270.

FOO, J. W., JOWETT, A. and STAFFORD, A. (1968) The effects of some β-adrenoceptor blocking drugs on the uptake and release of noradrenaline by the heart. *Brit. J. Pharmacol.* **34**, 141-147.

FORSSMAN, W. G. (1964) Studien über den Feinbau des Ganglion cervicale superius der Ratte. I. Normale Struktur. *Acta anat. (Basel).* **59**, 106.

FREDHOLM, B. and HEDQVIST, P. (1973) Increased release of noradrenaline from stimulated guinea pig vas deferens after indomethacin treatment. *Acta. physiol. scand.* **87**, 570-573.

FREEMAN, N. E. (1935) The effect of temperature on the rate of blood flow in the normal and in the sympathectomized hand. *Amer. J. Physiol.* **113**, 384-398.

FRIEDMAN, W. F., POOL, P. E., JACOBOWITZ, D., SEAGREN, S. C. and BRAUNWALD, E. (1968) Sympathetic innervation of the developing rabbit heart. *Circulation Res.* **23**, 25-32.

FROLKIS, V. V., BEZRUKOV, V. V., BOGATSKAYA, L. N., VERKHRATSKY, N. S., ZAMOSTIAN, V. P., SHEUTCHUK, V. G. and SHTCHEGOLEVA, I. V. (1970) Catecholamines in the metabolism and functions regulation in aging. *Gerontologia.* **16**, 129-140.

REFERENCES

FURCHGOTT, R. F. (1952) Effect of dibenamine on response of arterial smooth muscle to epinephrine, nor-epinephrine. electrical stimulation and anoxia. *Fed. Proc.* **11**, 217.

FURCHGOTT, R. F. (1972) A classification of adrenoceptors (adrenergic receptors). An evaluation from the stand-point of receptor theory. *In* 'Catecholamines'. H. Blaschko and J. E. Muscholl, eds. Springer-Verlag, Berlin, Heidelberg, New York, pp. 283–335.

FURNESS, J. B. (1969) An electrophysiological study of the innervation of the smooth muscle of the colon. *J. Physiol. (Lond.).* **205**, 549–562.

FURNESS, J. B. (1970a) The excitatory input to a single smooth muscle cell. *Pflüg. Arch. ges. Physiol.* **314**, 1–13.

FURNESS, J. B. (1970b) The effect of external potassium ion concentration on autonomic neuro-muscular transmission. *Pflüg. Arch. ges. Physiol.* **317**, 310–326.

FURNESS, J. B. (1971a) Some actions of 6-hydroxydopamine which affect autonomic neuromuscular transmission. *In* '6-Hydroxydopamine and Catecholamine Neurons'. T. Malmfors and H. Thoenen, eds. North-Holland Publishing Company, Amsterdam–London, pp. 205–214.

FURNESS, J. B. (1971b) The degeneration release of noradrenaline studied *in vitro*. *Proc. Aust. Physiol. Pharmacol. Soc.* **2**, 40–41.

FURNESS, J. B. (1974) Transmission to the longitudinal muscle of the guinea-pig vas deferens: the effect of pretreatment with guanethidine. *Brit. J. Pharmacol.* **50**, 63–68.

FURNESS, J. B. and BURNSTOCK, G. (1969) A comparative study of spike potentials in response to nerve stimulation in the vas deferens of the mouse, rat and guinea-pig. *Comp. Biochem. Physiol.* **31**, 337–345.

FURNESS, J. B. and BURNSTOCK, G. (1975) Role of circulating catecholamines in the gastrointestinal tract. *In* 'Handbook of Physiology. Endocrinology.' H. Blaschko and A. D. Smith, eds. *American Physiol. Soc.*, (Washington). (In press.)

FURNESS, J. B., CAMPBELL, G. R., GILLARD, S. M., MALMFORS, T., COBB, J. L. S. and BURNSTOCK, G. (1970) Cellular studies of sympathetic denervation produced by 6-hydroxydopamine in the vas deferens. *J. Pharmacol. exp. Ther.* **174**, 111–122.

FURNESS, J. B. and COSTA, M. (1971a) Morphology and distribution of intrinsic adrenergic neurones in the proximal colon of the guinea-pig. *Z. Zellforsch. Mikrosk. Anat.* **120**, 346–363.

FURNESS, J. B. and COSTA, M. (1971b) Monoamine oxidase histochemistry of enteric neurones in the guinea-pig. *Histochemie.* **28**, 324–336.

FURNESS, J. B. and COSTA, M. (1973) The nervous release and the action of substances which affect intestinal muscle through neither adrenoreceptors nor cholinoreceptors. *Phil. Trans. Roy. Soc. Lond. Ser. B.* **265**, 123–133.

FURNESS, J. B. and COSTA, M. (1974) Adrenergic innervation of the gastrointestinal tract. *Ergebn. Physiol.* **69**, 1-51.
FURNESS, J. B. and IWAYAMA, T. (1971) Terminal axons ensheathed in smooth muscle cells of the vas deferens. *Z. Zellforsch. Mikrosk. Anat.* **113**, 259-270.
FURNESS, J. B. and IWAYAMA, T. (1972) The arrangement and identification of axons innervating the vas deferens of the guinea-pig. *J. Anat.* **113**, 179-196.
FURNESS, J. B., MCLEAN, J. R. and BURNSTOCK, G. (1970) Distribution of adrenergic nerves and changes in neuromuscular transmission in the mouse vas deferens during post natal development. *Dev. Biol.* **21**, 491-505.
FUXE, K., GOLDSTEIN, M., HÖKFELT, T. and JOH, T. H. (1971) Cellular localization of dopamine-β-hydroxylase and phenylethanolamine-N-methyltransferase as revealed by immunochemistry. *Progr. Brain Res.* **34**, 127-138.
FUXE, K. and JONSSON, G. (1973) The histochemical fluorescence method for the demonstration of catecholamines: Theory, practice and application. *J. Histochem. Cytochem.* **21**, 293-311.
FUXE, K. and NILSSON, B. Y. (1965) Mechanoreceptors and adrenergic nerve terminals. *Experientia (Basel).* **21**, 641-642.
GABELLA, G. (1969) Taste buds and adrenergic fibres. *J. Neurol. Sci.* **9**, 237-242.
GABELLA, G. (1972) Fine structure of the myenteric plexus in the guinea-pig ileum. *J. Anat.* **111**, 69-97.
GABELLA, G. (1973) Fine structure of smooth muscle. *Phil. Trans. R. Soc. Lond.* Ser. B. **265**, 7-16.
GABELLA, G. and COSTA, M. (1969) Distribution of adrenergic fibers in the small intestine of the newborn guinea pig. *Panminerva Med.* **11**, 407-409.
GANNON, B. J., IWAYAMA, T., BURNSTOCK, G., GERKENS, I. and MASHFORD, M. L. (1971) Prolonged effects of chronic guanethidine treatment on the sympathetic innervation of the genitalia of male rats. *Med. J. Aust.* **2**, 207-208.
GANONG, W. F. (1972) Sympathetic effects on renin secretion: mechanism and physiological role. *In* 'Control of Renin Secretion'. Advances in Experimental Medicine and Biology. Tatiana A. Assaykeen, ed. Vol. **17** Plenum Press, New York, pp. 17-32.
GARRETT, J. R. (1971) Changes in autonomic nerves of salivary glands on degeneration and regeneration. *Progr. Brain Res.* **34**, 475-488.
GASKELL, W. H. (1886) On the structure, distribution and function of the nerves which innervate the visceral and vascular systems. *J. Physiol. (Lond.).* **7**, 1-80.
GASKELL, W. H. (1916) 'The Involuntary Nervous System.' Longmans, Green & Co., London.
GEFFEN, L. B. and HUGHES, C. C. (1972) Degeneration of sympathetic nerves *in vitro* and development of smooth muscle supersensitivity to noradrenaline. *J. Physiol. (Lond.).* **221**, 71-84.

GEFFEN, L. B., HUNTER, C. and RUSH, R. A. (1969) Is there bidirectional transport of noradrenaline in sympathetic nerves? *J. Neurochem.* **16**, 469–474.
GEFFEN, L. B. and LIVETT, B. G. (1971) Synaptic vesicles in sympathetic neurons. *Physiol. Rev.* **51**, 98–157.
GEFFEN, L. B. and OSTBERG, A. (1969) Distribution of granular vesicles in normal and constricted sympathetic neurones. *J. Physiol. (Lond.).* **204**, 583–592.
GEFFEN, L. B. and RUSH, R. A. (1968) Transport of noradrenaline in sympathetic nerves and the effect of nerve impulses on its contribution to transmitter stores. *J. Neurochem.* **15**, 925–930.
GEFFEN, L. B., LIVETT, B. G. and RUSH, R. A. (1970) Transmitter economy of sympathetic neurones. *Circulation Res.* **36 & 37**, Suppl. 2, 33–39.
GEFFEN, L. B., RUSH, R. A., LOUIS, W. J. and DOYLE, A. E. (1973a) Plasma dopamine-β-hydroxylase and noradrenaline amounts in essential hypertension. *Clinical Science.* **44**, 617–620.
GEFFEN, L. B., RUSH, R. A., LOUIS, W. J. and DOYLE, A. E. (1973b) Plasma catecholamine and dopamine-β-hydroxylase amounts in phaeochromo-cytoma. *Clinical Science.* **44**, 421–424.
GENNSER, G. and STUDNITZ, W. VON (1969) Monoamine oxidase, catechol-*o*-methyltransferase and phenylethanolamine–N–methyltransferase activity in para-aortic tissue of the human fetus. *Scand. J. clin. Lab. Invest.* **24**, 169–171.
GERO, J. and GEROVA, M. (1971) *In vivo* studies of sympathetic control of vessels of different function. *In* 'Physiology and Pharmacology of Vascular Neuroeffector Systems. Bevan, Furchgott, Maxwell and Somlyo, eds. S. Karger, Basel, Munich, Paris, London, New York, Sydney, pp. 86–94.
GERSHON, M. D. (1967) Inhibition of gastrointestinal movement by sympathetic nerve stimulation: the site of action. *J. Physiol. (Lond.).* **189**, 317–327.
GERSHON, M. D. and THOMPSON, E. B. (1973) The maturation of neuromuscular function in a multiply innervated structure: development of the longitudinal smooth muscle of the foetal mammalian gut and its cholinergic excitatory, adrenergic inhibitory and non-adrenergic inhibitory innervation. *J. Physiol. (Lond.).* **234**, 257–277.
GERTNER, S. B. and ROMANO, A. (1961) Action of guanethidine and bretylium of ganglionic transmission. *Fed. Proc.* **20**, 319.
GEWIRTZ, G. P. and Kopin, I. J. (1970) Effect of intermittent nerve stimulation on norepinephrine synthesis and mobilization in the perfused cat spleen. *J. Pharmacol. exp. Ther.* **175**, 514–520.
GIACOBINI, E., KARJALAINEN, K., KERPEL-FRONIUS, S. and RITZÉN, M. (1970) Monoamines and monoamine oxidase in denervated sympathetic ganglia of the cat. *Neuropharmacol.* **9**, 59–66.
GIACOBINI, E. and KERPEL-FRONIUS, S. (1970) Histochemical and biochemical correlations of monoamine oxidase activity in autonomic and sensory ganglia of the cat. *Acta physiol. scand.* **78**, 522–528.

GIBSON, A. and GILLESPIE, J. S. (1973) The effect of immunosympathectomy and of 6-hydroxydopamine on the response of the rat anococcygeus to nerve stimulation and to some drugs. *Brit. J. Pharmacol.* **47**, 261–267.

GIBSON, W. C. (1940) Degeneration and regeneration of sympathetic nerves. *J. Neurophysiol.* **3**, 237–247.

GILL, E. W. and VAUGHAN WILLIAMS, E. M. (1964) Local anaesthetic activity of the β-receptor antagonist, pronethanol. *Nature (Lond.).* **201**, 199.

GILLESPIE, J. S. (1962) Spontaneous mechanical and electrical activity of stretched and unstretched intestinal smooth muscle cells and their response to sympathetic nerve stimulation. *J. Physiol. (Lond.).* **162**, 54–75.

GILLESPIE, J. S., CREED, K. E. and MUIRE, T. C. (1973) The mechanism of action of neurotransmitters. Electrical changes underlying excitation and inhibition in intestinal and related smooth muscle. *Phil. Trans. Roy. Soc. Lond. Ser. B.* **265**, 95–106.

GILLESPIE, J. S., HAMILTON, D. N. H. and HOSIE, R. J. A. (1970) The extraneuronal uptake and localization of noradrenaline in the cat spleen and the effect on this of some drugs of cold and of denervation. *J. Physiol. (Lond.).* **206**, 563–590.

GILLESPIE, J. S. and KIRPEKAR, S. M. (1966) The uptake and release of radio-active noradrenaline by the splenic nerves of cats. *J. Physiol. (Lond.).* **187**, 51–68.

GILLESPIE, J. S. and MCGRATH, J. C. (1974) The effect of pithing and of nerve stimulation on the depletion of noradrenaline by reserpine in the rat anococcygeus muscle and vas deferens. *Brit. J. Pharmacol.* **52**, 585–590.

GILLESPIE, J. S. and MUIRE, T. C. (1970) Species and tissue variation in extraneuronal and neuronal accumulation of noradrenaline. *J. Physiol. (Lond.).* **206**, 591–604.

GILLESPIE, J. S. and RAE, R. M. (1970) Response of arteries to nerve stimulation and to noradrenaline and the relationship of this to innervation density and wall thickness. *J. Physiol. (Lond.).* **208**, 60–61P.

GILLESPIE, J. S. and TOWART, R. (1973) Uptake kinetics and ion requirements for extraneuronal uptake of NA by arterial smooth muscle and collagen. *Brit. J. Pharmacol.* **47**, 556–567.

GILLIS, C. N. (1971) Inactivation of norepinephrine released by electrical stimulation of rabbit aorta in a gaseous medium. *In* 'Symposium on the Physiology and Pharmacology of Vascular Neuroeffector Systems'. Interlaken 1969. J. A. Bevan, R. F. Furchgott, R. A. Maxwell and A. P. Somlyo, eds. S. Karger, Basel, pp. 47–52.

GILLIS, C. N. and SCHNEIDER, F. H. (1967) Frequency dependence potentiation by various drugs of the chronotropic response of isolated cat atria to sympathetic nerve stimulation. *Brit. J. Pharmacol.* **30**, 541–553.

GILLIS, C. N., SCHNEIDER, F. H., VAN ORDEN, L. S. and GIARMAN, N. H. (1966) Biochemical and microfluorometric studies of norepinephrine redistribution accompanying sympathetic nerve stimulation. *J. Pharmacol. exp. Ther.* **151**, 46–54.

GILLIS, C. N. and YATES, C. M. (1960) Effect of reserpine on the responses of isolated aorta to electrical stimulation. *Fed. Proc.* **19**, 105.

GLOVER, A. B. (1970) Effects of desmethylimipramine (DMI) and cocaine on sympathetic responses in the pithed rat. *J. Pharmac. Pharmacol.* **22**, 789–790

GLOWINSKI, J. and AXELROD, J. (1964) Inhibition of uptake of tritiated noradrenaline in the intact rat brain by imipramine and structurally related compounds. *Nature (Lond.).*, **204**, 1318–1319.

GLOWINSKI, J., AXELROD, J., KOPIN, I. and WURTMAN, R. (1964) Physiological disposition of H^3-norepinephrine in the developing rat. *J. Pharmacol. exp. Ther.* **146**, 48–53.

GOLDBERG, L. I., DA COSTA, F. M. and OZAKI, M. (1960) Actions of the decarboxylase inhibitor, α–methyl-3,4-dihydroxyphenylalanine, in the dog. *Nature (Lond.).* **188**, 502–504.

GOLDMAN, H. and JACOBOWITZ, D. (1971) Correlation of norepinephrine content with observations of adrenergic nerves after a single dose of 6-hydroxydopamine in the rat. *J. Pharmacol. exp. Ther.* **176**, 119–133.

GOLDSTEIN, M. N. (1967) Incorporation and release of H^3-catecholamines by cultured fetal human sympathetic nerve cells and neuroblastoma cells. *Proc. Soc. exp. Biol. Med.* **125**, 993–996.

GOLDSTEIN, M., FUXE, K. and HÖKFELT, T. (1972) Characterization and tissue localization of catecholamine synthesizing enzymes. *Pharmacol. Rev.* **24**, 293–309.

GOODALL, MCC. and KIRSHNER, N. (1958) Biosynthesis of epinephrine and norepinephrine by sympathetic nerves and ganglia. *Circulation.* **17**, 366–371.

GOODMAN, L. S. and GILMAN, A. (1971) 'The Pharmacological Basis of Therapeutics.' 4th ed. Collier-Macmillan Co., London, Toronto.

GORDON, R., REID, J. V. O., SJOERDSMA, A. and UDENFRIEND, S. (1966) Increased synthesis of norepinephrine in the rat heart on electrical stimulation of the stellate ganglia. *Mol. Pharmacol.* **2**, 610–613.

GORDON, R., SPECTOR, S., SJOERDSMA, A. and UDENFRIEND, S. (1966) Increased synthesis of norepinephrine and epinephrine in the intact rat during exercise and exposure to cold. *J. Pharmacol. exp. Ther.* **153**, 440–447.

GORIDIS, C. and NEFF, N. H. (1971a) Evidence for a specific monoamine oxidase associated with sympathetic nerves. *Neuropharmacol.* **10**, 557–564.

GORIDIS, C. and NEFF, N. H. (1971b) Monoamine oxidase in sympathetic nerves: a transmitter specific enzyme type. *Brit. J. Pharmacol.* **43**, 814–818.

GORKIN, V. Z. (1966) Monoamine oxidases. *Pharmacol. Rev.* **18**, 115–120.

GOVIER, W. C., SUGRUE, M. F. and SHORE, P. A. (1969) On the inability to produce supersensitivity to catecholamines in intestinal smooth muscle. *J. Pharmacol. exp. Ther.* **165**, 71–77.

GRAEFE, K. H., STEFANS, F. J. E. and LANGER, S. Z. (1973) Preferential metabolism of (—)-³H-norepinephrine through the deaminated glycol in the rat vas deferens. *Biochem. Pharmacol.* **22**, 1147–1160.

GRAHAM, J. D. P., LEVER, J. D. and SPRIGGS, T. L. B. (1968) An examination of adrenergic axons around pancreatic arterioles of the cat for the presence of acetylcholinesterase by high resolution autoradiographic and histochemical methods. *Brit. J. Pharmacol.* **33**, 15–20.

GRAHAM, J. M. and KEATINGE, W. R. (1972) Differences in sensitivity to vasoconstrictor drugs within the wall of the sheep carotid artery. *J. Physiol. (Lond.).* **221**, 477–492.

GRAHAM, M. H., ABBOUD, F. M. and ECKSTEIN, J. W. (1965) Effect of cocaine on cardiovascular responses in intact dogs. *J. Pharmacol. exp. Ther.* **150**, 46–52.

GRAY, E. G. and WHITTAKER, V. P. (1962) The isolation of nerve endings from brain: an electronmicroscope study of cell fragments derived by homogenization and centrifugation. *J. Anat.* **96**, 79–88.

GREEN, J. H. and HEFFRON, P. F. (1966) Simultaneous recording of sympathetic activity in different regions. *J. Physiol. (Lond.).* **185**, 48–50P.

GREEN, J. H. and HEFFRON, P. F. (1968) Studies upon patterns of activity in single post-ganglionic sympathetic fibres. *Arch. int. Pharmacodyn. Ther.* **173**, 232–243.

GREEN, R. D., DALE, M. M. and HAYLETT, D. G. (1972) Effect of adrenergic amines on the membrane potential of guinea-pig liver parenchymal cells in short term tissue culture. *Experientia (Basel).* **28**, 1073–1074.

GREENBERG, S., LONG, J. P., BURKE, J. P., CHAPNICK, B. and VAN ORDEN, L. S. (1973) Decreased contractility and NA content of guinea-pig seminal vesicles after chronic treatment with testosterone. *J. Pharmacol. exp. Ther.* **184**, 56–66.

GREENGARD, P., NATHASON, J. A. and KEBABIAN, J. W. (1973) Dopamine- octapamine- and serotonine-sensitive adenylate cyclases: possible receptors in aminergic neurotransmission. *In* 'Frontiers in Catecholamine Research'. E. Usdin and S. Snyder, eds. Pergamon Press, Oxford, pp. 377–382.

GREER, C. M., PINKSTON, J. O., BAXTER, J. H. JR. and BRANNON, E. S. (1938) Norepinephrine (β-(3-4-dihydroxyphenyl)-β-hydroxy-ethylamine) as a possible mediator in the sympathetic division of the autonomic nervous system. *J. Pharmacol. exp. Ther.* **62**, 189–227.

GRIGORÉVA, T. A. (1962) 'The Innervation of Blood Vessels.' Pergamon Press, New York, Oxford, London, Paris.

GRILLO, M. A. (1966) Electron microscopy of sympathetic tissues. *Pharmacol. Rev.* **18**, 387–399.

GRYNSZPAN-WINOGRAD, O. (1971) Morphological aspects of exocytosis in the adrenal medulla. *Phil. Trans. Roy. Soc. Lond. Ser. B.* **261**, 291–292.

GUTH, L. (1956) Regeneration in the mammalian peripheral nervous system. *Physiol. Rev.* **36**, 441–478.

GUTH, L. (1968) 'Trophic' influences of nerve on muscle. *Physiol. Rev.* **48**, 645–687.

GUTMAN, Y. and SEGAL, J. (1972) Effect of calcium, sodium and potassium on adrenal tyrosine hydroxylase activity *in vitro*. *Biochem. Pharmacol.* **21**, 2664–2666.

HAEFELY, W. (1970) Slow synaptic potentials in the cat superior cervical ganglion (SCG) in situ. *Experientia (Basel)*. **26**, 690.

HAEFELY, W. (1972) Electrophysiology of the adrenergic neuron. In 'Catecholamines'. H. Blaschko and E. Muscholl, eds. Springer-Verlag, Berlin, Heidelberg, New York, pp. 661–725.

HAEFELY, W., HÜRLIMANN, A. and THOENEN, H. (1964) A quantitative study of the effect of cocaine on the response of the cat nictitating membrane to nerve stimulation and to injected noradrenaline. *Brit. J. Pharmacol.* **22**, 5–21.

HAEFELY, W., HÜRLIMANN, A. and THOENEN, H. (1965) Relation between the rate of stimulation and the quantity of noradrenaline liberated from sympathetic nerve endings in the isolated perfused spleen of the cat. *J. Physiol. (Lond.).* **181**, 48–58.

HAEFELY, W., HÜRLIMANN, A. and THOENEN, H. (1966) The effect of stimulation of sympathetic nerves in the cat treated with reserpine, α-methyldopa and α-methyl-metatyrosine. *Brit. J. Pharmacol.* **26**, 172–185.

HAEFELY, W., HÜRLIMANN, A. and THOENEN, H. (1967) Adrenergic transmitter changes and response to sympathetic nerve stimulation after differing pretreatment with α-methyldopa. *Brit. J. Pharmacol.* **31**, 105–119.

HAEUSLER, G. (1972) Differential effect of verapamil on excitation-contraction coupling in smooth muscle and on excitation-secretion coupling in adrenergic nerves. *J. Pharmacol. exp. Ther.* **180**, 672–682.

HAEUSLER, G., HAEFELY, W. and HÜRLIMANN, A. (1969) On the mechanism of the adrenergic nerve blocking action of bretylium. *Naunyn-Schmied. Arch. exp. Path. Pharmak.* **265**, 260–277.

HAEUSLER, G., THOENEN, H., HAEFELY, W. and HÜRLIMANN, A. (1698) Electrical events in cardiac adrenergic nerves and noradrenaline release from the heart induced by acetylcholine and KCl. *Naunyn Schmied. Arch. exp. Path. Pharmak.* **261**, 389–411.

HÄGGENDAL, J. (1970) Some further aspects on the release of the adrenergic transmitter. In 'Bayer-Symposium II'. Springer-Verlag, Berlin, Heidelberg, New York. 100–109.

HÄGGENDAL, J. and DAHLSTRÖM, A. (1970) Uptake and retention of ^{3}H-noradrenaline in adrenergic nerve terminals after reserpine and axotomy. *Europ. J. Pharmacol.* **10**, 411–415.

HÄGGENDAL, J. and MALMFORS, T. (1969) The effect of nerve stimulation on the uptake of noradrenaline into the adrenergic nerve terminals. *Acta physiol. scand.* **75**, 28–32.

HÄGGENDAL, J., JOHANSSON, B., JONASON, J. and LJUNG, B. (1970) Correlation between noradrenaline release and effector response to nerve stimulation in the rat portal vein *in vitro*. *Acta physiol. scand. Suppl.* **349**, 17-32.

HÅKANSON, R. (1970) New aspects of the formation and function of histamine, 5-hydroxytryptamine and dopamine in gastric mucosa. *Acta physiol. scand. Suppl.* **340**, 1-134.

HÅKANSON, R., LOMBARD DES GOITTES, M. N. and OWMAN, C. (1967) Activities of trytophan hydroxylase, dopa decarboxylase and monoamine oxidase as correlated with the appearance of monoamines in developing rat pineal gland. *Life Sci.* **6**, 2577-2585.

HALSTEAD, D. C. and LARRABEE, M. G. (1972) Early effects of antiserum to the nerve growth factors on metabolism and transmission in superior cervical ganglion of mice. In 'Immunosympathectomy'. G. Steiner and C. Schönbaum, eds. Elsevier Publ. Co., Amsterdam, London, New York, pp. 221-236.

HAMBERGER, B. and NORBERG, K-A. (1963) Monoamines in sympathetic ganglia studied with fluorescence microscopy. *Experientia (Basel).* **19**, 580-581.

HAMBERGER, B. and NORBERG, K-A. (1965) Adrenergic synaptic terminals and nerve cells in bladder ganglia of the cat. *Int. J. Neuropharmacol.* **2**, 279-282.

HAMBERGER, B., NORBERG, K-A. and SJÖQVIST, F. (1965) Correlated studies of monoamines and acetyl-cholinesterase in sympathetic ganglia, illustrating the relative distribution of adrenergic and cholinergic neurons. In 'Pharmacology of Cholinergic and Adrenergic Transmission'. G. B. Koelle, W. W. Douglas and A. Carl, eds. Proc. Sec. Internat. Pharmacol. Meeting. Prague, 1963. Vol. 3, 41-53. Pergamon Press, Oxford.

HÁMORI, J., LANG, E. and SIMON, L. (1968) Experimental degeneration of the preganglionic fibers in the superior cervical ganglion of the cat. *Z. Zellforsch. Mikrosk. Anat.* **90**, 37-52.

HAND, A. R. (1970) Nerve-acinar cell relationships in the rat parotid gland. *J. Cell. Biol.* **47**, 540-543.

HAND, A. R. (1972) Adrenergic and cholinergic nerve terminals in the rat parotid gland. Electron microscopic observations on permanganate-fixed glands. *Anat. Rec.* **173**, 131-139.

HÄRKÖNEN, M. (1964) Carboxylic esteraics, oxidative enzymes and catecholamines in the superior cervical ganglion of the rat and the effect of pre- and post-ganglionic nerve division. *Acta physiol. scand. Suppl.* **237**, 1-94.

HARTMAN, B. K. (1973) Immunofluorescence of dopamine-β-hydroxylase: application of improved methodology to the localization of the peripheral and central noradrenergic nervous system. *J. Histochem. Cytochem.* **21**, 312-332.

HARTMAN, B. K. and UDENFRIEND, S. (1972) The application of immunological techniques to the study of enzymes regulating catecholamine synthesis and degradation. *Pharmacol. Rev.* **24**, 311-330.

HARTMAN, F. A., MCCORDOCK, H. A. and LODER, M. M. (1923) Conditions determining adrenal secretion. *Amer. J. Physiol.* **64**, 1-34.

HEATH, J. W., EVANS, B. K. and BURNSTOCK, G. (1973) Axon retraction following guanethidine treatment. Studies of sympathetic neurons *in vivo*. *Z. Zellforsch. Mikrosk. Anat.* **146**, 439-451.

HEATH, J. W., HILL, C. E. and BURNSTOCK, G. (1974) Axon retraction following guanethidine treatment. Studies of sympathetic neurons in tissue culture. *J. Neurocytol.* **3**, 263-276.

HEATH, J. W., EVANS, B. K., GANNON, B. J., BURNSTOCK, G. and JAMES, V. B. (1972) Degeneration of adrenergic neurons following guanethidine treatment: an ultrastructural study. *Virchows. Arch. Abt. B. Zell. Path.* **11**, 182-197.

HEDQVIST, P. (1970) Studies on the effect of prostaglandins E_1 and E_2 on the sympathetic neuromuscular transmission in some animal tissues. *Acta physiol. scand. Suppl.* **345**, 1-40.

HEDQVIST, P. (1973a) Prostaglandin as a tool for local control of transmitter release from sympathetic nerves. *Brain Res.* **62**, 483-488.

HEDQVIST, P. (1973b) Dissociation of prostaglandin and α-receptor mediated control of adrenergic transmitter release. *Acta physiol. scand.* **87**, 42A-43A.

HEDQVIST, P. and STJÄRNE, L. (1969) The relative role of recapture and of de novo synthesis for maintenance of neurotransmitter homeostasis in noradrenergic nerves. *Acta physiol. scand.* **76**, 270-283.

HENDRY, I. A. and IVERSEN, L. L. (1971) Effect of nerve growth factor and its antiserum on tyrosine hydroxylase activity in mouse superior cervical sympathetic ganglion. *Brain Res.* **29**, 159-162

HENDRY, I. A., IVERSEN, L. L. and BLACK, I. B. (1973) A comparison of the neural regulation of tyrosine hydroxylase activity in sympathetic ganglia of adult mice and rats. *J. Neurochem.* **20**, 1683-1689.

HENDRY, I. A., STÖCKEL, K., THOENEN, H. and IVERSEN, L. L. (1974) The retrograde axonal transport of nerve growth factor. *Brain Res.* **68**, 500-504.

HENNING, M. (1969) Studies on the mode of action of α-methyldopa. *Acta physiol. scand. Suppl.* **322**, 1-37.

HENNING, M. and SVENSSON, L. (1968) Adrenergic nerve function in the anaesthetized rat after treatment with α-methyldopa. *Acta pharmacol. (Kbh).* **26**, 425-436.

HERBERT, J. (1971) The role of the pineal gland in the control by light of the reproductive cycle of the ferret. In 'The Pineal Gland'. Ciba Foundation Symposium. G. E. W. Wolstenholme and J. Knight, eds. Churchill & Livingstone, Edinburgh-London, pp. 303-320.

HERMAN, H. (1952) Zusammenfassende Ergebnisse über Altersuaränderung am peripheren Nervensystem. *Z. Alternsfrorsch.* **6**, 197.

HERTTING, G. and AXELROD, J. (1961) Fate of tritiated noradrenaline at the sympathetic nerve-endings. *Nature. (Lond.).* **192**, 172-173.

HERTTING, G., POTTER, L. T. and AXELROD, J. (1962) Effect of decentralization and ganglion blocking agents on the spontaneous release of H³-norepinephrine. *J. Pharmacol. exp. Ther.* **136**, 289–292.

HERTZLER, E. C. (1961) 5-Hydroxytryptamine and transmission in sympathetic ganglia. *Brit. J. Pharmacol.* **17**, 406–413.

HERVONEN, A. (1971) Development of catecholamine storing cells in human fetal paraganglia and adrenal medulla. *Acta physiol. scand. Suppl.* **368**, 1–94.

HERVONEN, A. and KANERVA, L. (1972a) Cell types of human fetal superior cervical ganglion. *Z. Anat. EntwGesch.* **137**, 257–269.

HERVONEN, A. and KANERVA, L. (1972b) Adrenergic and noradrenergic axons of the rabbit uterus and oviduct. *Acta physiol. scand.* **85**, 139–141.

HERVONEN, A., KANERVA, L., LIETZEN, R. and PARTANEN, S. (1972) Ultrastructural changes induced by estrogen in the adrenergic nerves of the rabbit myometrium. *Acta physiol. scand.* **85**, 283–285.

HILL, C. E., MARK, G. E., ERÄNKÖ, O., ERÄNKÖ, L. and BURNSTOCK, G. (1973) Use of tissue culture to examine the actions of guanethidine and 6-hydroxydopamine. *Eur. J. Pharmacol.* **23**, 162–174.

HILLARP, N-Å. (1946) Structure of the synapse and the peripheral innervation apparatus of the autonomic nervous system. *Acta Anat. Suppl.* **4**, 1–153.

HILLARP, N-Å. (1959) The construction and functional organization of the autonomic innervation apparatus. *Acta physiol. scand. Suppl.* **157, 46**, 1–38.

HIMMS-HAGEN, J. (1972) Effect of catecholamines on metabolism. *In* 'Catecholamines'. H. Blaschko and E. Muscholl, eds. Springer-Verlag, Berlin, Heidelberg, New York, pp. 363–462.

HINES, M. and TOWER, S. S. (1928) Studies on the innervation of skeletal muscles. II. Of muscle spindles in certain muscles of the kitten. *Johns Hopk. Hosp. Bull.* **42**, 264–295.

HIRST, G. D. S. and MCKIRDY, H. C. (1974) Personal communication.

HOCH-LIGITI, C. and CAMP, J. L. III. (1959) Catecholamine production in tissue cultures of human adrenal medullary tumors and of adrenal medulla. *Proc. Soc. exp. Biol. Med.* **102**, 692–693.

HÖKFELT, T. (1967) Ultrastructural studies on adrenergic nerve terminals in the albino rat iris after pharmacological and experimental treatment. *Acta physiol. scand.* **69**, 125–126.

HÖKFELT, T. (1968) *In vitro* studies on central and peripheral monoamine neurons at the ultrastructural level. *Z. Zellforsch. Mikrosk. Anat.* **91**, 1–74.

HÖKFELT, T. (1969) Distribution of noradrenaline storing particles in peripheral adrenergic neurons as revealed by electron microscopy. *Acta physiol. scand.* **76**, 427–440.

HÖKFELT, T. (1971) Ultrastructural localization of intraneuronal monoamines – some aspects on methodology. *Progr. Brain Res.* **34**, 213–222.

HÖKFELT, T. (1973a) On the origin of small adrenergic storage vesicles: evidence for local formation in nerve endings after chronic reserpine treatment. *Experientia (Basel)* **29**, 580–582.

HÖKFELT, T. (1973b) Localization of catecholamines with special reference to synaptic vesicles. *Life Science.* **13**, 73–74.

HÖKFELT, T. and DAHLSTRÖM, A. (1971) Effect of two mitosis inhibitors (colchicine and vinblastine) on the distribution and axonal transport of noradrenaline storage particles, studied by fluorescence and electron microscopy. *Z. Zellforsch. Mikrosk. Anat.* **119**, 460–482.

HÖKFELT, T. and JONSSON, G. (1968) Studies on reaction and binding of monoamines after fixation and processing for electron microscopy with special reference to fixation with potassium permanganate. *Histochemie.* **16**, 45–67.

HÖKFELT, T. and LJUNGDAHL, A. (1972) Application of cytochemical techniques to the study of suspected transmitter substance in the nervous system. *In* 'Studies of Neurotransmitters at the Synaptic Level'. E. Costa, L. L. Iversen and R. Paoletti, eds. Advances in Biochemical Psychopharmacology **6**, pp. 1–36.

HÖKFELT, T., JONSSON, G. and SACHS, CH. (1972) Fine structure and fluorescence morphology of adrenergic nerves after 6–hydroxydopamine *in vivo* and *in vitro*. *Z. Zellforsch. Mikrosk. Anat.* **131**, 529–543.

HÖKFELT, T., FUXE, K., GOLDSTEIN, M. and JOH, T. H. (1973) Immunohistochemical localization of three catecholamine synthesizing enzymes: aspects on methodology. *Histochemie.* **33**, 231–254.

HOLMAN, M. E. (1967) Some electrophysiological aspects of transmission from noradrenergic nerves to smooth muscle. *Circulation Res.* **21**, Suppl. III, 71–81.

HOLMAN, M. E. (1969) Electrophysiology of vascular smooth muscle. *Ergebn. Physiol.* **61**, 137–177.

HOLMAN, M. E. (1970) Junction potentials in smooth muscle. *In* 'Smooth Muscle'. E. Bülbring, A. F. Brading, A. W. Jones and T. Tomita, eds. Edward Arnold, London, pp. 244–288.

HOLMAN, M. E. and JOWETT, A. (1964) Some actions of catecholamines on the smooth muscle of the guinea-pig vas deferens. *Aust. J. exp. Biol. med. Sci.* **42**, 40–53.

HOLTZ, P. (1939) Dopadecarboxylase. *Naturwissenschaften.* **27**, 724–725.

HOLTZ, P., HEISE, R. and LÜDTKE, K. (1938) Fermentativer Abbau von L-Dioxyphenylalanin (dopa) durch Niere. *Naunyn. Schmied. Arch. exp. Path. Pharmak.* **191**, 87–118.

HOLTZMAN, E. (1971) Cytochemical studies of protein transport in the nervous system. *Phil. Trans. Roy. Soc. Lond. Ser. B.* **261**, 407–421.

HOLZBAUER, M. and SHARMAN, D. F. (1972) The distribution of catecholamines in vertebrates. *In* 'Catecholamines'. H. Blaschko and E. Muscholl, eds. Springer-Verlag, Berlin, Heidelberg, New York, pp. 110–185.

HOPWOOD, D. (1971) The histochemistry and electron histochemistry of chromaffin tissue. *Progr. Histochem. Cytochem.* **3**, 1-66.

HONIG, C. R. and STAM, A. C. JR. (1967) The influence of adrenergic mediators and their structural analogs on cardiac actomyosin systems. *Ann. N.Y. Acad. Sci.* **139**, 724-740.

HORTON, E. W. (1973) Prostaglandins at adrenergic nerve-endings. *Brit. med. Bull.* **29**, 148-151.

HÖRTNAGL, H., HÖRTNAGL, H. and WINKLER, H. (1969) Bovine splenic nerve: characterization of noradrenaline-containing vesicles and other cell organelles by density gradient centrifugation. *J. Physiol. (Lond.).* **205**, 103-114.

HOTTA, Y. (1969) Some properties of the junctional and extrajunctional receptors in the vas deferens of the guinea-pig. *Agents and Actions.* **1**, 13-21.

HOWE, A. and NEIL, E. (1972) Arterial chemoreceptors. *In* 'Enteroceptors'. E. Neil, ed. Springer-Verlag, Berlin, Heidelberg, New York, pp. 47-80.

HRDINA, P. D. and LING, G. M. (1970) Studies on the mechanism of the inhibitory effect of desipramine (DMI) on vascular smooth muscle contraction. *J. Pharmacol. exp. Ther.* **173**, 407-415.

HUGHES, I. E., KIRK, J. A., KNEEN, B. and LARGE, B. J. (1973) An assessment of sympathetic function in isolated tissues from mice given nerve-growth-factor antiserum and 6-hydroxydopamine. *Brit. J. Pharmacol.* **47**, 748-759.

HUGHES, J. (1972) Evaluation of mechanisms controlling the release and inactivation of the adrenergic transmitter in the rabbit portal vein and vas deferens. *Brit. J. Pharmacol.* **44**, 472-491.

HUGHES, J. and ROTH, R. H. (1971) Evidence that angiotensin enhances transmitter release during sympathetic nerve stimulation. *Brit. J. Phrmacol.* **41**, 239-255.

HUKOVIĆ, S. (1961) Responses of the isolated sympathetic nerve-ductus deferens preparation of the guinea pig. *Brit. J. Pharmacol.* **16**, 188-194.

HUKOVIĆ, S. and MUSCHOLL, E. (1962) Die Noradrenalin-Abgabe aus dem isolierten Kaninchenherzen bei sympathischer Nervenreizung und ihre pharmakologische Beeinflussung. *Naunyn. Schmied. Arch. exp. Path. Pharmak.* **244**, 81-96.

HUME, W. R., DE LA LANDE, I. S. and WATERSON, J. G. (1972) Effect of acetylcholine on the response of the isolated rabbit ear artery to stimulation of the perivascular sympathetic nerves. *Europ. J. Pharmacol.* **17**, 227-233.

HUNT, C. C. (1960) The effect of sympathetic stimulation of mammalian muscle spindles. *J. Physiol. (Lond.).* **151**, 332-341.

IGGO, A. and VOGT, M. (1960) Preganglionic sympathetic activity in normal and in reserpine-treated cats. *J. Physiol. (Lond.).* **150**, 114-133.

IIZUKA, T., MARK, A. L., WENDLING, M. G., SCHMID, P. P. and ECKSTEIN, J. W. (1970) Differences in responses of saphenous and mesenteric veins to reflex stimuli. *Am. J. Physiol.* **219**, 1066-1070.

IKEDA, M., FAHIEN, L. A. and UDENFRIEND, S. (1966) A kinetic study of bovine adrenal tyrosine hydroxylase. *J. biol. Chem.* **241**, 4452-4456.

IMAGAWA, N. (1969) The ciliary muscle and the autonomic nervous system. Report 1. The activity of acetylcholine esterase in the cat ciliary muscle in electron microscope preparations. *Acta Societatis ophthalmologiscae japonicae.* **73**, 2152-2159.

IRIKI, M. and SIMON, E. (1973) Different autonomic control of regional circulating reflexes evoked by thermal stimulation and by hypoxia. *Austr. J. exp. biol. Med. Sci.* **51**, 283-293.

IRIKI, M., WALTHER, O. E., PLESCHKA, K. and SIMONE, E. (1971) Regional cutaneous and visceral sympathetic activity during asphyxia in the anesthetized rabbit. *Pflügers Arch. ges. Physiol.* **322**, 167-182.

IRISAWA, H., NINOMIYA, I. and WOOLEY, G. (1973) Efferent activity in renal and intestinal nerves during circulatory reflexes. *Jap. J. Physiol.* **23**, 657-666.

IVANOV, D. P. (1971) Connexions neuro-musculaires specialisées dans le canal déférent du rat. *J. Neuro-Visc. Relat.* **32**, 143.

IVERSEN, L. L. (1967) 'The Uptake and Storage of Noradrenaline in Sympathetic Nerves.' Cambridge University Press.

IVERSEN, L. L. (1971) Role of transmitter uptake mechanisms in synaptic neurotransmission. *Brit. J. Pharmacol.* **41**, 571-591.

IVERSEN, L. L. (1973) Catecholamine uptake processes. *Brit. med. Bull.* **29**, 130-135.

IVERSEN, L. L. and CALLINGHAM, B. A. (1970) Adrenergic transmission. In 'Synaptic Vesicles, Specific Granules, Autopharmacology', pp. 253-305. Reprinted from 'Fundamentals of Biochemical Pharmacology', Z. M. Bacq, ed. Pergamon Press, Oxford and New York.

IVERSEN, L. L. and SALT, P. J. (1970) Inhibition of catecholamine uptake$_2$ by steroids in the isolated rat heart. *Brit. J. Phrmacol.* **40**, 528-530.

IVERSEN, L. L., DECHAMPLAIN, J., GLOWINSKI, J. and AXELROD, J. (1967) Uptake, storage and metabolism of norepinephrine in tissues of the developing rat. *J. Pharmacol. exp. Ther.* **157**, 509-516.

IVERSEN, L. L., GLOWINSKI, J. and AXELROD, J. (1966) The physiologic disposition and metabolism of norepinephrine in immunosympathectomized animals. *J. Pharmacol. exp. Ther.* **151**, 273-284.

IWAYAMA, T. (1970) Ultrastructural changes in the nerves innervating the cerebral artery after sympathectomy. *Z. Zellforsch. Mikrosk. Anat.* **109**, 465-480.

IWAYAMA, T. and FURNESS, J. B. (1971) Enhancement of the granulation of adrenergic storage vesicles in drug-free solution. *J. Cell Biol.* **48**, 699-703.

IWAYAMA, T., FURNESS, J. B. and BURNSTOCK, G. (1970) Dual adrenergic and cholinergic innervation of the cerebral arteries of the rat. *Circulation Res.* **26**, 635-646.

JACOBOWITZ, D. (1965) Histochemical studies of the autonomic innervation of the gut. *J. Pharmacol. exp. Ther.* **149**, 358-364.

JACOBOWITZ, D. (1967) Histochemical studies of the relationship of chromaffin cells and adrenergic nerve fibers to the cardiac ganglia of several species. *J. Pharmacol. exp. Ther.* **158**, 227-240.

JACOBOWITZ, D. (1970) Catecholamine fluorescence studies of adrenergic neurons and chromaffin cells in sympathetic ganglia. *Fed Proc.* **29**, 1929-1944.

JACOBOWITZ, D. and BRUS, R. (1971) A study of extraneuronal uptake of norepinephrine in the perfused heart of the guinea-pig. *Eur. J. Pharmacol.* **15**, 274-284.

JACOBOWITZ, D. and WOODWARD, J. K. (1968) Adrenergic neurons in the cat superior cervical ganglion and cervical sympathetic nerve trunk. A histochemical study. *J. Pharmacol. exp. Ther.* **162**, 213-226.

JACOBOWITZ, D., KENT, K. M., FLEISCH, J. H. and COOPER, T. S. (1973) Histofluorescent study of catecholamine-containing elements in cholinergic ganglia from the calf and dog lung. *Proc. Soc. exp. Biol. Med.* **144**, 464-466.

JAIM-ETCHEVERRY, G. and ZIEHER, L. M. (1971a) Ultrastructural aspects of neurotransmitter storage in adrenergic nerves. *In* 'Advances in Cytopharmacology.' Vol. **1**. 1st Int. Symposium on Cell Biology and Cytopharmacology. Raven Press, New York.

JAIM-ETCHEVERRY, G. and ZIEHER, L. M. (1971b) Permanent depletion of peripheral norepinephrine in rats treated at birth with 6-hydroxydopamine. *Eur. J. Pharmacol.* **13**, 272-276.

JAJU, B. P. (1969) Burn and Rand's Hypothesis. *Ind. J. Physiol and Pharmacol.* **13**, 1-27.

JÄNIG, W. and SCHMIDT, R. F. (1970) Single unit responses in the cervical sympathetic trunk upon somatic nerve stimulation. *Pflüg. Arch. ges. Physiol.* **314**, 199-216.

JANKOWSKA, E., LUBINSKA, L. and NIEMIERKO, S. (1969) Translocation of AChE - containing particles in the axoplasm during nerve activity. *Comp. Biochem. Physiol.* **28**, 907-913.

JANSSON, G. and MARTINSON, J. (1966) Studies on the ganglionic site of action of sympathetic outflow to the stomach. *Acta physiol. scand.* **68**, 184-192.

JARROTT, B. (1970) Uptake and metabolism of catecholamines in the perfused hearts of different species. *Brit. J. Pharmacol.* **38**, 810-821.

JARROTT, B. (1971a) Occurrence and properties of monoamine oxidase in adrenergic neurons. *J. Neurochem.* **18**, 7-16.

JARROTT, B. (1971b). Occurrence and properties of catechol-*o*-methyl transferase in adrenergic neurons. *J. Neurochem.* **18**, 17-27.

JARROTT, B. and IVERSEN, L. L. (1968) Subcellular distribution of monoamine oxidase activity in rat liver and vas deferens. *Biochem. Pharmacol.* **17**, 1619-1625.

JARROTT, B. and IVERSEN, L. L. (1971) Noradrenaline metabolizing enzymes in normal and sympathetically denervated vas deferens. *J. Neurochem.* **18**, 1–6.

JARROTT, B. and LANGER, S. Z. (1971) Changes in monoamine oxidase and catechol-*o*-methyl transferase activities after denervation of the nictitating membrane of the cat. *J. Physiol. (Lond.).* **212**, 549–559.

JENKINSON, D. H. (1973) Classification and properties of peripheral adrenergic receptors. *Brit. med. Bull.* **29**, 142–147.

JENKINSON, D. H. and MORTON, I. K. M. (1967) The role of α- and β-adrenergic receptors in some actions of catecholamines on intestinal smooth muscle. *J. Physiol. (Lond.).* **188**, 387–402.

JENSEN-HOLM, J. and JUUL, P. (1971) Ultrastructural changes in the rat superior cervical ganglion following prolonged guanethidine administration. *Acta pharmacol. toxicol.* **30**, 308–320.

JOH, T. H., KAPIT, R. and GOLDSTEIN, M. (1969) A kinetic study of particulate bovine adrenal tyrosine hydroxylase. *Biochem. biophys. Acta.* **171**, 378–380.

JOHANSSON, B., JOHANSSON, S. R., LJUNG, B. and STAGE, L. (1972) A receptor kinetic model of a vascular neuroeffector. *J. Pharmacol. exp. Ther.* **180**, 636–646.

JOHNSON, D. G. (1973) Modulatory control of sympathetic neurotransmission by prostaglandins. *Trans. Amer. Soc. Neurochem.* **4** (1), p. 56.

JOHNSON, D. F., SILBERSTEIN, S. D., HANBAUER, I. and KOPIN, I. J. (1972) The role of nerve growth factor in the ramification of sympathetic nerve fibres into the rat iris in organ culture. *J. Neurochem.* **19**, 2025–2029.

JOHNSON, D. G., THOA, N. B., WEINSHILBOUM, R., AXELROD, J. and KOPIN, I. J. (1971) Enhanced release of dopamine-β-hydroxylase from sympathetic nerves by calcium and phenoxybenzamine and its reversal by prostaglandins. *Proc. nat. Acad. Sci. U.S.A.* **68**, 2227–2230.

JOHNSON, G. and RITZÉN, M. (1966) Microspectrofluorometric identification of metaraminol in sympathetic adrenergic neurons. *Acta physiol. scand.* **67**, 505–513.

JONSSON, G. and SACHS, ch. (1970) Effects of 6-Hydroxydopamine on the uptake and storage of noradrenaline in sympathetic adrenergic neurons. *Eur. J. Pharmacol.* **9**, 141–155.

JOO, F., LEVER, J. D., IVENS, C., MOTTRAM, D. R. and PRESLEY, R. (1971) A fine structural and electron histochemical study of axon terminals in the rat superior cervical ganglion after acute and chronic preganglionic denervation. *J. Anat.* **110**, 181–189.

JUNSTAD, M. and WENNMALM, A. (1973) On the release of prostaglandin E_2 from the rabbit heart following infusion of noradrenaline. *Acta physiol. scand.* **87**, 573–574.

JUUL, P. and ISAAC, R. L. (1973) The effect of guanethidine on the noradrenaline content of the adult rat superior cervical ganglion. *Acta pharmacol. toxicol.* **32**, 382–389.

KADOWITZ, P. J. (1972) Effect of prostaglandins E_1, E_2 and A_2 on vascular resistance and responses to noradrenaline, nerve stimulation and angiotension in the dog hind limb. *Brit. J. Pharmacol.* **46**, 395–400.

KADOWITZ, P. J., SWEET, C. S. and BRODY, M. J. (1971) Potentiation of adrenergic venomotor responses by angiotensin prostaglandin $F_{2\alpha}$ and cocaine. *J. Pharmacol. exp. Ther.* **176**, 167–173.

KADOWITZ, P. J., SWEET, C. S. and BRODY, M. J. (1972) Enhancement of sympathetic neurotransmission by prostaglandin $F_{2\alpha}$ in the cutaneous vascular bed of the dog. *Eur. J. Pharmacol.* **18**, 189–194.

KAJIMOTO, N., KIRPEKAR, S. M. and WAKADE, A. R. (1972) An investigation of spontaneous potentials recorded from the smooth muscle cells of the guinea-pig seminal vesicle. *J. Physiol. (Lond.).* **224**, 105–119.

KALSNER, S. (1969) Steroid potentiation of responses to sympathomimetic amines in aortic strips. *Brit. J. Pharmacol.* **36**, 582–593.

KALSNER, S. (1971) The mechanism of potentiation of contractile responses to catecholamines by methylxanthines in aortic strips. *Brit. J. Pharmacol.* **43**, 379–388.

KALSNER, S. (1972) Effects of the inhibition of noradrenaline uptake and synthesis on the maintenance of the response to continuous nerve stimulation in the central artery of the rabbit ear. *Brit. J. Pharmacol.* **45**, 1–12.

KALSNER, S. and NICKERSON, M. (1968) A method for the study of drug disposition in smooth muscle. *Canad. J. Physiol. Pharmacol.* **46**, 59–70.

KALSNER, S. and NICKERSON, M. (1969a) Mechanism of cocaine potentiation of responses to amines. *Brit. J. Pharmacol.* **35**, 428–439.

KALSNER, S. and NICKERSON, M. (1969b) Disposition of norepinephrine and epinephrine in vascular tissue, determined by the technique of oil immersion. *J. Pharm. exp. Ther.* **165**, 152–165.

KAMIJO, K., KOELLE, G. B. and WAGNER, H. H. (1956) Modification of the effects of sympathomimetic amines and of adrenergic nerve stimulation by 1-isonicotinyl-2-isopropylhydrazine (IIN) and isonicotinic acid hydrazide (INH). *J. Pharmacol. exp. Ther.* **117**, 213–227.

KANERVA, L. (1972) Development, histochemistry and connections of the paracervical (Frankenhäuser) ganglion of the rat uterus. A light and electron microscopic study. *Acta Instituti Anatomici Universitatis Helsinkiensis Supplementum 2.*

KAPELLER, K. and MAYOR, D. (1967) The accumulation of noradrenaline in constricted sympathetic nerves as studied by fluorescence and electron microscopy. *Proc. Roy. Soc. Lond. Ser. B.* **167**, 282–292.

KAPELLER, K. and MAYOR, D. (1969a) An electron microscopic study of the early changes proximal to a constriction in sympathetic nerves. *Proc. Roy. Soc. Lond. Ser. B.* **172**, 39–51.

KAPELLER, K. and MAYOR, D. (1969b) An electron microscopic study of the early changes distal to a construction in sympathetic nerves. *Proc. Roy. Soc. Lond. Ser. B.* **172**, 53–63.

KARIM, S. M., HILLIER, M. and DEVLIN, J. (1968) Distribution of prostaglandins E_1, E_2, $F_{1\alpha}$ and $F_{2\alpha}$ in some animal tissues. *J. Pharmac. Pharmacol.* **20**, 749–753.

KATZ, B. (1962) The transmission of impulses from nerve to muscle, and the subcellular unit of synaptic action. *Proc. Roy. Soc. Lond. Ser. B.* **155**, 455–477.

KATZ, B. (1969) 'The release of neural transmitter substances', Liverpool University Press.

KATZ, B. (1971). Quantal mechanism of neural transmitter release. *Science.* **173**, 123–126.

KAUFMAN, S. (1966) Coenzymes and hydroxylases: ascorbate and dopamine-β-hydroxylase; tetrahydropteridines and phenylalanine and tyrosine hydroxylases. *Pharmacol. Rev.* **18**, 61–70.

KAUFMAN, S. and FRIEDMAN, S. (1965) Dopamine-β-hydroxylase. *Pharmacol. Rev.* **17**, 71–100.

KAUMANN, A., BASSO, N. and ARAMEDIA, P. (1965) The cardiovascular effects of N-(α-methylaminopropyliminodibenzyl)-HCl (Desmethylimipramine) and guanethidine. *J. Pharmacol. exp. Ther.* **147**, 54–64.

KEBABIAN, J. W. and GREENGARD, P. (1971) Dopamine-sensitive adenyl cyclase: possible role in synaptic transmission. *Science.* **174**, 1346–1349.

KEEN, P. and LIVINGSTON, A. (1970) Effect of intravenous vinblastine on intraneuronal transport of noradrenaline in the rat. *Acta physiol. scand. Suppl.* **357**, 13.

KEENE, M. F. and HEWER, E. E. (1927) Observations on the development of the human suprarenal gland. *J. Anat.* **61**, 302–324.

KENDRICK, E., ÖBERG, B. and WENNERGREN, G. (1972) Vasoconstrictor fibre discharge to skeletal muscle kidney, intestine and skin at varying levels of arterial baroreceptor activity in the cat. *Acta physiol. scand.* **85**, 464–476.

KERKUT, G. A., SHAPIRA, A. and WALKER, R. J. (1967) The transport of ^{14}C-labelled material from CNS ⇌ muscle along a nerve trunk. *Comp. Biochem. Physiol.* **23**, 729–748.

KILBINGER, H., LINDMAN, R. LÖFFELHOLZ, K., MUSCHOLL, E. and PATIL, P. N. (1971) Storage and release of false transmitters after infusion of (+) and (−)-α-methyldopamine. *Naunyn. Schmied. Arch. exp. Path. Pharmak.* **271**, 234–248.

KIRAN, B. K. and KHAIRALLAH, P. A. (1969) Angiotensin and norepinephrine efflux. *Eur. J. Pharmacol.* **6**, 102–108.

KIRPEKAR, S. M. and CERVONI, P. (1963) Effect of cocaine, phenoxybenzamine and phentolamine on the catecholamine output from spleen and adrenal medulla. *J. Pharmacol. exp. Ther.* **142**, 59–70.

KIRPEKAR, S. M. and FURCHGOTT, R. F. (1964) The sympathomimetic action of bretylium on isolated atria and aortic smooth muscle. *J. Pharmacol. exp. Ther.* **143**, 64–76.

KIRPEKAR, S. M. and MISU, Y. (1967) Release of noradrenaline by splenic nerve stimulation and its dependence on calcium. *J. Physiol. (Lond.).* **188**, 219–235.

KIRPEKAR. S. M. and PUIG, M. (1971) Effect of flow-stop on noradrenaline release from normal spleens and spleens treated with cocaine, phentolamine or phenoxybenzamine. *Brit. J. Pharmacol.* **43**, 359–369.

KIRPEKAR, S. M., FURCHGOTT, F., WAKADE, A. R. and PRAT, J. C. (1973) Inhibition by sympathomimetic amines of the release of norepinephrine evoked by nerve stimulation in the cat spleen. *J. Pharmacol. exp. Ther.* **187**, 529–538.

KIRPEKAR, S. M., PRAT, J. C., PUIG, M. and WAKADE, A. R. (1972) Modification of the evoked release of noradrenaline from the perfused cat spleen by various ions and agents. *J. Physiol. (Lond.).* **221**, 601–615.

KIRPEKAR, S. M., WAKADE, A. R., DIXON, W. and PRAT, J. C. (1969) Effect of cocaine, phenoxybenzamine and calcium on the inhibition of norepinephrine output from the cat spleen by guanethidine. *J. Pharmacol. exp. Ther.* **165**, 166–175.

KIRSHNER, N. and KIRSHNER, A. G. (1971) Chromogranin A, dopamine β-hydroxylase and secretion from the adrenal medulla. *Phil. Trans. Roy. Soc. Lond. Ser. B.* **261**, 279–289.

KIRSHNER, N., SCHANBERG, S. M. and KIRPEKAR, R. M. (1972) Molecular aspects of the storage and uptake of catecholamines. *In* 'Advances in Drug Research'. N. J. Harper and A. B. Simmonds, eds. Academic Press, London, pp. 121–156.

KISIN, I. E. (1967) Der Einfluss von Reserpin, α-Methyldopa und Bretylium auf die Erregungsübertragung von den Sympathischen Nerven auf die Gefässe. *Vehr. dtsch. Ges. exp. Med.* **19**, 228–236.

KLEIN, R. L. and THURESON-KLEIN, Å. (1971) An electron microscopic study of noradrenaline storage vesicles isolated from bovine splenic nerve trunk. *J. Ultrastruct. Res.* **34**, 473–491.

KLINGMAN, G. I. and KLINGMAN, J. D. (1972a) Immunosympathectomy as an ontogenic tool. *In* 'Immunosympathectomy'. G. Steiner and E. Schönbaum, eds. Elsevier Publ. Co., Amsterdam, London, New York, pp. 91–95.

KLINGMAN, G. I. and KLINGMAN, J. D. (1972b) Immunosympathectomy and biogenic amines. *In* 'Immunosympathectomy'. G. Steiner and E. Schönbaum, eds. Elsevier Publ. Co., Amsterdam, London, New York, pp. 121–130.

KNOCHE, H. and TERWORT, H. (1973) Elektron mikroskopischer Beitrag zur Kenntnis von Degenerationsformer der vegetativen Endstrecke Nach Dunchscheidung prosganglionären Fasern. *Z. Zellforsch. Mikrosk. Anat.* **141**, 181–202.

KNOEFEL, P. and DAVIS, H. (1933) The frequency of impulses in the nerve fibers to the nictitating membranes of the cat. *Amer. J. Physiol.* **104**, 81–89.

KNYIHÁR, E., RISTOVSKY, K., KÓLMON, G. and CSILLIK, B. (1969) Chemical sympathectomy: Histochemical and submicroscopical consequences of 6-hydroxydopamine treatment in the rat iris. *Experientia (Basel).* **25**, 518–520.

KOBAYASHI, S. (1971) Comparative cytological studies of the carotid body. 1. Demonstration of monoamine-storing cells by correlated chromaffin reaction and fluorescence histochemistry. *Arch. hist. Jap.* **33**, 319–339.

KOELLE, G. B. (1963) Cytological distributions and physiological functions of cholinesterases. *In* 'Choninesterases and Anticholinesterase Agents; Handbuch der Experimentellen Pharmakologie'. G. B. Koelle, ed. Suppl. 15, Springer-Verlag, Berlin. pp. 187–298.

KOERKER, R. L. and MORÓN, N. C. (1971) An evaluation of the inability of cocaine to potentiate the responses to cardiac sympathetic nerve stimulation in the dog. *J. Pharmacol. exp. Ther.* **178**, 482–496.

KÖHN, A. (1903) Die Paraganglien. *Arch mikr. Anat.* **62**, 263–365.

KOIZUMI, K. and SUDA, I. (1963) Induced modulations in autonomic efferent neuron activity. *Amer. J. Physiol.* **205**, 738–744.

KÖLLIKER, A. (1879) 'Entwickelungesgesichte des Menschen und der Höheren Tiere'. Leipzig.

KOPIN, I. J. (1964) Storage and metabolism of catecholamines: the role of monoamine oxidase. *Pharmacol Rev.* **16**, 179–191.

KOPIN, I. J. (1968a) False adrenergic transmitters. *Ann. Rev. Pharmacol.* **8**, 377–394.

KOPIN, I. J. (1968b) The influence of false adrenergic transmitters on adrenergic neurotransmission. *In* 'Adrenergic Neurotransmission', Ciba Foundation Study Group No. 33, G. E. W. Wolstenholme and M. O'Connor, eds. J. & A. Churchill Ltd., London, pp. 95–104.

KOPIN, I. J. (1972) Metabolic degradation of catecholamines. The relative importance of different pathways under physiological conditions and after administration of drugs. *In* 'Catecholamines'. H. Blaschko and E. Muscholl, eds. Springer-Verlag, Berlin, Heidelberg, New York, pp. 270–282.

KOPIN, I. J. and SILBERSTEIN, S. D. (1972) Axons of sympathetic neurons: transport of enzymes *in vivo* and properties of axonal sprouts *in vitro*. *Pharmacol. Rev.* **24**, 245–254.

KOPIN, I. J., BREESE, G. R., KRAUSS, K. R. and WEISE, V. K. (1968) Selective release of newly synthesized norepinephrine from the cat spleen during sympathetic nerve stimulation. *J. Pharmacol. exp. Ther.* **161**, 271–278.

KOPIN, I. J. and GORDON, E. K. (1962) Metabolism of norepinephrine-H³ released by tyramine and reserpine. *J. Pharmacol. exp. Ther.* **138**, 351–359.

KOPIN, I. J. and GORDON, E. K. (1963) Metabolism of administered and drug-released norepinephrine-7-H^3 in the rat. *J. Pharmacol. exp. Ther.* **140**, 207–216.

KOROCHKIN, L. I. and KOROCHKINA, L. S. (1970) Hormonal influence on the differentiation of nerve cells of sympathetic and parasympathetic nervous system. *Z. Mikrosk. Anat. Forsch.* **82**, 293–321.

KOSTERLITZ, H. W. and LEES, G. M. (1961) Action of the bretylium on the isolated guinea-pig ileum. *Brit. J. Pharmacol.* **17**, 82–86.

KOSTERLITZ, H. W. and LEES, G. M. (1972) Interrelationships between adrenergic and cholinergic mechanisms. *In* 'Catecholamines'. H. Blaschko and E. Muscholl, eds. Springer-Verlag, Berlin, Heidelberg, New York. pp. 762–812.

KRAYER, O. and FUENTES, T. (1958) Changes of heart rate caused by direct cardiac action of reserpine. *J. Pharmacol. exp. Ther.* **123**, 145–152.

KUBO, T., MISHIO, H. and MISU, Y. (1970) The acute partial failure of neuromuscular transmission by reserpine in the isolated perfused rabbit's heart. *Jap. J. Pharmacol.* **20**, 442–445.

KUNTZ, A. (1910) The development of the sympathetic nervous system in mammals. *J. comp. Neurol.* **20**, 211–258.

KUNTZ, A. (1912) The development of the adrenals in the turtle. *Amer. J. Anat.* **13**, 71–89.

KUNTZ, A. (1920–21) The development of the sympathetic nervous system in man. *J. comp. Neurol.* **32**, 173–229.

KUNTZ, A. (1953) 'The Autonomic Nervous System'. Lea & Febiger, Philadelphia.

KUNTZMANN, R., COSTA, E., GESSA, G. L. and BRODIE, B. B. (1962) Reserpine and guanethidine action on peripheral stores of catecholamines. *Life Sci.* **3**, 65–74.

KUPFERMAN, A., GILLIS, C. N. and ROTH, R. H. (1970) Influence of sympathetic nerve stimulation on conversion of H^3-tyrosine to H^3-catecholamine and on H^3-norepinephrine disposition in rabbit pulmonary artery. *J. Pharmacol. exp. Ther.* **171**, 214–222.

KURIYAMA, H. (1963) Electrophysiological observations on the motor innervation of the smooth muscle cells in the guinea-pig vas deferens. *J. Physiol. (Lond.).* **169**, 213–228.

KURIYAMA, H. (1970) Effects of ions and drugs on the electrical activity of smooth muscle. *In* 'Smooth Muscle'. E. Bülbring, A. F. Brading, A. W. Jones and T. Tomita, eds. Edward Arnold, London, pp. 366–395.

KVETNANSKÝ, R., GEWIRTZ, G. P., WEISE, V. K. and KOPIN, I. J. (1971) Enhanced synthesis of adrenal dopamine β-hydroxylase induced by repeated immobilization in rats. *Mol. Pharmacol.* **7**, 81–86.

LABATE, J. S. (1941) Influence of cocaine on uterine reactions induced by adrenaline and hypogastric nerve stimulation. *J. Pharmacol. exp. Ther.* **72**, 370–382.

LADURON, P. (1970) Differential accumulation of various enzymes in constricted sciatic nerves. *Arch. int. Pharmacodyn. Therap.* **185**, 200–203.
LADURON, P. and BELPAIRE, F. (1968a) Transport of noradrenaline and dopamine-β-hydroxylase in sympathetic nerves. *Life Sci.* **7**, 1–7.
LADURON, P. and BELPAIRE, F. (1968b) Tissue fractionation and catecholamines. II. Intracellular distribution patterns of tyrosine hydroxylase, dopa decarboxylase, dopamine-α-hydroxylase, phenylethanolamine N-methyltransferase and monoamine oxidase in adrenal medulla. *Biochem. Pharmacol.* **17**, 1127–1140.
LAGERCRANTZ, H. (1971) Isolation and characterization of sympathetic nerve trunk vesicles. *Acta physiol. scand. Suppl.* **366**, 1–44.
LAGERSPETZ, K. Y. H. and HISSA, R. (1968) Catecholamines and medullary volume in the adrenal glands of young mice. *Acta Endocr.* **57**, 473–477.
LANDS, A. M., ARNOLD, A., MCAULIFF, J. P., LUDUENA, F. P. and BROWN, T. G. (1967) Differentiation of receptor systems activated by sympathomimetic amines. *Nature (Lond.).* **214**, 597–598.
LANE, B. P. and RHODIN, J. A. G. (1964) Cellular interrelationships and electrical activity in two types of smooth muscle. *J. Ultrastruct. Res.* **10**, 470–488.
LANGENDORFF, O. (1900) Zur Deutung der 'Paradoxen'. Pupillenerweiterung. *Klin. Mbl. Augenheilk.* **38**, 823–827.
LANGER, S. Z. (1966a) The degeneration contraction of the nictitating membrane in the unanesthetized cat. *J. Pharmacol. exp. Ther.* **151**, 66–72.
LANGER, S. Z. (1966b) Presence of tone in the denervated and in the decentralized nictitating membrane of the spinal cat and its influence on determination of supersensitivity. *J. Pharmacol. exp. Ther.* **154**, 14–34.
LANGER, S. Z. (1970) The metabolism of ³H-noradrenaline-released by electrical stimulation from the isolated nictitating membrane of the cat and from the vas deferens of the rat. *J. Physiol. (Lond.).* **208**, 515–546.
LANGER, S. Z. (1973a) The regulation of transmitter release elicited by nerve stimulation through a presynaptic feed-back mechanism. *In* 'Frontiers in catecholamine Research'. E. Usdin and S. Snyder, eds. Pergamon Press, Oxford, pp. 543–550.
LANGER, S. Z. (1973b) The role of the alpha-adrenergic receptor in regulating transmitter release elicited by nerve stimulation. *Acta Physiol. Latino. Amer.* **23**, 236–237.
LANGER, S. Z., STEFANO, F. J. E. and ENERO, M. A. (1972) Pre- and post-synaptic origin of the norepinephrine metabolites formed during transmitter release elicited by nerve stimulation. *J. Pharmacol. exp. Ther.* **183**, 90–102.
LANGER, S. Z. and VOGT, M. (1969) Release of noradrenaline from the isolated nictitating membranes of the cat by electrical stimulation. Abstracts of the 4th Int. Congr. Pharmacol., p. 141. Basel, Switzerland.
LANGER, S. Z. and VOGT, M. (1971) Noradrenaline release from isolated muscle of the nictitating membrane of the cat. *J. Physiol. (Lond.).* **214**, 159–171.

LANGER, S. Z., ADLER, E., ENERO, M. A. and STEFANO, F. J. E. (1971) The role of the alpha receptor in regulating noradrenaline overflow by nerve stimulation. XXVth Int. Congr. Physiol. Sciences, p. 335.

LANGLEY, J. N. (1897/98) On the regeneration of pre-ganglionic and of post-ganglionic visceral nerve fibres. *J. Physiol. (Lond.).* **22**, 215–230.

LANGLEY, J. N. (1901) Observations on the physiological action of extracts of the supra-renal bodies. *J. Physiol. (Lond.).* **27**, 237–256.

LANGLEY, J. N. (1905) On the reaction of cells and of nerve endings to certain poisons; chiefly as regards the reaction of striated muscle to nicotine and to curare. *J. Physiol. (Lond.).* **33**, 374–413.

LANGLEY, J. N. (1921). The Autonomic Nervous System. Part I. Heffer, Cambridge.

LANGLEY, J. N. and ANDERSON, H. K. (1904) On the union of the fifth cervical nerve with the superior cervical ganglion. *J. Physiol. (Lond.).* **31**, 439–442.

LARRABEE, M. G. (1970) Metabolism of adult and embryonic sympathetic ganglia. *Fed. Proc.* **29**, 1919–1928.

LAWRENTJEW, B. I. (1925) Über die Erscheinungen der Degeneration und Regeneration in sympathischen Nervensystem. *Z. Mikrosk. Anat. Forsch.* **2**, 201–223.

LEADERS, F. E. (1963) Local cholinergic-adrenergic interaction: mechanism for the biphasic chronotropic response to nerve stimulation. *J. Pharmacol. exp. Ther.* **142**, 31–38.

LE COMPTE, P. M. (1941) Observations on the return of vascular tone after sympathectomy. *Amer. J. Physiol.* **135**, 43–57.

LEE, F. L. (1967) The relation between norepinephrine content and response to sympathetic nerve stimulation of various organs of cats pretreated with reserpine. *J. Pharmacol. exp. Ther.* **156**, 137–141.

LEMPINEN, M. (1964) Extra-adrenal chromaffin tissue of the rat and the effect of cortical hormones on it. *Acta physiol. scand. Suppl.* **231**, 1–91.

LEVER, J. D. and ESTERHUIZEN, A. C. (1961) Fine structure of the arteriolar nerves in the guinea-pig pancreas. *Nature (Lond.).* **192**, 566–567.

LEVER, J. D., GRAHAM, J. D. P., IRVINE, G. and CHICK, W. J. (1965) The vesiculated axons in relation to arteriolar smooth muscle in the pancreas. A fine structural and quantitative study. *J. Anat.* **99**, 299–313.

LEVER, J. D., SPRIGGS, T. L. B. and GRAHAM, J. D. P. (1968) A formol-fluorescence, fine-structural and autoradiographic study of the adrenergic innervation of the vascular tree in the intact and sympathectomized pancreas of the cat. *J. Anat.* **103**, 15–34.

LEVI, G. (1925) Die strukturelle Grundlage der Körpergrösse bei Vollausgebildeten und im Wachstum begriffenen Tieren. *Ergebn. Anat. Entw. Gesch.* **86**, 87–342.

LEVI, G. (1946). Accrescimento e senescenza. La 'Nuova Italia', Florence, 1946.

LEVI-MONTALCINI, R. and BOOKER, B. (1960) Excessive growth of the sympathetic ganglia evoked by a protein isolated from mouse salivary glands. *Proc. nat. Acad. Sci., U.S.A.* **46**, 373–384.

LEVI-MONTALCINI, R. (1971) The Nerve Growth Factor. *In* 'Immunosympathectomy'. G. Steiner and E. Schönbaum, eds. Elsevier Publ. Co., Amsterdam, London, New York, pp. 25–36.

LEVI-MONTALCINI, R. and ANGELETTI, P. U. (1961) Growth control of the sympathetic system by a specific protein factor. *Quart. Rev. Biol.* **36**, 99–108.

LEVI-MONTALCINI, R. and ANGELETTI, P. U. (1966) Immunosympathectomy. *Pharmacol. Rev.* **18**, 619–628.

LEVI-MONTALCINI, R. and ANGELETTI, P. U. (1968) Nerve growth factor. *Physiol. Rev.* **48**, 534–569.

LEVI-MONTALCINI, R., MEYER, H. and HAMBURGER, V. (1954) *In vitro* experiments on the effects of Mouse sarcomas 180 and 37 on the spinal and sympathetic ganglia of the chick embryo. *Cancer Res.* **14**, 49–57.

LEVIN, J. A. and FURCHGOTT, R. F. (1970) Interactions between potentiating agents of adrenergic amines in rabbit aortic strips. *J. Pharmacol. exp. Ther.* **172**, 320–331.

LEVITT, M., SPECTOR, S., SJOERDSMA, A. and UDENFRIEND, S. (1965) Elucidation of the rate-limiting step in norepinephrine biosynthesis in the perfused guinea-pig heart. *J. Pharmacol. exp. Ther.* **148**, 1–8.

LEWANDOWSKY, M. (1900) Wirkung des Neberinierenextractes auf die glatten Muskeln der Hant. *Zbl. Physiol.* **14**, 433.

LIBET, B. (1964) Slow synaptic responses and excitatory changes in sympathetic ganglia. *J. Physiol. (Lond.).* **174**, 1–25.

LIBET, B. (1970) Generation of slow inhibitory and excitatory postsynaptic potentials. *Fed. Proc.* **29**, 1945–1956.

LIBET, B. and TOSAKA, T. (1969) Slow inhibitory and excitatory postsynaptic responses in single cells of mammalian sympathetic ganglia. *J. Neurophysiol.* **32**, 43–50.

LIBET, B. and TOSAKA, T. (1970) Dopamine as a synaptic transmitter and modulator in sympathetic ganglia: A different mode of synaptic action. *Proc. Nat. Acad. Sci. U.S.A.* **67**, 667–673.

LIGHTMAN, S. L. and IVERSEN, L. L. (1969) The role of Uptake$_2$ in the extra neuronal metabolism of catecholamines in the isolated rat heart. *Brit. J. Pharmacol.* **37**, 638–649.

LINDMAR, R., MUSCHOLL, E. and SPRENGER, E. (1967) Funktionelle Bedeutung der Freisetzung von Dihydroxyephedrin und Dihydroxypseudoephedrin als falschen sympathischen Überträgerstoffen am Hersen. *Naunyn. Schmied. Arch. exp. Path. Pharmak.* **256**, 1–25.

LIU, A. C. and ROSENBLUETH, A. (1935) Reflex liberation of circulating sympathin. *Amer. J. Physiol.* **113**, 555–559.

LIVETT, B. G., GEFFEN, L. B. and AUSTIN, L. (1968) Proximodistal transport of ^{14}C-noradrenaline and protein in sympathetic nerves. *J. Neurochem.* **15**, 931–939.

LIVETT, B. G., GEFFEN, L. B. and RUSH, R. A. (1971) Immunochemical methods for demonstrating macromolecules in sympathetic neurons. *Phil. Trans. Roy. Soc. Lond. Ser. B.* **261**, 359–361.

LJUNG, B. (1970) Nervous and myogenic mechanisms in the control of a vascular neuroeffector system. *Acta Physiol. Scand. Suppl.* **349**, 33–68.

LOEWENSTEIN, W. R. and ALTAMIRANO-ORREGO, R. (1956) Enhancement of activity in pacinian corpuscle by sympathomimetic agents. *Nature (Lond.).* **178**, 1292–1293.

LOEWI, O. (1921) Uber humorale Ubertragbarkeit der Herznervenwirkung. *Pflüg. Arch. ges. Physiol.* **189**, 239–242.

LÖFFELHOLZ, K. and MUSCHOLL, E. (1970) Inhibition by parasympathetic nerve stimulation of the release of the adrenergic transmitter. *Naunyn. Schmied. Arch. exp. Path. Pharmak.* **267**, 181–184.

LOVENBERG, W., WEISSBACH, H. and UDENFRIEND, S. (1962) Aromatic L-amino acid decarboxylase. *J. biol. Chem.* **237**, 89–93.

LUBÍNSKA, L. and NIERMIERKO, S. (1971) Velocity and intensity of bidirectional migration of acetylcholinesterase in transected nerves. *Brain Res.* **27**, 329–342.

LUNDBERG, D. (1969) Adrenergic neuron blockers and transmitter release after sympathetic denervation studied in the conscious rat. *Acta physiol. scand.* **75**, 415–426.

LUNDBERG, D. (1971) Studies on 6-hydroxydopamine in conscious rats. In '6-Hydroxydopamine and Catecholamine Neurons'. T. Malmfors and H. Thoenen, eds. North Holland Publ. Co., Amsterdam-London, pp. 225–237.

LUNDBORG, P. (1967) Studies on the uptake and subcellular distribution of catecholamines and their α-methylated analogues. *Acta physiol. scand. Suppl.* **302**, 3–34.

LUNDBORG, P. and STITZEL, R. E. (1968) Studies on the dual action of guanethidine in sympathetic nerves. *Acta physiol. scand.* **72**, 100–107.

LUTZ, G. (1968) Die Entwicklung des Halssympathicus und des Nervus vertebralis. *Z. Anat. EntwGesch.* **127**, 187–200.

MCAFEE, D. A. and GREENGARD, P. (1972) Adenosine 3′, 5′-Monophosphate: Electrophysiological evidence for a role in synaptic transmission. *Science.* **178**, 310–312.

MCCULLOCH, M. W., RAND, M. J. and STORY, D. F. (1972) Inhibition of ^{3}H-noradrenaline release from sympathetic nerves of guinea-pig atria by a presynaptic α-adrenoceptor mechanisms. *Brit. J. Pharmacol.* **46**, 523–524P.

MACHADO, A. B. M. (1971) Electron microscopy of developing sympathetic fibres in the rat pineal body. The formation of granular vesicles. *Progr. Brain Res.* **34**, 171–185.

MACHADO, C. R. S., WRAGG, L. E. and MACHADO, A. B. M. (1968) A histochemical study of sympathetic innervation of 5-hydroxytryptamine in the developing pineal body of the rat. *Brain Res.* **8**, 310–318.

MACHIDA, K. (1929) Observations on the degeneration and renegeration of postganglionic nerve fibers. *Bull Johns. Hopk. Hosp.* **45**, 247–263.

MACHOVA, J. and KRISTOFOVA, A. (1973) The effect of dibutyryl-cyclic AMP, dopamine and aminophylline on ganglionic surface potential and transmission. *Life Sci.* **13**, 525–536.

MCLEAN, J. R. and BURNSTOCK, G. (1972) Axoplasmic flow of adrenaline and monoamine oxidase in amphibian sympathetic nerves. *Z. Zellforsch. Mikrosk. Anat.* **124**, 44–56.

MCMAHAN, U. J. and KUFFLER, S. W. (1971) Visual identification of synaptic boutons on living ganglion cells and of varicosities in postganglionic axons in the heart of the frog. *Proc. Roy. Soc. Lond. Ser. B.* **177**, 485–508.

MACMILLAN, W. H. (1959) A hypothesis concerning the effect of cocaine on the action of sympathomimetic amines. *Brit. J. Pharmacol.* **14**, 385–391.

MAINLAND, J. F. and SHAW, F. H. (1952) Comparison of ganglionic blocking agents. *Nature (Lond.).* **170**, 418–419.

MAITRE, L. and STAEHELIN, M. (1967) On the norepinephrine replacement of α-methyl-norepinephrine in the rat heart after treatment with α-methyl-DOPA. *Experientia (Basel).* **23**, 810–811.

MALAMED, S., POISNER, A. M., TRIFARÓ, J. M. and DOUGLAS, W. W. (1968) The fate of the chromaffin granule during catecholamine release from the adrenal medulla. III. Recovery of a purified fraction of electron translucent granules. *Biochem. Pharmacol.* **17**, 241–246.

MALIK, K. U. and LING, G. M. (1969) Modification by acetylcholine of the response of rat mesenteric arteries to sympathetic stimulation. *Circulation Res.* **25**, 1–9.

MALIK, K. U. and MUSCHOLL, E. (1969) The effect of α-methyldopa on the vasoconstrictor responses of the rat mesenteric artery preparation to nerve stimulation. *Arzneim. Forsch* **19**, 1111–1113.

MALMFORS, T. (1965) Studies on adrenergic nerves. The use of rat and mouse iris for direct observations on their physiology and pharmacology at cellular and subcellular levels. *Acta physiol. scand. Suppl.* **248**, 1–93.

MALMFORS, T. (1969) Histochemical studies on the release of the adrenergic transmitter by nerve impulses in combination with drugs especially adrenergic neuron blocking agents. *Pharmacol. Rev.* **2**, 138–150.

MALMFORS, T. (1971) The effects of 6-hydroxydopamine on the adrenergic nerves as revealed by the fluorescence histochemical method. *In* '6-Hydroxydopamine and Catecholamine Neurons'. T. Malmfors and H. Thoenen, eds. North-Holland Publ. Co., Amsterdam, London, pp. 37–58.

MALMFORS, T., FURNESS, J. B., CAMPBELL, G. R. and BURNSTOCK, G. (1971)

Re-innervation of smooth muscle of the vas deferens transplanted into the interior chamber of the eye. *J. Neurobiol.* **2**, 193–207.

MALMFORS, T. and OLSON, L. (1967) Adrenergic reinnervation of anterior chamber transplants. *Acta physiol. scand.* **71**, 401–402.

MALMFORS, T. and SACHS, ch. (1965a) Direct demonstration of the system of terminals belonging to an individual adrenergic neuron and their distribution in the rat iris. *Acta physiol. scand.* **64**, 377–382.

MALMFORS, T. and SACHS, ch. (1965b) Direct studies on the disappearance of the transmitter and changes in the uptake-storage mechanisms of degenerating adrenergic nerves. *Acta physiol. scand.* **64**, 211–223.

MALMFORS, T. and SACHS, ch. (1968) Degeneration of adrenergic nerves produced by 6-hydroxydopamine. *Eur. J. Pharmacol.* **3**, 89–92.

MALMFORS, T. and THOENEN, H. (1971) 6-Hydroxydopamine and Catecholamine Neurons. T. Malmfors and H. Thoenen, eds. North-Holland Publ. Co., Amsterdam, London.

MANNARD, A. and POLOSA, C. (1973) Analysis of background firing of single sympathetic preganglionic neurons of cat cervical nerve. *J. Neurophysiol.* **36**, 398–408.

MARK, G. E., CHAMLEY, J. H. and BURNSTOCK, G. (1973) Interactions between autonomic nerves and smooth and cardiac muscle cells in tissue culture. *Dev. Biol.* **32**, 194–200.

MASCETTI, G. G. (1972) An attempt to localize the cervical sympathetic action on the retina of the cat. *Brain Res.* **41**, 221–224.

MATSUDA, H. (1970) Ultrastructural localization of cholinesterase activity in the nervous system of the rabbit iris dilator. *Jap. J. Opthalmol.* **14**, 21–28.

MATSUMOTO, T. (1920) The granules, vacuoles and mitochondria in the sympathetic nerve-fibers cultivated *in vitro*. *Johns Hopk. Hosp. Bull.* **31**, 91–93.

MATTHEWS, M. R. (1972) Evidence suggesting translocation of small dense cored vesicles from the cell bodies to regenerating axon tips of adrenergic neurons. *J. Anat.* **111**, 508–510.

MATTHEWS, M. R. and RAISMAN, G. (1969) The ultrastructure and somatic efferent synapses of small granule-containing cells in the superior cervical ganglion. *J. Anat.* **105**, 255–282.

MATTHEWS, M. R. and RAISMAN, G. (1972) A light and electron microscopic study of the cellular response to axonal injury in the superior cervical ganglion of the rat. *Proc. Roy. Soc. Lond. Ser. B.* **181**, 43–79.

MAXWELL, R. A., PLUMMER, A. J., OSBORNE, M. W. and ROSS, S. D. (1956) Evidence for a peripheral action of reserpine. *J. Pharmacol. exp. Ther.* **116**, 42.

MAXWELL, R. A., PLUMMER, A. J., SCHNEIDER, F., POUALSKI, H. and DANIEL, A. I. (1960) Pharmacology of 2-(octahydro-1-azocinyl)-ethyl-guanidine-sulfate (SU-5864). *J. Pharmac. exp. Ther.* **128**, 22–29.

MAYOR, D., BANKS, P., TOMLINSON, D. R. and GRIGARIS, R. (1971) Noradrenaline transport in sympathetic nerves. *Progr. Brain Res.* **34**, 489–498.

MECKEL, J. F. (1751) Observation anatomique avec l'examen physiologique du veritable usage des noeuds, ou ganglions des nerfs. *Mem. Acad. Roy. d. Soc.* **5**, 84.

MEKATA, F. (1973) Electrophysiological studies of the smooth muscle cell membrane of the rabbit common carotid artery. *J. Gen. Physiol.* **57**, 738–751.

MELVILLE, K. I. (1937) The antisympathomimetic action of dioxane compounds (F883 and F933) with special reference to the vascular responses to dihydroxyphenyl ethanolamine (arterenol) and nerve stimulation. *J. Pharmacol. exp. Ther.* **59**, 317–327.

MERRILLEES, N. C. R. (1968) The nervous environment of individual smooth muscle cells of the guinea-pig vas deferens. *J. Cell Biol.* **37**, 794–817.

MERRILLEES, N. C. R., BUENSTOCK, G. and HOLMAN, M. E. (1963) Correlation of fine structure and physiology of the innervation of smooth muscle in the guinea-pig vas deferens. *J. Cell Biol.* **19**, 529–550.

MILLS, E. (1968) Activity of aortic chemoreceptors during electrical stimulation of the stellate ganglion in the cat. *J. Physiol. (Lond.).* **199**, 103–114.

MIRKIN, B. L. (1972) Ontogenesis of the adrenergic nervous system: functional and pharmacologic implications. *Fed. Proc.* **31**, 65–73.

MOHAMMED, S., GAFFNEY, T. E., YARD, A. C. and GOMEZ, H. (1968) Effect of methyldopa, reserpine and guanethidine on hindleg vascular resistance. *J. Pharmacol. exp. Ther.* **160**, 300–307.

MOLINOFF, C. and AXELROD, J. (1969) Octopamine: normal occurrence in sympathetic nerves of rats. *Science.* **164**, 428–429.

MOLINOFF, P. B. and AXELROD, J. (1971) Biochemistry of catecholamines. *Ann. Rev. Biochem.* **40**, 465–500.

MOLINOFF, P. B., BRIMIJOIN, S., WEINSHILBOUM, R. and AXELROD, J. (1970) Neurally mediated increase in dopamine-β-hydroxylase activity. *Proc. nat. Acad. Sci. U.S.A.* **66**, 453–458.

MONTANARI, R., COSTA, E., BEAVEN, M. A. and BRODIE, B. B. (1963) Turnover of norepinephrine in heart of intact mice, rats and guinea-pigs using tritiated norepinephrine. *Life Sci.* **2**, 232–240.

MOORE, J. I. (1966) Potentiation of the cardiac and pressor responses to electrical stimulation of the cardiac sympathetic nerves by cocaine in open chest of dogs. *J. Pharmacol. exp. Ther.* **153**, 218–224.

MORALES-AGUILERA, A. and VAUGHAN WILLIAMS, E. M. (1965) The effects on cardiac muscle of β-receptor antagonists in relation to their activity as local anaesthetics. *Brit. J. Pharmacol.* **24**, 332–338.

MORAN, N. C. (1967) The development of β-adrenergic blocking drugs: A retrospective and prospective evaluation. *Ann. N.Y. Acad. Sci.* **139**, 649–660.

MOTT, J. C. (1961) The stability of the cardiovascular system. *In* 'Somatic

Stability in the Newly Born'. A Ciba Foundation Symposium. G. E. W. Wolstenholme and M. O'Connor, eds. Boston, Little, Brown Co., pp. 192-214.

MUELLER, R. A., THOENEN, H. and AXELROD, J. (1969a) (1) Compensatory increase in adrenal tyrosine hydroxylase activity after chemical sympathectomy. *Science*. **163**, 468-469.

MUELLER, R. A., THOENEN, H. and AXELROD, J. (1969b) Inhibition of trans-synaptically increased tyrosine hydroxylase activity by cycloheximide and actinomycin D. *Mol. Pharmacol*. **5**, 463-469.

MUELLER, R. A., THOENEN, H. and AXELROD, J. (1969c) Increase in tyrosine hydroxylase activity after reserpine administration. *J. Pharmacol. exp. Ther.* **169**, 74-79.

MURRAY, M. R. (1965) Nervous tissues *in vitro*. In 'Cells and Tissues in Culture Methods. Biology and Physiology., 2'. E. N. Willmer, ed. Academic Press, London, N.Y., pp. 373-455.

MURRAY, M. R. and STOUT, A. P. (1947) Adult human sympathetic ganglion cells cultivated *in vitro*. *Amer. J. Anat.* **80**, 225-273.

MUSACCHIO, J. M., FISCHER, J. E. and KOPIN, I. J. (1966) Subcellular distribution and release by sympathetic nerve stimulation of dopamine and methyldopamine. *J. Pharmacol. exp. Ther.* **152**, 51-55.

MUSACCHIO, J. M. and GOLDSTEIN, M. (1963) Biosynthesis of norepinephrine and norsynephrine in the perfused rabbit heart. *Biochem. Pharmacol.* **12**, 1061-1063.

MUSACCHIO, J. M. and WEISE, V. K. (1965) Effects of decentralization on norepinephrine biosynthesis from tyrosine, dopa and dopamine. *Pharmacologist*. **7**, 156.

MUSCHOLL, E. (1961) Effect of cocaine and related drugs on the uptake of noradrenaline by heart and spleen. *Brit. J. Pharmacol.* **16**, 352-359.

MUSCHOLL, E. (1966) Indirectly acting sympathomimetic amines. *Pharmacol. Rev.* **18**, 551-559.

MUSCHOLL, E. (1970) Cholinomimetic Drugs and Release of the Adrenergic Transmitter. *In* Bayer Symposium II; 'New Aspects of Storage and Release Mechanisms of Catecholamines'. H. J. Shümann and G. Kroneburg, eds. Springer-Verlag, Berlin, Heidelberg, New York, pp. 168-186.

MUSCHOLL, E. (1972) Adrenergic false transmitters. *In* 'Catecholamines'. H. Blaschko and E. Muscholl, eds. Springer-Verlag, Berlin, Heidelberg, New York, pp. 618-660.

MUSCHOLL, E. (1973) Muscarinic inhibition of the norepinephrine release from peripheral sympathetic fibres. *In* 'Pharmacology and the Future of Man'. *Proc. 5th Int. Congr. Pharmacol. San Francisco*, 1972, Vol. **4**, 440-457. Karger, Basel.

MUSCHOLL, E., LINDMAR, R., LÖFFELHOLZ, K. and FOZARD, J. R., (1973) Muscarinic inhibition of the release of the adrenergic transmitter from peripheral sympathetic fibres. *Acta Physiol. Polonica.* **24**, 177-184.

MUSCHOLL, E. and SPRENGER, E. (1966) Vergleichende Untersuchung der Blutdruckwirkung, Aufnahme und Speicherung von Dihydroxyephedrin (α-methyladrenalin) und Dihydroxypseudoephedrin. *Naunyn Schmied. Arch. exp. Path. Pharmak.* **254**, 109–124.

MUSCHOLL, E. and VOGT, M. (1958) The action of reserpine on the peripheral sympathetic system. *J. Physiol. (Lond.).* **141**, 132–155.

MUSCHOLL, E. and VOGT, M. (1964) Secretory responses of extra medullary chromaffin tissue. *Brit. J. Pharmacol.* **22**, 193–203.

MUSICH, J. and HUBBART, J. I. (1972) Biological sciences. Release of protein from mouse motor nerve terminal. *Nature (Lond.).* **237**, 279–281.

MUSTONEN, T. and TERÄVÄINEN, H. (1971) Synaptic connections of the paracervical (Frankenhäuser) ganglion of the rat uterus examined with the electron microscope after division of the sympathetic and sacral parasympathetic nerves. *Acta physiol. scand.* **82**, 264–267.

MYLECHARANE, E. J. and RAPER, C. (1973) Further studies on the adrenergic neuron blocking activity of some β-adrenoreceptor antagonists and guanethidine. *J. Pharmac. Pharmacol.* **25**, 213–220.

NAGASAWA, J. and MITO, S. (1967) Electron microscopic observations on the innervation of the smooth muscle. *Tohoku J. exp. Med.* **91**, 277–293.

NAGATSU, T., LEVITT, M. and UDENFRIEND, S. (1964) Conversion of L-tyrosine to 3,4-dihydroxyphenylalanine by cell-free preparations of brain and sympathetically innervated tissues. *Biochem, Biophys. Res. Comm.* **14**, 543–549.

NAGATSU, T. and UDENFRIEND, S. (1972) Photometric assay of dopamine-β-hydroxylase activity in human blood. *Clinic. Chem.* **18**, 980–983.

NAKAZATO, Y. and OHGA, A. (1972) Acute inhibitory effects on the sympathetically induced response of dog stomach. *Jap. J. Pharmacol.* **22**, 167–174.

NEFF, N. H. and GORIDIS, C. (1972) Neuronal monoamine oxidase; specific enzyme types and their rates of formation. In 'Monoamineoxidases New Vistas'. Advances in Biochemical Pharmacology 5. E. Costa and M. Sandler, eds. Raven Press, N.Y., pp. 307–323.

NELSON, E. and RENNELS, M. (1970) Innervation of intracranial arteries. *Brain.* **93**, 475–490.

NICKEL, E. and POTTER, L. T. (1970) Synaptic vesicles in freeze-etched electric tissue of *Torpedo*. *Brain Res.* **23**, 95–100.

NICKEL, E. and POTTER, L. T. (1971) Synaptic vesicles in freeze-etched electric tissue of *Torpedo*. *Phil. Trans. Roy. Soc. Lond. Ser. B.* **261**, 383–385.

NIELSEN, K. C., OWMAN, ch, and SPORRONG, B. (1971) Ultrastructure of the autonomic innervation apparatus in the main pial arteries of rats and cats. *Brain Res.* **27**, 25–32.

NILSSON, B. Y. (1972) Effects of sympathetic stimulation on mechanoreceptors of cat vibrissae. *Acta physiol. scand.* **85**, 390–397.

NINOMIYA, I., IRISAWA, A. and NISIMARU, N. (1973) Non-uniformity of sympathetic nerve activity to the skin and kidney. *Amer. J. Physiol.* **224**, 256–264.

NORBERG, K.-A. (1964) Adrenergic innervation of the intestinal wall studied by fluorescence microscopy. *Int. J. Neuropharmacol.* **3**, 379–382.

NORBERG, K.-A. (1967) Transmitter histochemistry of the sympathetic adrenergic nervous system. *Brain Res.* **5**, 125–170.

NORBERG, K.-A. and HAMBERGER, B. (1964) The sympathetic adrenergic neuron. *Acta physiol. scand. Suppl.* **238**, 1–42.

NORBERG, K.-A., RITZÉN, M. and UNGERSTEDT, U. (1966) Histochemical studies on a special catecholamine-containing cell type in sympathetic ganglia. *Acta physiol. scand.* **67**, 260–270.

NORBERG, K.-A. and SJÖQVIST, F. (1966) New possibilities for adrenergic modulation of ganglionic transmission. *Pharmacol. Rev.* **18**, 743–751.

NORDENFELDT, I. (1965) Choline acetylase in salivary glands of the cat after sympathetic denervation. *Quart. J. exp. Physiol.* **50**, 57–64.

NORMANN, T. C. (1965) The neurosecretory system of the adult *Calliphora Erythrocephala*. I. The fine structure of the corpus cardiacum with some observations on adjacent organs. *Z. Zellforsch. Mikrosk. Anat.* **67**, 461–501.

O'BRIEN, R. A., DA PRADA, M. and PLETSCHER, A. (1972) The ontogenesis of catecholamines and adenosine – 5'-triphosphate in the adrenal medulla. *Life Sci.* **11**, 749–759.

OCHI, J., KONISHI, M., YOSHIKAWA, H. and SANO, Y. (1968) Fluorescence and electron microscopic evidence for the dual innervation of the iris sphincter muscle of the rabbit. *Z. Zellforsch. Mikrosk. Anat.* **91**, 90–95.

OCHOA, E., FISZER DE PLAZAS, S. and DE ROBERTIS, E. (1972) Conductance changes produced by L-norepinephrine on lipid membrane containing a proteolipid from the bovine spleen capsule. *Mol. Pharmacol.* **8**, 215–221.

O'DONNELL, S. R. and SAAR, N. (1973) A histochemical study of extraneuronal accumulation of NA in the guinea-pig trachea. *Brit. J. Pharmacol.* **49**, 267–278.

OHLIN, P. and STRÖMBLAD, B. C. R. (1963) Observations on the isolated vas deferens. *Brit. J. Pharmacol.* **20**, 299–306.

OLIVER, G. and SCHAFER, E. A. (1895) Physiological effects of extracts of the suprarenal capsules. *J. Physiol. (Lond.).* **18**, 230–279.

OLIVERIO, A. and STJÄRNE, L. (1965) Acceleration of noradrenaline turnover in the mouse heart by cold exposure. *Life Sci.* **4**, 2339–2343.

OLSON, L. (1967) Outgrowth of sympathetic adrenergic neurons in mice treated with a nerve-growth factor (N.G.F.). *Z. Zellforsch. Mikrosk. Anat.* **81**, 155–173.

OLSON, L. (1969) Intact and regenerating sympathetic noradrenaline axons in the rat sciatic nerve. *Histochemie.* **17**, 349–367.

OLSON, L. (1970) Fluorescence histochemical evidence for axonal growth and secretion from transplanted adrenal medullary tissue. *Histochemie.* **22**, 1–7.

OLSON, L. and MALMFORS, T. (1970) Growth characteristics of adrenergic nerves in the adult rat. *Acta physiol. scand. Suppl.* **348**, 1–112.

ORLOV, R. S. (1962) On impulse transmission from motor sympathetic nerve to smooth muscle. *Fiziol. Zh. SSSR.* **48**, 342–348.

OSSWALD, W., GUIMARÃES, S. and COIMBRA, A. (1971) The termination of action of catecholamines in the isolated venous tissue of the dog. *Naunyn Schmied. Arch. exp. Path. Pharmak.* **265**, 67–80.

OTTEN, U., PARAVICINI, U., OESCH, F. and THOENEN, H. (1973) Time requirement for the single steps of trans-synaptic induction of tyrosine hydroxylase in the peripheral sympathetic nervous system. *Naunyn Schmied. Arch. exp. Path. Pharmak.* **280**, 117–127.

OWMAN, Ch. (1964) Sympathetic nerves probably storing two types of monoamines in the rat pineal gland. *Int. J. Neuropharmacol.* **3**, 105–112.

OWMAN, Ch. (1965) Localization of neuronal and parenchymal monoamines under normal and experimental conditions in the mammalian pineal gland. *Progr. Brain Res.* **10**, 423–453.

OWMAN, Ch., EDVINSSON, L. and NIELSEN, K. C. (1974) Autonomic neuroreceptor mechanisms in brain vessels. *Blood Vessels.* **11**, 2–31.

OWMAN, Ch., SJÖBERG, N.-O., SJÖSTRAND, N.-O. and SWEDIN, G. (1970) Effect of high doses of oestrogen and progesterone on the noradrenaline content of the short adrenergic neurons innervating the male genital tract of the rat. *Acta endocr., Copenhagen.* **64**, 459–465.

OWMAN, Ch., SJÖBERG, N.-O. and SWEDIN, G. (1971) Histochemical and chemical studies on pre- and post-natal development of the different systems of 'short' and 'long' adrenergic neurons in peripheral organs of the rat. *Z. Zellforsch. Mikrosk. Anat.* **116**, 319–341.

OWMAN, Ch. and SJÖSTRAND, N.-O. (1965) Short adrenergic neurons and catecholamine-containing cells in vas deferens and accessory male genital glands of different mammals. *Z. Zellforsch. Mikrosk. Anat.* **66**, 300–320.

OWMAN, Ch., SJÖBERG, N.-O. and SJÖSTRAND, N.-O. (1974) Short adrenergic neurons, a peripheral neuroendocrine mechanism. *In* 'Amine Fluorescence Histochemistry'. M. Fujiwara and C. Tanaka, eds. Igaku Shoin Ltd., Tokyo. pp. 47–66.

PADDLE, B. M., and BURNSTOCK, G. (1974) Unpublished Results.

PAINTAL, A. S. (1964). Effects of drugs on vertebrate mechanoreceptors. *Pharmacol. Rev.* **16**, 341–380.

PALAIC, D. and PANISSET, J. C. (1969) Inhibition of the noradrenaline uptake in guinea-pig vas deferens by continuous nerve stimulation. *J. Pharm. Pharmacol.* **21**, 328–329.

PAPKA, R. E. (1972) Ultrastructural and fluorescence histochemical studies of developing sympathetic ganglia in the rabbit. *Amer. J. Anat.* **134**, 337–364.

PAPKA, R. E. (1973) The ultrastructure of adrenergic neurons in sympathetic ganglia of the newborn rabbit after treatment with 6-hydroxydopamine. *Amer. J. Anat.* **137**, 447–465.

PARTINGTON, P. P. (1936) The production of sympathin in response to physiological stimuli in the anesthetized animal. *Amer. J. Physiol.* **117**, 55–58.

PATERSON, A. M. (1890) Development of the sympathetic nervous system in mammals. *Phil. Trans. Roy. Soc. Lond. Ser. B.* **181**, 159–186.

PATERSON, G. (1965) The response to transmural stimulation of isolated arterial strip and its modifications by drugs. *J. Pharmac. Pharmacol.* **17**, 341–349.

PATIL, P. N. and JACOBOWITZ, D. (1968) Steric aspects of adrenergic drugs. IX. Pharmacologic and histochemical studies on isomers of cobefrin (α-methylnorepinephrine). *J. Pharmacol. exp. Ther.* **161**, 279–295.

PATON, W. D. M. and VIZI, E. S. (1969) The inhibitory action of noradrenaline and adrenaline on acetylcholine output by guinea-pig ileum longitudinal muscle strip. *Brit. J. Pharmacol.* **35**, 10–28.

PEIPER, U., GRIEBEL, L. and WENDE, W. (1971) Activation of vascular smooth muscle of rat aorta by noradrenaline and depolarization: two different mechanisms. *Pflüg. Arch. ges Physiol.* **330**, 74–89.

PELLEGRINO DE IRALDI, A. and DE ROBERTIS, E. (1968) The neurotubular system of the axon and the origin of granulated and non-granulated vesicles in regenerating nerves. *Z. Zellforsch. Mikrosk. Anat.* **87**, 330–344.

PELLEGRINO DE IRALDI, A. and DE ROBERTIS, E. (1970) Studies on the origin of the granulated and non-granulated vesicle. *In* 'New Aspects of Storage and Release Mechanisms of Catecholamines. H. J. Schümann and G. Kroneberg eds. Springer-Verlag, Berlin, Heidelberg, New York, pp. 4–19.

PELLEGRINO DE IRALDI, A. and SABURO, A. M. (1971) Two compartments in the granulated vesicles of the pineal nerves. *In* 'The Pineal Gland'. Ciba Foundation Symposium. G. E. W. Wolstenholme and Julie Knight, eds. J. & A. Churchill, London, pp. 177–195.

PERRI, V., SACCHI, O. and CASELLA, C. (1970a) Electrical properties and synaptic connections of the sympathetic neurons in the rat and guinea-pig superior cervical ganglion. *Pflüg. Arch. ges. Physiol.* **314**, 40–54.

PERRI, V., SACCHI, O. and CASELLA, C. (1970b) Synaptically mediated potentials elicited by the stimulation of postganglionic trunks in the guinea-pig superior cervical ganglion. *Pflüg. Arch. ges. Physiol.* **314**, 55–67.

PETRAS, J. M. and CUMMINGS, J. F. (1972) Autonomic neurons in the spinal cord of the rhesus monkey: a correlation of the findings of cytoarchitectonics and sympathectomy with fiber degeneration following dorsal rhizotomy. *J. comp. Neurol.* **146**, 189–218.

PFENNIGER, K., AKERT, K., MOOR, H. and SANDRI, C. (1971) Freeze-fracturing of presynaptic membranes in the central nervous system. *Phil. Trans. Roy. Soc. Lond. Ser. B.* **261**, 387.

PICK, J. (1970) 'The Autonomic Nervous System'. J. B. Lippincott Co., Philadelphia, Toronto.

PICK, J., GERDIN, C. and DELEMOS, C. (1964) An electron microscopical study of developing sympathetic neurons in man. *Z. Zellforsch. Mikrosk. Anat.* **62**, 402–415.

PILAR, G. and LANDMESSER, L. (1972) Axotomy mimicked by localized colchicine application. *Science.* **177**, 1116–1118.

PITTS, R. F., LARRABEE, M. G. and BRONK, D. W. (1941) An analysis of hypothalamic cardiovascular control. *Amer. J. Physiol.* **134**, 359–383.

POISNER, A. M., TRIFARÓ, J. M. and DOUGLAS, W. W. (1967) The fate of the chromaffin granule during catecholamine release from the adrenal medulla. II. Loss of protein and retention of lipid in subcellular fractions. *Biochem. Pharmacol.* **16**, 2101–2108.

POHORECKY, L. A. and WURTMAN, R. J. (1971) Adrenocortical control of epinephrine synthesis. *Pharmacol. Rev.* **23**, 1–35.

POLOSA, C. (1967) The silent period of sympathetic preganglionic neurons. *Canad. J. Physiol. Pharmacol.* **45**, 1033–1045.

POLOSA, C. (1968) Spontaneous activity of sympathetic preganglionic neurons. *Canad. J. Physiol. Pharmacol.* **46**, 887–896.

POTTER, L. T. (1966) Storage of norepinephrine in sympathetic nerves. *Pharmacol. Rev.* **18**, 439–451.

POTTER, L. T. and AXELROD, J. (1963) Studies on the storage of norepinephrine and the effect of drugs. *J. Pharmacol. exp. Ther.* **140**, 199–206.

POTTER, L. T., COOPER, T., WILLMAN, V. L. and WOLFE, D. E. (1965) Synthesis, binding, release, and metabolism of norepinephrine in normal and transplanted dog hearts. *Circulation Res.* **16**, 468–481.

POWELL, C. E. and SLATER, I. H. (1958) Blocking of inhibitory adrenergic receptors by a dichloro analog of isoproterenol. *J. Pharmacol. exp. Ther.* **122**, 480–488.

POWELL, J. R. and BRODY, M. J. (1973) Peripheral facilitation of reflex vasoconstriction by prostaglandin $F_2\alpha$. *J. Pharmacol. exp. Ther.* **187**, 495–500.

PURVES, R. D. (1974) Muscarinic excitation: a microelectrophoretic study on cultured smooth muscles. *Brit. J. Pharmacol.* **52**, 77–86.

PURVES, R. D., HILL, C. E., CHAMLEY, J., MARK, G. E., FRY, D. M. and BURNSTOCK, G. (1974) Functional autonomic neuromuscular junctions in tissue culture. *Pflüg. Arch. ges Physiol.* **350**, 1–7.

RAND, M. J. and VARMA, B. (1970) The effects of cholinomimetic drugs on responses to sympathetic nerve stimulation and noradrenaline in the rabbit ear artery. *Brit. J. Pharmacol.* **38**, 758–770.

RAND, M. J. and WILSON, J. (1967) The relationship between adrenergic neurone blocking activity and local anesthetic activity in a series of guanethidine derivatives. *Eur. J. Pharmacol.* **1**, 200–209.

RANSON, S. W. and BILLINGSLEY, P. R. (1918) The superior cervical ganglion and the cervical portion of the sympathetic trunk. *J. comp. Neurol.* **29**, 313-358.

RAPER, C. and MCCULLOCH, M. W. (1971) Adrenoreceptor classification. *Med. J. Aust.* **2**, 1331-1335.

READ, J. B. and BURNSTOCK, G. (1969a) A method for the localization of adrenergic nerves during early development. *Histochemie.* **20**, 197-200.

READ, J. B. and BURNSTOCK, G. (1969b) Adrenergic innervation of the gut musculature in vertebrates. *Histochemie.* **17**, 263-272.

READ, J. B. and BURNSTOCK, G. (1970) Development of the adrenergic innervation and chromaffin cells in the human fetal gut. *Dev. Biol.* **22**, 513-534.

REIFFESTEIN, R. J. (1968) Effects of cocaine on the rate of contraction to noradrenaline in the cat spleen strip: mode of action of cocaine. *Brit. J. Pharmacol.* **32**, 591-597.

REISS, M. (1955) The development of endocrine gland function in relation to brain function. *In* 'Biochemistry of the Developing Nervous System.' Heinrich Waelsch, ed. Academic Press, New York, pp. 450-458.

RESTI, M. (1962) Relievi istomorfologici ed istochimici in gangli simpatici lombari di conigli trattati con forti dosi di cortisone. *Acta chir. Ital.* **18**, 627-637.

RICHARDSON, K. C. (1962) The fine structure of autonomic nerve endings in smooth muscle of the rat vas deferens. *J. Anat.* **96**, 427-442.

RICHARDSON, K. C. (1964) The fine structure of the albino rabbit iris with special reference to the identification of adrenergic and cholinergic nerves and nerve endings in its intrinsic muscles. *Amer. J. Anat.* **114**, 173-205.

ROBERTSON, J. D. (1956) The ultrastructure of a reptilian myoneural junction. *J. biophys. biochem. Cytol.* **2**, 381-394.

ROBINSON, P. M. (1969) A cholinergic component of the innervation of the longitudinal smooth muscle of the guinea-pig vas deferens: The fine structural localization of cholinesterase. *J. Cell. Biol.* **41**, 462-476.

ROBINSON, P. M. (1971) The demonstration of acetylcholinesterase in autonomic axons with the electronmicroscope. *Progr. Brain Res.* **34**, 371-376.

ROBINSON, P. M. and BELL, C. (1967) The localization of acetylcholinesterase at the autonomic neuromuscular junction. *J. Cell. Biol.* **33**, 93-102.

ROBISON, G. A., BUTCHER, R. W. and SUTHERLAND, E. W. (1970) The catecholamines. *In* Biochemical Actions of Hormones', 2nd ed. G. Litwack, ed. Academic Press, New York, pp. 81-111.

ROFFI, J. (1968a) Evolution des quantités d'adrénaline et de noradrénaline dans les surrénales des foetus et des nouveaunés de rat et de lapin. *Ann. Endocr., (Paris).* **29**, 277-300.

ROFFI, J. (1968b) Influence des corticosurrénales sur la synthèse d'adrénaline chez le foetus et le nouveau-né de rat et de lapin. *J. Physiol. (Paris).* **60**, 455-494.

ROGERS, D. C. (1972) Cell contacts and smooth muscle bundle formation in tissue transplants into the anterior eye chamber. *Z. Zellforsch. Mikrosk. Anat.* **133**, 21–33.

ROGERS, D. C. and BURNSTOCK, G. (1966a) The interstitial cell and its place in the concept of the autonomic ground plexus. *J. comp. Neurol.* **126**, 255–284.

ROGERS, D. C. and BURNSTOCK, G. (1966b) Multiaxonal autonomic junctions in intestinal smooth muscle of the toad (*Bufo marinus*). *J. comp. Neurol.* **126**, 625–652.

ROGERS, L. A., ATKINSON, A. and LONG, J. P. (1966) Effects of various autonomic drugs on isolated perfused mesenteric arteries. *J. Pharmacol. exp. Ther.* **151**, 313–320.

ROSELL, S., KOPIN, I. J. and AXELROD, J. (1963) Fate of H^3-noradrenaline in skeletal muscle before and following sympathetic stimulation. *Amer. J. Physiol.* **205**, 317–321.

ROSENGREN, E. and SJÖBERG, N.-O. (1968) Changes in the amount of adrenergic transmitter in the female genital tract of rabbit during pregnancy. *Acta physiol. scand.* **72**, 412–424.

ROSENBLUETH, A. (1950) Transmission of nerve impulses of neuroeffector junctions and peripheral synapses. I. Transmission at autonomic neuroeffector junctions. Chapman and Hall, London.

ROSENBLUETH, A. and RIOCH, K. D. MCK. (1933) The nature of the response of smooth muscle to adrenin and the augmentor action of cocaine for sympathetic stimuli. *Amer. J. Physiol.* **103**, 681–685.

ROSZKOWSKI, A. P. and KOELLE, G. B. (1960) Enhancement of inhibitory and excitatory effects of CA. *J. Pharmacol. exp. Ther.* **128**, 227–232.

ROTH, R. H. and HUGHES, J. (1972) Acceleration of protein synthesis by angiotensin-correlation with angiotensin's effect on catecholamine biosynthesis. *Biochem. Pharmacol.* **21**, 3182–3187.

ROTH, R. H., STJÄRNE, L. and VON EULER, U. S. (1967a) Factors influencing the rate of norepinephrine biosynthesis in nerve tissue. *J. Pharmacol. exp. Ther.* **158**, 373–377.

ROTH, R. H., STJÄRNE, L. and VON EULER, U. S. (1967b) Acceleration of noradrenaline biosynthesis by nerve stimulation. *Life Sci.* **5**, 1071–1075.

ROTH, R. H., STJÄRNE, L., LEVINE, R. L. and GIARMAN, N. J. (1968) Abnormal regulation of catecholamine synthesis in pheochromocytoma. *J. Lab. clin. Med.* **3**, 397–403.

RUBIN, R. P. (1970) The role of calcium in the release of neurotransmitter substances and hormones. *Pharmacol. Rev.* **22**, 389–428.

RUBIO, M. C. and LANGER, S. Z. (1973) Effects of the NA metabolites on tyrosine hydroxylase activity in guinea pig atria. *Naunyn Schmied. Arch. exp. Path. Pharmak.* **280**, 315–380.

RUSH, R. A. and GEFFEN, L. B. (1972) Radioimmunoassay and clearance of circulating dopamine-β-hydroxylase. *Circulation Res.* **31**, 444–452.

RYALL, R. W. (1961) Effects of cocaine and antidepressant drugs on the nictitating membrane of the cat. *Brit. J. Pharmacol.* **17**, 339–357.

SACHS, ch. (1971) Effect of 6-hydroxydopamine *in vitro* and on the adrenergic neuron. In '6-Hydroxydopamine and Catecholamine Neurons'. T. Malmfors and H. Thoenen, eds. North Holland Publ. Co., Amsterdam, London, pp. 59–74.

SACHS, ch., DE CHAMPLAIN, J., MALMFORS, T. and OLSON, L. (1970) The postnatal development of noradrenaline uptake in the adrenergic nerves of different tissues from the rat. *Eur. J. Pharmacol.* **9**, 67–79.

SACHS, ch. and JONSSON, G. (1973) Quantitative microfluorimetric and neurochemical studies on degenerating adrenergic nerves. *J. Histochem. Cytochem.* **21**, 902–911.

SALA, L. (1893) Sur la fine anatomie des ganglions du sympathique. *Arch. Biol. Ital.* **18**, 439–458.

SALMON, G. K. and IRESON, J. D. (1970) A correlation between the hypotensive action of methyldopa and its depression of peripheral sympathetic function. *Arch. int. Pharmacodyn. Ther.* **183**, 60–64.

SALT, P. J. (1972) Inhibition of noradrenaline uptake$_2$ in the isolated rat heart by steroids, clonidine and methoxylated phenylethylamines. *Eur. J. Pharmacol.* **20**, 329–340.

SALT, P. J. and IVERSEN, L. L. (1972) Inhibition of extraneuronal uptake of catecholamine in the isolated rat heart by cholesterol. *Nature: New Biol.* **238**, 91–92.

SAMUEL, E. P. (1953) Chromidial studies on the superior cervical ganglion of the rabbit (A) Caudally projected postganglionic axons (B) Intercalary 'commissural' neurons. *J. comp. Neurol.* **98**, 93–111.

SAMPSON, S. R. (1972) Mechanism of efferent inhibition of carotid body chemoreceptors in the cat. *Brain Res.* **45**, 266–270.

SANO, Y., ODAKE, G. and YONEZAWA, T. (1967) Fluorescence microscopic observations of catecholamines in cultures of the sympathetic chains. *Z. Zellforsch. Mikrosk. Anat.* **80**, 345–352.

SANTINI, M. (1969) New fibers of sympathetic nature in the inner core region of Pacinian corpuscles. *Brain Res.* **16**, 535–538.

SAUM, W. R. and DE GROAT, W. C. (1972) Parasympathetic ganglia: Activation of an adrenergic inhibitory mechanism by cholinomimetic agents. *Science.* **175**, 659–661.

SCHMITT, F. O. (1968) The molecular biology of neuronal fibrous proteins. *Neurosci. Res. Progr. Bull.* **6**, 119–144.

SCHMITT, H. and PÉTILLOT, N. (1970) Influence du remplacement de la noradrénaline par des faux médiateurs et de l'inhibition de la synthèse sur l'excitabilité sympathique. *J. Pharmacol. (Paris).* **1**, 183–194.

SCHNAITMAN, C., ERWIN, V. G. and GREENAWALT, J. W. (1967) The submitochondrial localization of monoamine oxidase. An enzymatic marker for the outer membrane of rat liver mitochondria. *J. Cell. Biol.* **32**, 719–735.
SCHNEIDER, F., SMITH, A. D. and WINKLER, H. (1967) Secretion from the adrenal medulla: biochemical evidence for exocytosis. *Brit. J. Pharmacol.* **31**, 94–104.
SCHOFIELD, R. M. (1952) The innervation of the cervix and cornu uterii in the rabbit. *J. Physiol. (Lond.).* **117**, 317–328.
SCHUCKER, F. (1972) Effects of NGF-antiserum in sympathetic neurons during early postnatal development. *Exp. Neurol.* **36**, 59–78.
SCHÜMANN, H.-J. (1969) Vergleichende Untersuchungen über das Wachstum von Ratten- und Mäuse- embryonen. *Wilhelm Roux'-Arch. Entwickl-Mech. Olg.* **163**, 325–333.
SCHÜMANN, H. J. (1970) Effect of angiotensin on noradrenaline release of the isolated rabbit heart. *In* 'New Aspects of Storage and Release Mechanisms of Catecholamines'. Bayer Symposium II. Springer-Verlag, Berlin, pp. 202–209.
SCHÜMANN, H. J., STARKE, K. and WERNER, U. (1970) Interactions of inhibitors of noradrenaline uptake and angiotensin on the sympathetic nerves of the isolated rabbit heart. *Brit. J. Pharmacol.* **39**, 390–397.
SCHWARTZ, H. G. (1934) Reflex activity within the sympathetic nervous system. *Amer. J. Physiol.* **109**, 593–604.
SCHWIELER, G. H., DOUGLAS, J. S. and BOUHUYS, A. (1970) Postnatal development of autonomic efferent innervation in the rabbit. *Amer. J. Physiol.* **219**, 391–397.
SCRIABINE, A. (1969) Some observations on the adrenergic blocking activity of desipramine and amitriptyline on aortic strips of rabbit aorta to various agents. *Experientia (Basel).* **25**, 164–165.
SEARS, M. L. and BÁRÁNY, E. H. (1960) Outflow resistance and adrenergic mechanisms. *Arch. Ophthal.* **64**, 839–848.
SEARS, M. L. and GILLIS, C. N. (1967) Mydriasis and the increase in outflow of the aqueous humor from the rabbit eye after cervical ganglionectomy in relation to the release of norepinephrine from the iris. *Biochem. Pharmacol.* **16**, 777–782.
SEDVALL, G. C. and KOPIN, I. J. (1967a) Influence of sympathetic denervation and nerve impulse activity of tyrosine hydroxylase in the rat submaxillary gland. *Biochem. Pharmacol.* **16**, 39–46.
SEDVALL, G. C. and KOPIN, I. J. (1967b) Acceleration of norepinephrine synthesis in the rat submaxillary gland *in vivo* during sympathetic nerve stimulation. *Life Sci.* **6**, 45–51.
SEDVALL, G. C. and THORSON, J. (1965) Adrenergic transmission at vaso-

constrictor nerve terminals partially depleted of noradrenaline. *Acta physiol. scand.* **64**, 251–258.

SEDVALL, G. C., WEISE, V. K. and KOPIN, I. J. (1968) The rate of norepinephrine synthesis measured *in vivo* during short intervals; influence of adrenergic nerve impulse activity. *J. Pharmacol. exp. Ther.* **159**, 274–282.

SETEKLEIV, J. (1970) Effects of drugs on ion distribution and flux in smooth muscle. In 'Smooth Muscle'. E. Bülbring, A. Brading, A. Jones and T. Tomita, eds. Publ. Edward Arnold, London, pp. 343–365.

SHARMAN, D. F. (1972) The catabolism of catecholamines, Recent studies. *Brit. med. Bull.* **29**, 110–119.

SHORE, P. A. (1962) Release of serotonin and catecholamines by drugs. *Pharmacol. Rev.* **14**, 531–550.

SHORE, P. A. (1966) The mechanism of norepinephrine depletion by reserpine, metaraminol and related agents. The role of monoamine oxidase. *Pharmacol. Rev.* **18**, 561–568.

SHORE, P. A. (1971) Transport and storage of biogenic amines. *Ann. Rev. Pharmacol.* **12**, 209–226.

SIEGRIST, G., DOLIVO, M., DUNANT, Y., FOROGLOU-KERAMEUS, C., RIBAU-PIERRE FR. DE, and ROUILLER, ch. (1968) Ultrastructure and function of the chromaffin cells in the superior cervical ganglion of the cat. *J. Ultrastruct. Res.* **25**, 381–407.

SIGG, E. B., SOFFER, L. and GYERMEK, L. (1963) Influence of imipramine and related psychoactive agents on the effect of 5-hydroxytryptamine and catecholamines on the cat nictitating membrane. *J. Pharmacol. exp. Ther.* **142**, 13–20.

SILBERSTEIN, S. D., BRIMIJOIN, S., MOLINOFF, P. B. and LEMBERGER, L. (1972) Induction of dopamine-β-hydroxylase in rat superior cervical ganglion in organ culture. *J. Neurochem.* **19**, 919–921.

SILBERSTEIN, S. D., JOHNSON, D. G., JACOBOWITZ, D. M. and KOPIN, I. J. (1971) Sympathetic reinnervation of the rat iris in organ culture. *Proc. nat. Acad. Sci. U.S.A.* **68**, 1121–1124.

SILINSKY, E. M. and HUBBART, J. I. (1973) Release of ATP from rat motor nerve terminals. *Nature (Lond.).* **243**, 404–405.

SIMEONE, F. A. (1937) The effect of regeneration of the nerve supply on the sensitivity of the denervated nictitating membrane to adrenine. *Amer. J. Physiol.* **120**, 466–474.

SIMPSON, L. L. (1968) The role of calcium in neurohumoral and neurohormonal extrusion processes. *J. Pharm. Pharmacol.* **20**, 889–910.

SIMPSON, F. O. and DEVINE, C. E. (1966) The fine structure of autonomic neuromuscular contacts in arterioles of sheep renal cortex. *J. Anat.* **100**, 127–137.

SJÖBERG, N.-O. (1967) The adrenergic transmitter of the female reproductive tract: Distribution and functional changes. *Acta physiol. scand. Suppl.* **305**, 5–32.

SJÖBERG, N.-O. (1968) Increase in transmitter content of adrenergic nerves in the reproductive tract of female rabbits after oestrogen treatment. *Acta endocr., Copenhagen.* **57**, 405–413.

SJÖSTRAND, N. O. (1965) The adrenergic innervation of the vas deferens and the accessory male genital glands. *Acta physiol. scand. Suppl.* **257**, 1–82.

SKOK, V. I. (1973) 'Physiology of Autonomic Ganglia'. Igaku Shoin Ltd., Tokyo.

SMITH, A. D. (1971a) Secretion of proteins (chromogranin A and dopamine β-hydroxylase) from a sympathetic neuron. *Phil. Trans. Roy. Soc. Lond. Ser. B.* **261**, 363–370.

SMITH, A. D. (1971b) Summing up: some implications of the neuron as a secreting cell. *Phil. Trans. Roy. Soc. Lond. Ser. B.* **261**, 423–440.

SMITH, A. D. (1972a) Cellular control of the uptake, storage and release of noradrenaline in sympathetic nerves. *Biochem. Soc. Symp.* **36**, 103–131.

SMITH, A. D. (1972b) Subcellular localization of noradrenaline in sympathetic neurons. *Pharmacol. Rev.* **24**, 435–437.

SMITH, A. D., DE POTTER, W. P., MOERMAN, E. J. and DE SCHAEPDRYVER, A. F. (1970) Release of dopamine β-hydroxylase and chromogranin A upon stimulation of the splenic nerve. *Tissue & Cell.* **2**, 547–568.

SMITH, A. D., DE POTTER, W. P. and DE SCHAEPDRYVER, A. F. (1969) Subcellular fractionation of bovine splenic nerves. *Arch. int. Pharmacodyn. Ther.* **179**, 495–496.

SMITH, A. D. and WINKLER, H. (1972) Fundamental mechanisms in the release of catecholamines. *In* 'Catecholamines'. H. Blaschko and E. Muscholl, eds. Springer-Verlag, Berlin, Heidelberg, New York, pp. 538–617.

SMITH, C. B. (1966) The role of monoamine oxidase in the intraneuronal metabolism of norepinephrine released by indirectly-acting sympathomimetic amines or by adrenergic nerve stimulation. *J. Pharmacol. exp. Ther.* **151**, 207–220.

SMITH, C. B., TRENDELENBURG, U., LANGER, S. Z. and TSAI, T. H. (1966). The relation of retention of norepinephrine-H^3 to the norepinephrine content of the nictitating membrane of the spinal cat during development of denervation supersensitivity. *J. Pharmacol. exp. Ther.* **151**, 87–94.

SMITTEN, N. A. (1963). Cytological analysis of catecholamine synthesis in the ontogenesis of vertebrates and problems of melanogenesis. *Gen. comp. Endocrin.* **3**, 362–377.

SNIDER, S. R., ALMGREN, O. and CARLSSON, A. (1973) The occurrence and functional significance of dopamine in some peripheral adrenergic nerves of the rat. *Naunyn Schmied. Arch. exp. Path. Pharmak.* **278**, 1–12.

SNYDER, S. H., FISCHER, J. and AXELROD, J. (1965) Evidence for the presence of monoamine oxidase in sympathetic nerve endings. *Biochem. Pharmacol.* **14**, 363–365.

SOMLYO, A. P. and SOMLYO, A. V. (1968) Vascular smooth muscle. I. Normal structure, pathology, biochemistry and biophysics. *Pharmacol. Rev.* **20,** 197-272.

SOURKES, T. L. (1966) Dopa decarboxylase: substrates, coenzyme, inhibitors. *Pharmacol. Rev.* **18,** 53-60.

SPECTOR, S. (1966) Inhibitors of endogenous catecholamine biosynthesis. *Pharmacol. Rev.* **18,** 599-609.

SPECTOR, S., SJOERDSMA, A., ZALTMAN-NIRENBERG, P. and UDENFRIEND, S. (1963) Norepinephrine synthesis from tyrosine-C^{14} in isolated perfused guinea pig heart. *Science.* **139,** 1299-1301.

SPECTOR, S., GORDON, R., SJOERDSMA, A. and UDENFRIEND, S. (1967) End-product inhibition of tyrosine hydroxylase as a possible mechanism for regulation of norepinephrine synthesis. *Mol. Pharmacol.* **3,** 549-555.

SPECTOR, S., MELMON, K. and SJOERDSMA, A. (1962) Evidence for rapid turnover of norepinephrine in rat heart and brain. *Proc. Soc. Exp. Biol. Med.* **111,** 79-81.

SPECTOR, S., SJOERDSMA, A. and UDENFRIEND, S. (1965) Blockade of endogenous norepinephrine synthesis by α-methyl-tyrosine, an inhibitor of tyrosine hydroxylase. *J. Pharmacol. exp. Ther.* **147,** 86-95.

SPECTOR, S., TARVER, J. and BERKOWITZ, B. (1972) Effect of drugs and physiological factors in the disposition of catecholamines in blood vessels. *Pharmacol. Rev.* **24,** 191-202.

SPEDEN, R. N. (1967) Adrenergic transmission in small arteries. *Nature (Lond.).* **216,** 289-290.

SPEDEN, R. N. (1970) Excitation of vascular smooth muscle. *In* 'Smooth Muscle'. E. Bülbring, A. Brading, A. Jones and T. Tomita, eds. Edward Arnold (Publ.) Ltd., London, pp. 558-588.

SPOENDLIN, H. and LICHTENSTEIGER, W. (1966) The adrenergic innervation of the labyrinth. *Acta otolaryng.* **61,** 423-434.

SPRATTO, G. R. and MILLER, J. W. (1968a) The effect of various estrogens on the weight, catecholamine content and rate of contractions of rat uteri. *J. Pharmacol. exp. Ther.* **161,** 1-6.

SPRATTO, G. R. and MILLER, J. W. (1968b) An investigation of the mechanism by which estradiol-17β elevates the epinephrine content of the rat uterus. *J. Pharmacol. exp. Ther.* **161,** 7-13.

SPRIGGS, T. L. B. (1966) Peripheral noradrenaline and adrenergic transmission in the rat. *Brit. J. Pharmacol.* **26,** 271-281.

STAFFORD, A. (1963) Potentiation of some catecholamines by phenoxybenzamine, guanethidine and cocaine. *Brit. J. Pharmacol.* **21,** 361-367.

STARKE, K. (1972a) Alpha sympathomimetic inhibition of adrenergic and cholinergic transmission in the rabbit heart. *Nauyn. Schmied. Arch. exp. Path. Pharmak.* **274,** 18-45.

STARKE, K. (1972b) Influence of extracellular noradrenaline on the stimulation evoked secretion of noradrenaline from sympathetic nerves: evidence for an α-receptor-mediated feed back inhibition of noradrenaline release. *Naunyn Schmied. Arch. exp. Path. Pharmak.* **275**, 11–23.

STARKE, K. and MONTEL, H. (1973) Sympathomimetic inhibition of NA release mediated by prostaglandins? *Naunyn Schmied. Arch. exp. Path. Pharmak.* **278**, 111.

STARKE, K., MONTEL, H. and SCHÜMANN, H. J. (1971) Influence of cocaine and phenoxybenzamine on noradrenaline uptake and release. *Naunyn. Schmied. Arch. exp. Path. Pharmak.* **270**, 210–214.

STARKE, K. and SCHÜMANN, H. J. (1972) Interactions of angiotensin, phenoxybenzamine and propanolol on noradrenaline release during sympathetic nerve stimulation. *Eur. J. Pharmacol.* **18**, 27–30.

STARKE, K., WAGNER, J. and SCHÜMANN, H. J. (1972) Adrenergic neuron blockade by clonidine: comparison with guanethidine and local anesthetics. *Arch. int. Pharmacodyn. Thér.* **195**, 191–272.

STEHLE, R. L. and ELLSWORTH, H. C. (1937) Dihydroxyphenylethanolamine (Arterenol) as a possible sympathetic hormone. *J. Pharmacol. exp. Ther.* **59**, 114–121.

STEINER, G. and SCHÖNBAUM, E. (1972) 'Immunosympathectomy'. Elsevier Publ. Co., Amsterdam, London, New York.

STEINSLAND, O. S., FURCHGOTT, R. F. and KIRPEKAR, S. M. (1973) Inhibition of adrenergic neurotransmission by parasympathomimetics in the rabbit ear artery. *J. Pharmacol. exp. Ther.* **184**, 346–356.

STJÄRNE, L. (1966) Storage particles in noradrenergic tissues. *Pharmacol. Rev.* **18**, 425–432.

STJÄRNE, L. (1972a) The synthesis, uptake and storage of catecholamines in the adrenal medulla. The effect of drugs. *In* 'Catecholamines'. H. Blaschko and E. Muscholl, eds. Springer-Verlag, Berlin, Heidelberg, New York, pp. 231–269.

STJÄRNE, L. (1972b) Alpha-adrenoceptor mediated feed-back control of sympathetic neurotransmitter secretion in guinea-pig vas deferens. *Nature: New Biol.* **241**, 190–191.

STJÄRNE, L. (1973a) Michaelis-Menten kinetics of secretion of sympathetic neurotransmitter as a function of external calcium: effect of graded α-receptor blockade. *Naunyn. Schmied. Arch. exp. Path. Pharmak.* **278**, 323–327.

STJÄRNE, L. (1973b) Kinetics of secretion of sympathetic neurotransmitter as a function of external calcium: mechanism of inhibitory effect of prostaglandin E. *Acta physiol. scand.* **87**, 428–430.

STJÄRNE, L. (1973c) Frequency dependence of dual negative feed-back control of secretion of sympathetic neurotransmitter in guinea-pig vas deferens. *Brit. J. Pharmacol.* **49**, 358–360.

STJÄRNE, L. (1973d) Lack of correlation between profiles of transmitter efflux and of muscular contractions in response to nerve stimulation in isolated guinea-pig vas deferens. *Acta physiol. scand.* **88**, 137–144.

STJÄRNE, L., HEDQVIST, P. and BYGDEMAN, S. (1969) Neurotransmitter quantum released from sympathetic nerves in cat's skeletal muscle. *Life Sci.* **8**, 189–196.

STJÄRNE, L., ROTH, R. H., BLOOM, F. and GIARMAN, N. J. (1970) Norepinephrine concentrating mechanisms in sympathetic nerve trunks. *J. Pharmacol. exp. Ther.* **171**, 70–79.

STJÄRNE, L. and WENNMALM, Å. (1970) Preferential secretion of newly formed noradrenaline in the perfused rabbit heart. *Acta physiol. scand.* **80**, 428–429.

STÖCKEL, K., PARAVICINI, U. and THOENEN, H. (1974) Specificity of the retrograde axonal transport of nerve growth factor. *Brain Res.* **76**, 413–421.

STOHR, P. JR. (1957) 'Hanbuch Der Mikroskopischen Anatomie des Menschen'. Bd. IV. Teil. J. ed. Springer-Verlag, Berlin.

STRÖMBLAD, B. C. R. (1956) Supersensitivity and amine oxidase activity in denervated salivary glands. *Acta physiol. scand.* **36**, 137–157.

STRÖMBLAD, B. C. R. and NICKERSON, M. (1961) Accumulation of epinephrine and norepinephrine by some rat tissues. *J. Pharmacol. exp. Ther.* **134**, 154–159.

SU, C. and BEVAN, J. A. (1970a) The release of H^3-norepinephrine in arterial strips studied by the technique of superfusion and transmural stimulation. *J. Pharmacol. exp. Ther.* **172**, 62–68.

SU, C. and BEVAN, J. A. (1970b) Blockade of the nicotine-induced norepinephrine released by cocaine, phenoxybenzamine and desipramine. *J. Pharmacol. exp. Ther.* **175**, 533–540.

SU, C. and BEVAN, J. A. (1971) Adrenergic transmitter release and distribution in blood vessels. *In* 'Physiology and Pharmacology of Vascular Neuroeffector Systems'. J. A. Bevan, R. F. Furchgott, R. A. Maxwell, and A. P. Somlyo, eds. S. Karger, Basel, Münich, Paris, London, New York, Sydney, pp. 13–21.

SU, C., BEVAN, J. and BURNSTOCK, G. (1971) [^{3}H]-Adenosine triphosphate: Release during stimulation of enteric nerves. *Sci. Amer.* **173**, 337–339.

SUGARMAN, S. R., MARGOLIUS, H. S., GAFFNEY, T. E. and MOHAMMED, S. (1968) Effect of methyldopa on chronotropic responses to cardioaccelerator nerve stimulation in dogs. *J. Pharmacol. exp. Ther.* **162**, 115–120.

SWEDIN, G. (1970) Comparison of the effects of α-methyltyrosine on the noradrenaline-stores of different peripheral organs of the rat. *Life Sci.* **9**, 1249–1259.

SWEDIN, G. (1971) Studies on neurotransmission mechanisms in the rat and guinea-pig vas deferens. *Acta physiol. scand. Suppl.* **369**, 1–34.

SWEDIN, G. (1972) Abolition of the nerve-induced responses of the isolated vas deferens of the young rat by agents blocking α-adrenoceptors. *Brit. J. Pharmacol.* **44**, 160–161.

TAFURI, W. L. (1964) Ultrastructure of the vesicular component in the intramural nervous system of the guinea-pig's intestine. *Z. Naturf.* **19B**, 622–625.

TAKAMINE, J. (1901) Adrenaline, the active principle of the suprarenal glands and its mode of preparation. *Amer. J. Pharm.* **73**, 523–531.

TAMARIND, D. L. and QUILLIAM, J. P. (1971) Synaptic organization and other ultrastructural features of the superior cervical ganglion of the rat, kitten and rabbit. *Micron.* **2**, 204–234.

TARLOV, S. R. and LANGER, S. Z. (1971) The fate of ^{3}H-norepinephrine released from isolated atria and vas deferens: Effect of field stimulation. *J. Pharmacol. exp. Ther.* **179**, 186–197.

TARVER, J. JR., BERKOWITZ, B. and SPECTOR, S. (1971) Alterations in tyrosine hydroxylase and monoamine oxidase activity in blood vessels. *Nature: New Biol.* **231**, 252–253.

TARVER, J. H. and SPECTOR, S. (1970) Catecholamine metabolic enzymes in the vasculature. *Fed. Proc.* **29**, 278.

TAXI, J. (1965) Contribution à l'étude des connexions des neurones moteurs du système nerveux autonome. *Ann. Sci. Naturelles (Zoologie) Ser. 12.* **7**, 413–674.

TAXI, J. and DROZ, B. (1969) Radioautographic study of the accumulation of some biogenic amines in the autonomic nervous system. *In* 'Cellular Dynamics of the Neuron'. Samuel H. Barondes, ed. *Symp. int. Soc. Cell Biol.* **8**, 175–190.

TAXI, J., GAUTRON, J. and L'HERMITE, P. (1969) Données ultrastructurales sur une éventuelle modulation adrénergique de l'activité du ganglion cervical supérieur du rat. *C.R. Acad. Sci., Paris.* **269**, 1281–1284.

TERNI, T. (1922) Ricerche sulla struttura e sull'evoluzione del simpatico dell'uomo. I. Le differenze nella costituzione dei gangli simpatici delle varie regioni. II. Le trasformazioni delle cellule simpatiche durante l'accrescimento fetale e postnatale fino alla senescenza. *Monitore Zool. ital.* **33**, 63–72.

THAEMERT, J. C. (1966) Ultrastructural interrelationships of nerve processes and smooth muscle cells in three dimensions. *J. Cell Biol.* **28**, 37–49.

THOA, N. B., JOHNSON, D. G., KOPIN, I. J. and WEINER, N. (1971) Acceleration of catecholamine formation in the guinea-pig vas deferens after hypogastric nerve stimulation: Roles of tyrosine hydroxylase and new protein synthesis. *J. Pharmacol. exp. Ther.* **178**, 442–449.

THOENEN, H. (1970) Induction of tyrosine hydroxylase in peripheral and central adrenergic neurons by cold-exposure in rats. *Nature (Lond.).* **228**, 861–862.

THOENEN, H. (1972) Surgical, immunological and chemical sympathectomy. Their application in the investigation of the physiology and pharmacology of the sympathetic nervous system. *In* 'Catecholamines'. H. Blaschko and E. Muscholl, eds. Springer-Verlag, Berlin, Heidelberg, New York, pp. 813–844.

THOENEN, H., ANGELETTI, P. U., LEVI-MONTALCINI, R. and KETTLER, R. (1971) Selective induction by nerve growth factor of tyrosine hydroxylase and dopamine-β-hydroxylase in the rat superior cervical ganglia. *Proc. nat. Acad. Sci., U.S.A.* **68**, 1598–1602.

THOENEN, H., HAEFELY, W., GEY, K. F. and HÜRLIMANN, A. (1965) Diminished effects of sympathetic nerve stimulation in cats pretreated with disulfiram. *Life Sci.* **4**, 2033–2038.

THOENEN, H., HAEFELY, W., GEY, K. F. and HÜRLIMANN, A. (1966) The effect of α-methyl-tyrosine on peripheral sympathetic transmission. *Life Sci.* **5**, 723–730.

THOENEN, H., HÜRLIMANN, A. and HAEFELY, W. (1964) The effect of sympathetic nerve stimulation on volume, vascular resistance and norepinephrine output in the isolated perfused spleen of the cat, and its modification by cocaine. *J. Pharmacol. exp. Ther.* **143**, 57–63.

THOENEN, H., HÜRLIMANN, A. and HAEFELY, W. (1966) Interaction of phenoxybenzamine with guanethidine and bretylium at the sympathetic nerve endings of the isolated perfused spleen of the cat. *J. Pharmac. exp. Ther.* **151**, 189–195.

THOENEN, H., HÜRLIMANN, A. and HAEFELY, W. (1969) Cation dependence of the noradrenaline-releasing action of tyramine. *Eur. J. Pharmacol.* **6**, 29–37.

THOENEN, H., KETTLER, R., BURKARD, W. and SANER, A. (1971) Neurally mediated control of enzymes involved in the synthesis of norepinephrine; are they regulated as an operational unit? *Nauyn. Schmeid. Arch. exp. Path. Pharmak.* **270**, 146–160.

THOENEN, H., MUELLER, R. A. and AXELROD, J. (1969) Transsynaptic induction of adrenal tyrosine hydroxylase. *J. Pharmacol. exp. Ther.* **169**, 249–254.

THOENEN, H., MUELLER, R. A. and AXELROD, J. (1970) Phase difference in the induction of tyrosine hydroxylase in cell body and nerve terminals of sympathetic neurones. *Proc. nat. Acad. Sci., U.S.A.* **65**, 58–62.

THOENEN, H., SANER, A., ANGELETTI, P. U. and LEVI-MONTALCINI, R. (1972a) Increased activity of choline acetyltransferase in sympathetic ganglia after prolonged administration of nerve growth factor. *Nature (Lond.).* **236**, 26–28.

THOENEN, H., SANER, A., KETTLER, R. and ANGELETTI, P. U. (1972b) Nerve growth factor and preganglionic cholinergic nerves; their relative importance to the development of the terminal adrenergic neuron. *Brain Res.* **44**, 593–603.

THOENEN, H. and TRANZER, J. P. (1968) Chemical sympathectomy by selective destruction of adrenergic nerve endings with 6-hydroxydopamine. *Nauyn. Schmeid. Arch. exp. Path. Pharmak.* **261**, 271–288.

THOENEN, H. and TRANZER, J. P. (1971) Functional importance of subcellular distribution of false adrenergic transmitters. *Progr. Brain Res.* **34**, 223–236.

THOENEN, H. and TRANZER, J. P. (1973) The pharmacology of 6-Hydroxydopamine. *Ann. Rev. Pharmacol.* **13**, 169–180.

THURESON-KLEIN, Å., KLEIN, R. L. and YEN, S. G. (1973) Ultrastructure of highly purified sympathetic nerve vesicles. Correlation between matrix density and norepinephrine content. *J. Ultrastruct. Res.* **43**, 18–35.

TIFFENEAU, M. M. (1939) Sur les effets hypertenseurs de quelques dérivés de la series de l'ephedrine et de l'adrénaline. – Considérations sur les determinisme de la formation de l'adrénaline dans l'organisme animal. *Ann. de Physiol.* **5**, 571–577.

TIPTON, K. F. (1967) The sub-mitochondrial localization of monoamine oxidase in rat liver and brain. *Biochim. biophys. Acta.* **135**, 910–920.

TIPTON, K. F. (1973) Biochemical aspects of monoamine oxidase. *Brit. med. Bull.* **29**, 116–119.

TJÄLVE, H. (1971) Catechol- and Indolamines in some endocrine cell systems. An autoradiographical, histochemical and radioimmunological study. *Acta physiol. scand. Suppl.* **360**, 1–122.

TODA, N. (1971) Influence of cocaine and desipramine on the contractile response of isolated rabbit pulmonary arteries and aortae to transmural stimulation. *J. Pharmacol. exp. Ther.* **179**, 198–206.

TOMITA, T. (1967a) Current spread in the smooth muscle of the guinea-pig vas deferens. *J. Physiol. (Lond.).* **189**, 163–176.

TOMITA, T. (1967b) Spike propagation in the smooth muscle of the guinea-pig taenia coli. *J. Physiol. (Lond.).* **191**, 517–527.

TOMITA, T. (1970) Electrical properties of mammalian smooth muscle. In 'Smooth Muscle'. E. Bülbring, A. Brading, A. Jones and T. Tomita, eds. Edward Arnold London, pp. 197–243.

TORCHIANA, M. L., PORTER, C. C., STONE, C. A. and HANSON, H. M. (1970) Some biochemical and pharmacological actions of α-methylphenylalanine. *Biochem. Pharmacol.* **19**, 1601–1614.

TRANZER, J. P. (1972) A new amine storing compartment in adrenergic axons. *Nature: New Biol.* **237**, 57–58.

TRANZER, J. P. (1973) New aspects of the localization of catecholamines in adrenergic neurons. In 'Frontiers in Catecholamine Research'. E. Usdin and S. Snyder, eds. Pergamon Press, Oxford, pp. 453–458.

TRANZER, J. P. and RICHARDS, J. G. (1971) Fine structural aspects of the effect of 6-hydroxydopamine on peripheral adrenergic neurons. In '6-Hydroxy-dopamine and Catecholamine Neurons'. T. Malmfors and H. Thoenen, eds. North Holland Publishing Co., Amsterdam, London, pp. 15–31.

TRANZER, J. P. and SNIPES, R. L. (1968) Fine structural localization of nora-drenaline in sympathetic nerve terminals: a critical study of the influence of fixation. In 'Proc. Europ. Reg. Conf. Elect. Micros'. 4th ed. Vol. **2**, pp. 519–520. D. S. Bocciarelli, Rome.

TRANZER, J. P. and THOENEN, H. (1967a) Significance of 'empty vesicles' in post-ganglionic sympathetic nerve terminals. *Experientia (Basel).* **23**, 123–124.

TRANZER, J. P. and THOENEN, H. (1967b) Ultramorphologische Veränderungen der sympathischen Nervenendigungen der Katze nach Vorbehandlung mit

5- und 6-Hydroxy–Dopamin. *Naunyn. Schmied. Arch. exp. Path. Pharmak.* **257**, 343–344.

TRANZER, J. P. and THOENEN, H. (1967c) Electronmicroscopic localization of 5-hydroxydopamine (3,4,5-trihydroxyphenyl-ethylamine) a new 'false' sympathetic transmitter. *Experimentia (Basel).* **23**, 743.

TRANZER, J. P. and THOENEN, H. (1968) An electron microscopic study of selective acute degeneration of sympathetic nerve terminals after administration of 6-hydroxydopamine. *Experientia (Basel).* **24**, 155–156.

TRANZER, J. P., THOENEN, H., SNIPES, R. and RICHARDS, J. G. (1969) Recent developments on the ultrastructural aspect of adrenergic nerve endings in various experimental conditions. *Progr. Brain Res.* **31**, 33–46.

TRENDELENBURG, U. (1959) The supersensitivity caused by cocaine. *J. Pharmacol. exp. Ther.* **125**, 55–65.

TRENDELENBURG, U. (1961) Pharmacology of autonomic ganglia. *Ann. Rev. Pharmac.* **1**, 219–238.

TRENDELENBURG, U. (1963) Supersensitivity and subsensitivity to sympathomimetic amines. *Pharmacol. Rev.* **15**, 225–276.

TRENDELENBURG, U. (1965) Supersensitivity by cocaine to dextrorotary isomers of norepinephrine. *J. Pharmacol. exp. Ther.* **148**, 329–338.

TRENDELENBURG, U. (1966) Supersensitivity to norepinephrine induced by continuous nerve stimulation. *J. Pharmacol. exp. Ther.* **151**, 95–102.

TRENDELENBURG, U. (1967) Some aspects of the pharmacology of autonomic ganglion cells. *Ergebn. Physiol.* **59**, 1–85.

TRENDELENBURG, U. (1972) Factors influencing the concentration of catecholamines. *In* 'Catecholamines'. H. Blaschko and E. Muscholl, eds. Springer-Verlag Berlin, Heidelberg, New York, pp. 726–761.

TRENDELENBURG, U. (1974) The relaxation of rabbit aortic strips after a preceding exposure to sympathomimetic amines. *Naunyn. Schmied. Arch. exp. Path. Pharmak.* **281**, 13–46.

TRENDELENBURG, U. and WAGNER, K. (1971) The effect of 6-hydroxydopamine on the cat's nictitating membrane; degeneration, contraction and sensitivity to (-)-noradrenaline. *In* '6-Hydroxodopamine and Catecholamine Neurons.' T. Malmfors and H. Thoenen, eds. North Holland Publ. Co., Amsterdam, London, pp. 215–223.

TRIGGLE, D. J. (1965) 'Chemical Aspects of the Autonomic Nervous System'. Academic Press, London and New York.

TRIGGLE, D. J. (1972) Adrenergic receptors. *Ann. Rev. Pharmacol.* **12**, 185–196.

TUCKETT, I. L. (1896) On the structure and degeneration of non-medullated nerve fibres. *J. Physiol. (Lond.).* **19**, 267–311.

TÜRKER, R. K. and KHAIRALLAH, P. A. (1967) Desmethylimipramine (desipramine), an adrenergic blocking agent. *Experientia (Basel).* **23**, 252.

UDENFRIEND, S. (1966a) Biosynthesis of the sympathetic neurotransmitter norepinephrine. *In* 'The Harvey Lectures'. **60**, 57–83. Academic Press, New York.

UDENFRIEND, S. (1966b) Tyrosine hydroxylase. *Pharmacol. Rev.* **18**, 43–52.

UDENFRIEND, S. (1968) Physiological regulation of noradrenaline biosynthesis. *In* 'Adrenergic neurotransmission'. Ciba Foundation. G. E. W. Wolstenholme and M. O'Connor, eds. J. & A. Churchill Ltd. London, pp. 3–11.

UDENFRIEND, S., ZALTZMAN-NIRENBERG, P., GORDON, R. and SPECTOR, S. (1966) Evaluation of the biochemical effects produced *in vivo* by inhibitors of the three enzymes involved in norepinephrine biosynthesis. *Mol. Pharmacol.* **2**, 95–105.

UEHARA, Y. and BURNSTOCK, G. (1971) Inclusion bodies in fibroblast-like cells in the mucosa of the guinea-pig ureter. *J. Ultrastruct. Res.* **34**, 175–180.

UEHARA, Y. and BURNSTOCK, G. (1972) Postsynaptic specialization of smooth muscle at close neuromuscular junctions in the guinea-pig sphincter pupillae. *J. Cell Biol.* **53**, 849–853.

UNGER, K. (1951) Über Altersveränderungen in den Grenzstrang-Ganglien der Ratte. *Anat. Anz.* **98**, 13–23.

UNGERSTED, U., BUTCHER, L. L., BUTCHER, S. G., ANDÉN, N.-E. and FUXE, K. (1969) Direct chemical stimulation of dopaminergic mechanisms in the neostriatum of the rat. *Brain Res.* **14**, 461–471.

UNGVÁRY, G. and LÉRÁNTH, C. (1970) Termination in the prevertebral abdominal sympathetic ganglia of axons arising from the local (terminal) vegetative plexus of visceral organs. *Z. Zellforsch. Mikrosk. Anat.* **110**, 185–191.

UVNÄS, B. (1973) An attempt to explain nervous transmitter release as due to nerve impulse-induced cation exchange. *Acta physiol. scand.* **87**, 168–175.

VANHOUTTE, P. M. and SHEPHERT, J. T. (1973) Venous relaxation caused by acetylcholine acting on the sympathetic nerves. *Circulation Res.* **32**, 259–267.

VAN ORDEN, III. L. S., BENSCH, K. G., LANGER, S. Z. and TRENDELENBURG, U. (1967) Histochemical and fine structural aspects of the onset of denervation supersensitivity in the nictitating membrane of the spinal cat. *J. Pharmacol. exp. Ther.* **157**, 274–283.

VAN ORDEN, III. L. S., BLOOM, F. E., BARRNETT, R. J. and GIARMAN, N. J. (1966) Histochemical and functional relationships of catecholamines in adrenergic nerve endings. I. Participation of granular vesicles. *J. Pharmacol. exp. Ther.* **154**, 185–199.

VAN ORDEN, III. L. S., BURKE, J. P., GEYER, M. and LODOEN, F. V. (1970) Localisation of depletion-sensitive and depletion-resistant norepinephrine storage sites in autonomic ganglia. *J. Pharmacol. exp. Ther.* **174**, 56–71.

VAN ORDEN, III. L. S., ROBB, B. J., BHATNAGAR, R. K. and BURKE, J. P. (1972) Lack of effect of electrical stimulation on norepinephrine stores of rat vas deferens. *Life Sci.* **11**, 1123–1133.

VARAGIC, V. (1956) An isolated rabbit hypogastric-nerve uterus preparation, with observations on the hypogastric transmitter. *J. Physiol. (Lond.).* **132**, 92–99.

VARMA, D. R. (1967) Antihypertensive effect of methyldopa in metacorticoid immunosympathectomized rats. *J. Pharm. Pharmacol.* **19**, 61–62.

VARMA, D. R. and BENFEY, B. G. (1963) Antagonism of reserpine-induced subsensitivity to tyramine by methyldopa. *J. Pharmacol. exp. Ther.* **141**, 310–313.

VASALLE, M., LEVINE, M. J. and STUCKEY, J. H. (1968) On the sympathetic control of ventricular automaticity. *Circulation Res.* **23**, 249–258.

VERITY, M. A. (1971) Morphologic studies of the vascular neuroeffector apparatus. *In* 'Physiology and Pharmacology of Vascular Neuroeffector Systems'. Bevan, Furchgott, Maxwell and Somlyo, eds. S. Karger, Basel, München, Paris, London, New York, Sydney, pp. 2–12.

VERITY, M. A. and BEVAN, J. A. (1968) Fine structural study of the terminal effector plexus, neuromuscular and intermuscular relationships in the pulmonary artery. *J. Anat.* **103**, 49–63.

VIVEROS, O. H., ARQUEROS, L. and KIRSHNER, N. (1969a) Quantal secretion from adrenal medulla: all or none release of storage vesicle content. *Science.* **165**, 911–913.

VIVEROS, O. H., ARQUEROS, L., CONNETT, R. J. and KIRSHNER, N. (1969b) Mechanism of secretion from the adrenal medulla. IV. Fate of the storage vesicles following insulin and reserpine administration. *Mol. Pharmacol.* **5**, 69–82.

VOGT, M. (1964) Sources of noradrenaline in the immunosympathectomized rat. *Nature (Lond.).* **204**, 1315–1316.

VOGT, M. (1973) Quantitative aspects of the release of noradrenaline from peripheral adrenergic neurons. *Acta physiol. polonica.* **24**, 165–169.

VOLLE, R. L. (1969) Ganglionic transmission. *Ann. Rev. Pharmacol.* **9**, 135–146.

VOLLE, R. L. and HANCOCK, J. C. (1970) Transmission in sympathetic ganglia. *Fed. Proc.* **29**, 1913–1918.

VON EULER, U. S. (1946) A specific sympathomimetic ergone in adrenergic nerve fibres (sympathin) and its relations to adrenaline and noradrenaline. *Acta physiol. scand.* **12**, 73–97.

VON EULER, U. S. (1956) 'Noradrenaline'. Charles C. Thomas, Springfield, Illinois, U.S.A.

VON EULER, U. S. (1966) Release and uptake of noradrenaline in adrenergic nerve granules. *Acta physiol. scand.* **67**, 430–440.

VON EULER, U. S. (1968) Some aspects of the mechanisms involved in adrenergic neurotransmission. *Perspect. Biol. Med.* **12**, 79–94.

VON EULER, U. S. (1969) Acute neuromuscular transmission failure in vas deferens after reserpine. *Acta physiol. scand.* **76**, 255–256.

VON EULER, U. S. (1972) Synthesis, uptake and storage of catecholamines in adrenergic nerves. The effect of drugs. *In* 'Catecholamines'. H. Blaschko and

E. Muscholl, eds. Springer-Verlag, Berlin, Heidelberg, New York, pp. 186–230.

VON EULER, U. S. and HILLARP, N.-Å. (1956) Evidence for the presence of noradrenaline in submicroscopic structures of adrenergic axons. *Nature (Lond.).* 177, 44–45.

VON EULER, U. S., and LISHAJKO, F. (1963) Effect of reserpine on the uptake of catecholamines in isolated nerve storage granules. *Int. J. Neuropharmacol.* 2, 127–134.

VON STUDNITZ, W. (1968) Catecholamine and adenosine triphoshate content of human fetal adrenals. *Scand. J. Clin. Lab. Invest.* 22, 185–188.

WAGNER, K. and TRENDELENBURG, U. (1971) Development of degeneration contraction and supersensitivity in the cat's nictitating membrane after 6-hydroxydopamine. *Naunyn. Schmied. Arch. exp. Path. Pharmak.* 270, 215–236.

WAKADE, A. R. and KIRPEKAR, S. M. (1973) 'Trophic' influences on the sympathetic nerves of the vas deferens and seminal vesicle of the guinea-pig. *J. Pharmacol. exp. Ther.* 186, 528–586.

WAKADE, A. R. and KRUSZ, J. (1972) Effect of reserpine, phenoxybenzamine and cocaine on neuromuscular transmission in the vas deferens of the guinea-pig. *J. Pharmacol. exp. Ther.* 181, 310–317.

WALSNER, M. (1971) Sodium Excretion. In 'The Kidney III', C. Rouiller and A. F. Muller, eds. Academic Press, New York, pp. 127–207.

WALTHER, D. E., IRIKI, M. and SIMON, E. (1970) Antagonistic changes of blood flow and sympathetic activity in different vascular beds following central thermal stimulation. II. Cutaneous and visceral sympathetic activity during spinal cord heating and cooling in anesthetized rabbits and cats. *Pflüg. Arch. ges Physiol.* 319, 162–182.

WARD, J. W. (1936) A histological study of transplanted sympathetic ganglia. *Amer. J. Anat.* 58, 147–179.

WARING, H. (1935–36) The development of the adrenal gland of the mouse. *Quart. J. Micr. Sci.* 78, 329–366.

WATANABE, H. (1969) Electron microscopic observations on the innervation of smooth muscle in the guinea pig vas deferens. *Acta anat. nippon.* 44, 189–202.

WATANABE, H. (1971) Adrenergic nerve elements in the hypogastric ganglion of the guinea-pig. *Amer. J. Anat.* 130, 305–329.

WATERSON, J. G. (1973) Comparison of constrictor responses of the rabbit ear artery to norepinephrine and to sympathetic stimulation. *Circulation Res.* 32, 323–328.

WATERSON, J. G., HUME, W. R. and DE LA LANDE, I. S. (1970) The distribution of cholinesterase in the rabbit ear artery. *J. Histochem. Cytochem.* 18, 211–216.

WEINER, N. (1970) Regulation of norepinephrine biosynthesis. *Annu. Rev. Pharmacol.* 10, 273–290.

WEINER, N. and BJUR, R. (1972) The role of intraneuronal monoamine oxidase in the regulation of norepinephrine synthesis. *In* 'Monoamine Oxidases, New Vistas'. Advances in Biochem. Psychopharmacol. 5. E. Costa and M. Sandler, eds. Raven Press, N.Y., pp. 409–419.

WEINER, N., CLOUTIER, G., BJUR, R. and PFEFFER, R. I. (1972) Modification of norepinephrine synthesis in intact tissue by drugs and during short-term adrenergic nerve stimulation. *Pharmacol. Rev.* **24**, 203–222.

WEINER, R. and KALEY, G. (1969) Influence of prostaglandin E_1 on the terminal vascular bed. *Amer. J. Physiol.* **217**, 563–566.

WEINER, N. and RABADJIJA, M. (1968a) The effect of nerve stimulation on the synthesis and metabolism of norepinephrine in the isolated guinea-pig hypogastric nerve – vas deferens preparation. *J. Pharmacol. exp. Ther.* **160**, 61–71.

WEINER, N. and RABADJIJA, M. (1968b) The regulation of norepinephrine synthesis. Effect of puromycin on the accelerated synthesis of norepinephrine associated with nerve stimulation. *J. Pharmacol. exp. Ther.* **164**, 103–114.

WEINSHILBOUM, R. M., THOA, N. B., JOHNSON, D. G., KOPIN, I. J. and AXELROD, J. (1971) Proportional release of norepinephrine and dopamine-β-hydroxylase from sympathetic nerves. *Science.* **174**, 1349–1351.

WENNMALM, A. (1971) Studies on mechanisms controlling the secretion of neurotransmitters in the rabbit heart. *Acta Physiol. scand. Suppl.* **365**, 1–35.

WEISS, P. (1961) The concept of perpetual neuronal growth and proximo-distal substance convection. *In* 'Regional Neurochemistry'. S. S. Kety and J. Ekles, eds. Pergamon Press, London, New York, pp. 220–240.

WEISS, P. and HISCOE, H. B. (1948) Experiments on the mechanism of nerve growth. *J. exp. Zool.* **107**, 315–395.

WEISS, P. and PILLAI, A. (1965) Convection and fate of mitochondria in nerve fibers: axonal flow as vehicle. *Proc. Nat. Acad. Sci. U.S.A.* **54**, 48–56.

WEKSTEIN, D. R. (1964) Sympathetic function and development of temperature regulation. *Amer. J. Physiol.* **206**, 823–826.

WEKSTEIN, D. R. (1965) Heart rate of the preweanling rat and its autonomic control. *Amer. J. Physiol.* **208**, 1259–1262.

WENZEL, B. M. (1972). Immunosympathectomy and behaviour. *In* 'Immunosympathectomy'. G. Steiner and E. Schönbaum, eds. Elsevier Publ. Co., Amsterdam, London, New York, pp. 199–219.

WERNER, U., STARKE, K. and SCHÜMANN, H. J. (1972) Actions of clonidine and 2-(2-methyl-6-ethyl-cyclohexylamino)-2-oxazolidine on postganglionic autonomic nerves. *Arch. int. Pharmacodyn. Ther.* **195**, 282–290.

WERNER, U., WAGNER, J. and SCHÜMANN, H. J. (1971) Effects of β-receptor blocking drugs on the output of noradrenaline from the isolated rabbit heart induced by sympathetic nerve stimulation. *Naunyn Schmied. Arch. exp. Path. Pharmak.* **268**, 102–113.

WESTFALL, D. P. (1973) Antagonism by protryptiline and desipramine of the response of the vas deferens of the rat to NE, ACh and K^+. *J. Pharmacol. exp. Ther.* **185**, 540–550.

WHITBY, L. G., AXELROD, J. and WEIL-MALHERBE, H. (1961) The fate of H^3-norepinephrine in animals. *J. Pharmacol. exp. Ther.* **132**, 193–201.

WHITE, J. C., OKELBERRY, A. M. and WHITELAW, G. P. (1936) Vasomotor tonus of the denervated artery, control of sympathectomized blood vessels by sympathomimetic hormones and its relation to the surgical treatment of patients with Raynaud's disease. *Arch. Neurol. Psychiat. (Chicago).* **36**, 1251–1276.

WHITE, J. C. and SMITHWICK, R. H. (1944) The Autonomic Nervous System. Henry-Kimpton, London.

WHITTAKER, V. P. and SHERIDAN, M. N. (1965) The morphology and acetylcholine content of isolated cerebral cortical synaptic vesicles. *J. Neurochem.* **12**, 363–372.

WIDDICOMBE, J. G. (1966) Action potentials in parasympathetic and sympathetic efferent fibres to the trachea and lungs of dogs and cats. *J. Physiol. (Lond.).* **186**, 56–88.

WILLIAMS, T. H. and PALAY, S. L. (1969) Ultrastructure of the small neurons in the superior cervical ganglion. *Brain Res.* **15**, 17–34.

WINCKLER, J. (1969) Über die adrenerger Herznerven bei Ratte und Meerschweinchen Entwicklung und Innervationsmuster. *Z. Zellforsch. Mikrosk. Anat.* **98**, 106–121.

WINICK, M. and GREENBERG, R. E. (1965) Appearance and localization of a nerve growth-promoting protein during development. *Pediatrics*, Springfield. **35**, 221–228.

WINKLER, H. (1971) The membrane of the chromaffin granule. *Phil. Trans. Roy. Soc. Lond. Ser. B.* **261**, 293–303.

WOLFE, D. E. (1965) The epiphyseal cell: an electron microscopic study of its intercellular relationships and intracellular morphology in the pineal body of the albino rat. *Progr. Brain Res.* **10**, 332–376.

WOOD, J. G. and BARRNETT, R. J. (1964) Histochemical demonstration of norepinephrine at a fine structural level. *J. Histochem. Cytochem.* **12**, 197–209.

WOODS, R. (1969) Acrylic aldehyde in sodium dichromate as a fixative for identifying catecholamine storage sites with the electron microscope. *J. Physiol. (Lond.).* **203**, 35–36P.

WOOTEN, G. F. and COYLE, J. T. (1973) Axonal transport of catecholamine synthesizing and metabolizing enzymes. *J. Neurochem.* **20**, 1361–1371.

WRETE, M. (1940) Beiträge zue Kenntris der Anatomie des Halssympathicus beim Kaninchen. *Z. Mikrosk. Anat. Forsch.* **49**, 317–332.

WURTMAN, R. J. (1973) Biogenic amines and endocrine function. *Fed. Proc.* **32**, 1769–1771.

WURTMAN, R. J. and AXELROD, J. (1966a) Control of enzymatic synthesis of adrenaline in the adrenal medulla by adrenal cortical steroids. *J. Biol. Chem.* **241**, 2301–2305.

WURTMAN, R. J. and AXELROD, J. (1966b) A 24-hour rhythm in the content of norepinephrine in the pineal and salivary glands of the rat. *Life Sci.* **5**, 665–669.

WURTMAN, R. J. and AXELROD, J. (1968) Daily rhythmic changes in tyrosine transaminase activity of the rat liver. *Proc. nat. Acad. Sci., U.S.A.* **57**, 1594–1598.

WURTMAN, R. J., AXELROD, J., SEDVALL, G. and MOORE, R. Y. (1971) Photic and neural control of the 24-hour norepinephrine rhythm in the rat pineal gland. *J. Pharmacol. exp. Ther.* **157**, 487–492.

WURTMAN, R. J., POHORECKY, L. A. and BALIGA, B. S. (1972) Adrenocortical control of the biosynthesis of epinephrine and proteins in the adrenal medulla. *Pharmacol. Rev.* **24**, 411–426.

WYLIC, D. W., ARCHER, S. and ARNOLD, A. (1960) Augmentation of pharmacological properties of catecholamines by *o*-methyltransferase inhibitors. *J. Pharmacol. exp. Ther.* **130**, 239–244.

YAMAMOTO, H. and KIRPEKAR, S. M. (1972) Effects of nerve stimulation on the uptake of norepinephrine by the perfused spleen of the cat. *Eur. J. Pharmacol.* **17**, 25–33.

YAMAUCHI, A. (1969) Innervation of the vertebrate heart as studied with the electron-microscope. *Arch. Histol. Jap.* **31**, 83–117.

YAMAUCHI, A. and BURNSTOCK, G. (1969) Post-natal development of the innervation of the mouse vas deferens. A fine structural study. *J. Anat.* **104**, 17–32.

YATES, C. M. and GILLIS, C. N. (1963) The response of rabbit vascular tissue to electrical and drug stimulation. *J. Pharmacol. exp. Ther.* **140**, 52–59.

YOKOTA, R. (1973) The granule-containing cell somata in the superior cervical ganglion of the rat, as studied by a serial sampling method for electron microscopy. *Z. Zellforsch. Mikrosk. Anat.* **141**, 331–345.

YOSHIKAWA, H. (1970) An experimental study on the sympathetic neuron chains using the fluorescence method for biogenic monoamines. *Arch. Hist Jap.* **31**, 495–509.

ZAIMIS, E. (1964) Pharmacology of the autonomic nervous system. *Ann. Rev. Pharmacol.* **4**, 365–400.

ZAIMIS, E. (1965) The immunosympathectomized animal: a valuable tool in physiological and pharmacological research. *J. Physiol. (Lond.).* **177**, 35–36P.

ZAIMIS, E. and BERK, L. (1965) Morphological, biochemical and functional changes in the sympathetic nervous system of rats treated with nerve growth factor-antiserum. *Nature (Lond.).* **206**, 1220–1222.

ZELANDER, T., EKHOLM, R. and EDLUND, Y. (1962) The ultrastructural organization of the rat oxocrine pancreas. III. Intralobular vessels and nerves. *J. Ultrastruct. Res.* **7**, 84–101.

ZIMMERMAN, B. G. and GISSLEN, J. (1968) Pattern of renal vasoconstriction and transmitter release during sympathetic stimulation in presence of angiotensin and cocaine. *J. Pharmacol. exp. Ther.* **163**, 320–329.

ZIMMERMAN, B. G. and WHITMORE, L. (1967) Transmitter release in skin and muscle blood vessels during sympathetic stimulation. *Amer. J. Physiol.* **212**, 1043–1054.

ZUCKERKANDL, E. (1912) The development of the chromaffin organs and of the suprarenal glands. *In* 'Keibal and Mall's Manual of Human Embryology'. U.S.A. Lippincott. Vol. 2, pp. 157–179.

INDEX

Acetylcholine (ACh), 14, 15, 59, 61, 75
Acetylcholinesterase (AChE), 33
Acoustic receptors, 13
Action potential, 11
Adenyl cyclase system, 66
Adipose tissue, 12
Adrenal glands, 1
Adrenaline, 1
  biosynthesis, 2
Adrenergic effects on cholinergic neurons, 16
Adrenergic inhibition, 10
Adrenergic mechanisms, 2
Adrenergic nerve, 1, 3
  dual role of, 100
  general organization and functions, 4–18
  terminal axons of, 18
Adrenergic nerve plexus, spatial arrangement of, 72
Adrenergic neuroeffector transmission, 14, 51–106
Adrenergic neurons
  and non-adrenergic neurons, 14
  bipolar, 19
  cell body, 20
  cholinergic effects on, 14
  collaterals of, 9
  components of, 20
  drug-induced, structural changes in, 122
  general functions of, 12
  general structure, 19
  input of, 5
  intraganglionic connections of, 7
  long, 20, 21
  monopolar, 19
  multipolar, 19
  output of, 11
  short, 20, 21
Adrenergic neurotransmission, onset of, 114
Adrenergic pharmacology, 3
Adrenergic response
  amplitude of, 96
  and neuromuscular geometry, 96
  termination of, 98
  time course of, 98
Adrenergic synapses, 11
Adrenergic terminal axons and muscle effectors, 51
  branching of, 129
Adrenergic terminal varicosities, 11
Adrenergic transmission, 3
  development of, 112
  electrophysiology of, 72
  excitatory, 72
  inhibitory, 76
  maintenance of, 90
  models of, 88
  pharmacology of, 76
  safety factor, 78
Adrenergic vesicles
  distribution of diameters of, 27
  life span, 34
  specific marker, for, 25
α-Adrenoceptor, 64, 65
β-Adrenoceptor, 64, 65
α-Adrenoceptor blocking drugs, 84, 114
β-Adrenoceptor blocking drugs, 85
Adult tissues, 108
Adventitial-medial border, 52
Ageing pigment, 122
Ageing sympathetic ganglia, 121
Amnion, 70
Amphetamine, 79

INDEX

Amplitude and duration of response, 83, 96
Anaesthetic action, 80
Angiotensin, 59, 60, 79, 115
Anococcygeus muscle, 125
Anterior eye chamber, 3
Anterior pelvic (hypogastric) ganglia, 132
Asphyxia, 118
ATP, 17, 56, 67, 118
Atrioventricular node, 52
Auerbach's plexus, 52, 104
Auricle, 108
Auricular arteries, 52
Autonomic ground plexus, 2, 19, 51
Autonomic nervous system, 121
Axon lengths, spectrum of, 21
Axoplasmic flow
   fast, 32
   slow, 32

Baroreceptors, 8, 13
Bethanedine, 80
Biogenic amines, fluorescence histochemical method for, 4
Bladder, 5
Blood pressure, 131
Blood stream, 88
Blood vessels, 36
Bovine splenic nerve, 23
Bretylium, 79, 124
Bronchial arteries, 52
Brown fat, 36
Butoxamine, 86

Calcium
   effect of NA on, 67–68
   in NA release, 58
Cardiac muscle, 12
Castration, 116
Catecholamines, 1, 134, 135
   circulating, 100, 103, 104
   turnover of, 37
Catechol-o-methyl transferase (COMT), 47, 48, 70, 71, 83, 89–90, 93, 98

Cation-exchange hypothesis, 57
Central nervous structures, 13
Chemical sympathectomy, 3
Chemoreceptors, 8, 13
Cholesterol, 29
Choline acetyltransferase, 14, 113, 116
Cholinergic effects on adrenergic neurons, 14
Cholinergic input from peripheral neurons, 8
Cholinergic link, 3, 15
Cholinergic nerves, 3
   adrenergic effects on, 16
   terminals of, 18
Chondroblasts, 70
Chromaffin cells, 35, 36, 109
Chromatolysis, 121
Chromogranin, 56
Chronic treatment, 124
Chronotropic effect, 86
Ciliary ganglion, 16
Clorgyline, 48
Close junctions, 52
Cocaine, 81–83
Cocaine-like supersensitivity, 126
Colchicine, 33, 57, 122
Cold exposure, 12, 131
Colon, 5
Concentration peaks, 63
Conduction, speeds of, 5
Convergence, 6
Cortical hormones, 36, 115, 118
Corticosteroids, 83
Corticotrophic hormones, 115
Cortisone, 120
Cyclic AMP, 66
Cytochalasin B, 57

Decentralization, 42, 44, 126
Degeneration, time course of, 121
Degeneration contraction, 125
Denervation
   supersensitivity, 125
   surgical or immunochemical, 48
Dense core vesicles, 23
Densely innervated tissues, 100

INDEX

Density of innervation, 51, 81, 108, 130
Descending inhibition, 17
Desmethylimipramine (DMI), 82
Dibenamine, 84
Dichloroisoproterenol (DCI), 85
Diethyldithio carbamate, 77
3,4-Dihydroxymandelic acid (DOMA), 50
5,6-Dihydroxytryptamine, 123
Dimethyl phenyl piperazinium (DMPP), 80
Dispensability of sympathetic system, 126
Disulfiram, 77
Divergence, 6
3,4-Dihydroxyphenylglycol (DOPEG), 50
Dog retractor penis, 68
α-*methyl*-dopa, 87
Dopa-decarboxylase, 32, 39
Dopa-decarboxylation, 40
Dopamine, 10, 59
  retrograde flow of, 32
Dopamine-β-hydroxylase (DβH), 28, 30, 31, 34, 38, 39, 41, 44, 57, 77, 91, 95
  radio immunoassay for, 57
  serum level of, 57
  trans-synaptic induction, 44
Ductus arteriosus, 111

Ear artery, 64
Effector cells, functional relations with, 108
Effector organs, 111, 112
Effector tissues, 2, 3, 108, 120
Effector unit, 101
Elastic lamellae, 73
Electrical stimulation, 33
Electronmicroscopic studies, 3
Electrotonic coupling, 54, 75
Embryological tissues, 110
Endocrine cells, 35
Endothelial cells, 70
Enteric ganglia, 115

Enteric system, 4
Enzyme acetylcholinesterase, 14
Epsilon-aminocaproic acid, 80
Estrogenic hormones, 115
Estrogens, 116
Excitatory junction potentials (EJP's), 72–75, 101, 103
Excitatory postsynaptic potential (EPSP), 9, 11
Explants of tissues, 108
Extracellular recordings, 9
Extracellular space, 88

False adrenergic transmitters, 86
Feed-back inhibition, 60
Fibroblasts, 70
Fluorescence histochemical method for biogenic amines, 4
Fluorescence microscopy, 3
Foetal neurons, 107
Freeze-etching technique, 56
Frequency-response curve, 96, 97
Frog heart, 2
Functional ganglionic transmission, 112
Functional transmission, 112, 119

Ganglion blocking effect, 80
Gastrointestinal tract, 12, 51
Genes, 114
Glands, 12
Glucocorticoids, 109, 115
Golgi region, 112
Granular vesicles, 23
  large, 24
  small, 24
Growth, 107
Growth cones, 119
Guanethidine, 79, 124
Guinea-pig intestine, 17
Guinea-pig pelvic ganglia, 7
Gut motility, 8

Heterogeneity
  of adrenergic neurons, 130
  or peripheral adrenergic system, 129

221

INDEX

Hibernating animals, 41
Homeostasis, 12, 134
Hormones, 115, 130
   cortical, 36, 115, 118
   sexual, 132
   thyrotrophic, 115
Humoral control, 104, 133
6-Hydroxydopamine (6-OHDA), 122–23, 125
Hyperpolarization, 67
Hypotension, 122

Immature tissues, 108
Immunofluorescence methods, 3
Immunosympathectomy, 3, 124
Imipramine, 82
Induction, 43, 45, 117
Inhibitory postsynaptic potential (IPSP), 9
Innervation
   asymmetrical, 81
   density of, 51, 81, 108, 130
   field of, 119
   pattern of, 119, 130
Inotropic effect, 86
Insulin, 115
Integrative properties, 8
Interactions, 9
Intermediolateral nuclei, 5
Interneurons, 9
Intestinal muscle, longitudinal, 104
Intracardiac ganglia, 16
Intracellular recordings, 9
Intraganglionic connections of adrenergic neurons, 7
Intraganglionic inhibitory inputs, 9
Intraganglionic input, 8
Isonicotinic acid, 83
Isoprenaline, 70

Junctional cleft, 52

Liver cells, 66

Marsilid, 83
Mast cells, 35
Maturation, 110, 114, 117, 118

Mecamylamine, 80
Mechanoreceptors, 13
Membrane depolarization, 97
Mesentery, 24
Methoxamine, 85
α-Methyl-p-tyrosine, 76
Michaelis–Menten kinetics, 69
Migratory ability, 107
Mineral oil bath, 100
Modulating factors, 130
Monamine oxidase (MAO), 32, 47, 50, 69, 70, 71, 83, 89–90, 93, 98
Muscarinic receptors, 10, 14, 16
Muscle effector bundles, 75
   and nexuses, 54
Muscle effectors and adrenergic terminal axons, 51
Muscle spindles, 13

Nerve activity, 41
   level of, 129
Nerve control, 104, 133
Nerve Growth Factor (NGF), 108, 114, 120, 124, 130
Neuroblastoma, 114
Neurofilaments, 20
Neuromuscular distance, 96, 97, 103
Neuromuscular junction
   dimensions of, 63, 78
   models of, 101
Neurotubules, 20
Nialamide, 83
Nicotinic drugs, 107
Nicotinic receptors, 10, 14, 80
Nictitating membrane, 51
Nissl substance, 115
Node crowding, 72
Non-terminal axons, 20
Noradrenaline, 1
   autoradiographic localization of, 2
   axonal flow of, 91
   axoplasmic transport of, 31
   biosynthesis, 39–47
   cytoplasmic, 88
   diffusion from neuromuscular junction into blood stream, 71

INDEX

Noradrenoline—*cont.*
  effect on membrane conductivity and calcium activity, 67
  enzymic inactivation, drugs affecting, 83
  extraneuronal uptake of, 69
  extravesicular storage of, 23
  in isolated granules, 28
  inactivation, 48, 68–72
  intraneuronal distribution of, 22
  metabolic degradation, 47–50, 71
  neuronal uptake of, 68
  peak concentration of, 96, 100
  postjunctional action, 64
  release of, 54–64, 90
    amount of, 61
    calcium in, 58
    drugs affecting, 79–80
    facilitation of, 60
    physiological control of, 58
    quantal hypothesis and quantitative estimation of, 61
    spontaneous nerve induced, 55
    spontaneously in packets, 55
  retrograde transport of, 32
  re-uptake of, 68, 91
  storage of, 90
    drugs affecting, 77
  storage particles of, 2
    heavy, 29
    light, 29
  storage sites, preservation of, 24
  synthesis of, 91
    drugs affecting, 76
    regulation in adrenergic neurons, 41
  synthesis site, 39
  to ATP, ratio of, 28
  to DβH, ratio of, 57
  transmitter mechanisms, 88
  uptake, drugs affecting, 81
  utilization rate, 77
  vesicles containing, electron microscopic identification, 23
Non-adrenergic neurons and adrenergic neurons, 14

Normetanephrine (NMN), 71

Oesophageal muscularis mucosae, 68
Olfactory receptors, 13
Overflows, 97

Pain receptors, 13
Paradoxical pupil dilatation, 125
Paraganglionic tissue, 35
Paravertebral ganglia, 4
Paravertebral sympathetic chain, 110
Pargyline, 83
Parotid gland, 52
Pathology, 121
Pelvic ganglia, 5
Pempidine, 80
Peripheral adrenergic system, heterogeneity of, 129
Peripheral neurons, cholinergic input from, 8
Peristaltic reflex, 17
Pheniprazine, 83
Phentolamine, 84
Phenoxybenzamine (PBA), 60, 61, 84
Phenylethanolamine-N-methyl transferase (PNMT), 117
Pheochromoblast, 110
Phospholipids, 29
Physiological roles, 113, 133
Pineal gland, 5, 13, 113
Pinealocytes, 5
Pituitary nervous lobe, 13
Portal vein, 64
Postganglionic discharge, 11
Postganglionic neuron, 4
Postjunctional membrane, 53
Postjunctional supersensitivity, 81
Postsynaptic receptors, drugs affecting, 84
Potentiation, 81
Precursors, 86
Preganglionic denervation, 120
Preganglionic nerve impulses, 130
Preganglionic neuron, 4
Pregnancy, 115
Prejunctional inhibitory mechanism, 98

INDEX

Prejunctional supersensitivity, 81
Prenylamine, 77
Presynaptic α-adrenoceptors, 59
Prevertebral ganglia, 4
Prevertebral sympathetic ganglia, 110
Primitive mammalian sympathetic cells, 110
Primitive stem cell, 114
Pronethanol, 86
Propranolol, 85, 86
Prostaglandins, 59, 60, 79
Pulmonary artery, 63, 97
Purinergic nerves, 3
Purinergic neurons, 16–17
Pyrogallol, 84

Rami communicantes, 5
Receptors, 13, 64
   chemical nature of, 66
   hypothetical types of, 10
   occupancy, 63
   sites, 2
Regenerating fibres, 119
Relaxation, rate of, 100
Repetitive nerve stimulation, 75
Reserpine, 55, 77
Response
   amplitude and duration of, 83, 96
   characteristics of, 129
Retraction of terminal axons, 124
Retrograde degeneration, 9

Saphenous vein, 99
Saturation, 69
Seminal vesicles, 13, 73
Senescence, 121
Sensory receptors, 12
Sexual hormones, 132
Sinoatrial node, 52
Skin vessels, 131
Smooth endoplasmic reticulum, 34, 112
Smooth muscle, 2, 12
   densely innervated, 81
   sparsely innervated, 81
   with dense adrenergic innervation and close neuromuscular distances, 103
   with low density of adrenergic nerves and large neuromuscular distances, 103
Sotalol, 85
Sparsely innervated smooth muscle, 100
Sperm, emission of, 132
Sphincter pupillae, 52
Sphincter urethrae, 124
Spike activity, 67
Splanchnic nerves, 5
Spleen capsule, 66
Spontaneous activity, 8
Spontaneous excitatory junction potentials (SEJP), 61
Sublingual gland, 120
Submandibular gland, 120
Subsurface cisternae, 54
Sucrose gradient centrifugation, 29
Sulphomucopolysaccharides (SMPS)-protein complexes, 57
Surgical sympathectomy, 121
Sympathectomy, 122
Sympathetic activity, fluctuation of, 43
Sympathetic ganglia, 44
Sympathetic neuroblast, 110
Sympathetic pathways, independent activation of, 131
Sympathetic system, dispensability of, 126
Sympatho-adrenal functions, dispensability of, 134
Sympatho-adrenal system, 1, 12, 109
   functional organisation of, 133
   generalized activation of, 132
Sympathomimetic effects, 79
Synapses 'en passage', 5
Synaptic vesicles, 56

Taenia coli, 53
Terminal axons, 21
Testosterone, 116, 120
Tetrabenazine, 77
Thermoregulation, 113, 131

INDEX

Thyrotrophic hormone, 115
Tissue culture, 107
Tonic activity, 8
Tonic pacemaker, 113
Transmural stimulation, 93
Tranylcypromine, 83
Tropolone, 84
Tubular reticulum, 91
Tyramine, 58, 79
Tyrosine, 39
Tyrosine hydroxylase (T-OH), 32, 39, 40, 41, 43, 44, 46, 50, 76
  rate-limiting step enzyme, 44
  trans-synaptic induction, 44
Tyrosine hydroxylation, 40

Umbilical artery, 70
Undesirable interaction between systems, 135
Ureter, 51
Uterine artery, 64
Uterus, 51, 115

Varicose terminal axons, 51–54
Vas deferens, 13, 16, 51, 52
Vagina, 115
Vasoconstriction, 103
Vesicle exocytosis, 55
Vinblastine, 33, 57
Visceral receptors, 8, 13